Lecture Notes in Computer Science 849

Edited by G. Goos and J. Hartmanis

Advisory Board: W. Brauer   D. Gries   J. Stoer

Reiner W. Hartenstein
Michal Z. Servít (Eds.)

Field-Programmable Logic

Architectures, Synthesis and Applications

4th International Workshop on Field-Programmable
Logic and Applications, FPL '94
Prague, Czech Republic, September 7–9, 1994
Proceedings

Springer-Verlag

Berlin Heidelberg New York
London Paris Tokyo
Hong Kong Barcelona
Budapest

Series Editors

Gerhard Goos
Universität Karlsruhe
Postfach 69 80
Vincenz-Priessnitz-Straße 1
D-76131 Karlsruhe, Germany

Juris Hartmanis
Cornell University
Department of Computer Science
4130 Upson Hall
Ithaca, NY 14853, USA

Volume Editors

Reiner W. Hartenstein
Fachbereich Informatik, Universität Kaiserslautern
Postfach 30 49, D-67653 Kaiserslautern, Germany

Michal Z. Servít
Department of Computers, Czech Technical University
Karlovo náměstí 13, 12135 Prague 2, Czech Republic

CR Subject Classification (1991): B.6-7, J.6

ISBN 3-540-58419-6 Springer-Verlag Berlin Heidelberg New York

CIP data applied for

© Springer-Verlag Berlin Heidelberg 1994
Printed in Germany

Typesetting: Camera-ready by author
SPIN: 10475493   45/3140-543210 - Printed on acid-free paper

Preface

This book contains papers first presented at the 4th International Workshop on Field-Programmable Logic and Applications (FPL'94), held in Prague, Czech Republic, September 7 - 9, 1994.

The FPL'94 workshop was organized by the Czech Technical University and the University of Kaiserslautern, in co-operation with IEEE Czechoslovakia Section and the University of Oxford (Dept. for Continuing Education), as a continuation of three already held workshops in Oxford (1991 and 1993) and in Vienna (1992).

The growing importance of field-programmable devices is demonstrated by the strongly increased number of submitted papers for FPL'94. For the workshop in Prague, 116 papers were submitted. It was pleasing to see the high quality of these papers and their international character with contributions from 27 countries. The list below shows the distribution of origins of the papers submitted to FPL'94 (some papers were written by an international team):

Austria:	7	Malaysia:	1
Belgium:	1	Norway:	2
Brazil:	1	Poland:	5
Canada:	1	Republic of Belarus:	5
Czech Republic:	5	Slovakia:	3
Finland:	2	South Africa:	1
France:	9	Spain:	4
Germany:	17	Sweden:	2
Greece:	2	Switzerland:	3
Hungary:	2	Syria:	1
India:	1	Turkey:	1
Japan:	2	United Kingdom:	18
Latvia:	1	USA:	19

The FPL'94 Technical Program offers an exciting array of regular presentations and posters covering a wide range of topics. From the 116 submitted papers the very best 40 regular papers and 24 high quality posters were selected. In order to give the industry a strong weight in the conference, there are 10 industrial papers among the 40 regular papers. All selected papers, except one, are included in this book.

We would like to thank the members of the Technical Program Committee for reviewing the papers submitted to the workshop. Our thanks go also to the authors who wrote the final papers for this issue.

We also gratefully acknowledge all the work done at Springer-Verlag in publishing this book.

July 1994

Reiner W. Hartenstein,

Michal Z. Servít

Program Committee

Jeffrey Arnold, IDA SRC, USA
Peter Athanas, Virginia Tech, USA
Stan Baker, PEP, USA
Klaus Buchenrieder, Siemens AG, FRG
Erik Brunvand, U. of Utah, USA
Pak Chan, U. of California (Santa Cruz), USA
Bernard Courtois, INPG, Grenoble, France
Keith Dimond, U. of Kent, UK
Barry Fagin, Dartmouth College, USA
Patrick Foulk, Heriot-Watt U., UK
Norbert Fristacky, Slovak Technical University, Slovakia
Manfred Glesner, TH Darmstadt, FRG
John Gray, Xilinx, UK
Herbert Gruenbacher, Vienna U., Austria
Reiner Hartenstein, U. Kaiserslautern, FRG
Sinan Kaptanoglu, Actel Corporation, USA
Andres Keevallik, Tallin Technical U., Estonia
Wayne Luk, Oxford U., UK
Amr Mohsen, Aptix, USA
Will Moore, Oxford U., UK
Klaus Müller-Glaser, U. Karlsruhe, FRG
Peter Noakes, U. of Essex, UK
Franco Pirri, U. of Firenze, Italy
Jonathan Rose, U. of Toronto, Canada
Zoran Salcic, Czech T. U., Czech Republic
Mariagiovanna Sami, Politechnico di Milano, Italy
Michal Servít, Czech T. U., Czech Republic
Mike Smith, U. of Hawaii, USA
Steve Trimberger, Xilinx, USA

Organizing Committee

Michal Servít, Czech Technical University, Czech Republic, General Chairman
Reiner Hartenstein, University of Kaiserslautern, Germany, Program Chairman

Table of Contents

Trade-Offs and Experience

Innovations and Smart Applications

FPGA-Based Computer Architectures

High Level Design

New Tools

CCMs and HW/SW Co-Design

Modelers

Educational Experience

Novel Architectures and Smart Applications

Applications and Educational Experience

Fault Modeling and Test Generation for FPGAs

Michael Hermann and Wolfgang Hoffmann

Institute of Electronic Design Automation,
Department of Electrical Engineering
Technical University of Munich, 80290 Munich, Germany

Abstract. This paper derives a fault model for one-time programmable
FPGAs from the general functional fault model and an algorithm to
perform test generation according to this model. The new model is characterized by the abstraction of functional faults from a set of possible
implementations of a circuit. In contrast to other functional-level test
generation procedures a fault coverage of 100% can be achieved regardless of the final implementation of the circuit.

1 Introduction

As new technologies for the implementation of digital designs are developed
there is also the need for adequate CAD tools to exploit new features or to
address new problems. One of the most important new technologies that have
been developed during the past few years is the *Field Programmable Gate Array*
(FPGA). A FPGA generally consists of an array of programmable *logic modules*
(LM) and a programmable interconnect area.

Although architecture and complexity of FPGAs are similar to those of small
gate arrays there are some important differences. In particular, test generation
is affected by the lack of knowledge about the final physical implementation of
the design.

This is due to the fact that the last step of a design visible to the designer is a
netlist of *feasible modules*. A *module* can be viewed as a black box with n inputs
and m outputs. The behaviour of any output is given by a completely specified
Boolean function of a subset of the n inputs. A module is called *feasible*, if it
can be implemented by a single LM of the FPGA.

Given a netlist containing only feasible modules a working FPGA can be obtained by the following two steps: *personalization* and *programming*. *Personalization* is the process of choosing one of several possible configurations of a
LM such that the LM performs the same Boolean functions as a given feasible
module. *Programming* denotes simply the process of implementing the choosen
personalization within the given FPGA. The actual personalization of the FPGA
is choosen by the vendor's design tool without interaction or notification of the
user. As the personalization is performed at the users's site a test of the personalized FPGA can be only performed by the user after personalization and
programming.

For some types of FPGAs the manufacturer can test the device right after it has
been manufactured. This way, the chance of a failure during programming can be

eliminated or at least significantly decreased. As an example, for reprogrammable FPGAs like the Xilinx-FPGAs a large number of different personalizations can be tested at the factory and thus almost any programming problem the user could encounter can be anticipated.

This method is not viable for one-time programmable FPGAs like the Actel-FPGAs [6]. In this case the manufacturer can test the circuitry of the device in its unprogrammed state only. Therefore programming faults can still occur [5].

Most programming faults can be detected on the fly by the programming equipment. However, some programming faults may still remain undetected that cause the device to malfunction. Therefore, a test set for a personalized FPGA is still required.

If a gate-level netlist of the circuit was available, a conventional gate-level testpattern generator could be used to generate the required test set. However, there is usually no information about the actual personalization available. Without this information it is not possible to generate a gate-level description of the personalized FPGA even though the internal structure of the LM may be known. Therefore a gate-level testpattern generator can not be used to generate a test set for a personalized FPGA.

This paper describes a method for deterministic testpattern generation in the absence of a gate-level description of the circuit. The paper is organized as follows:

In section 2 we will describe the fault model used for testpattern generation. Section 3 explains a method to generate the testpatterns according to this model. In section 4 we show some results of this approach and compare it to other approaches.

2 Fault Model

2.1 Previous Work

Several approaches for test generation have been developed that deal with the lack of an exact gate-level description of a circuit [1] [2] [3] [5]. We will shortly discuss the application of those approaches to a netlist containing feasible modules.

In [2] each function g associated with a module is described in a two-level representation. A two-level AND/OR (or OR/AND) implementation of g is then generated. At this point a fictive gate-level description of the circuit is available. The stuck-at fault model is applied to all signals of the netlist and a gate-level test generator can be used to generate a test set. The reported results indicate that the test set derived from this fictive gate-level netlist yields a very high fault coverage for varying implementations of the module. However, this approach can not provide 100% fault coverage in a deterministic way.

The approach described in [3] starts from the existing representation of g. Then a function f is derived by replacing some operations in the representation of g by other operations. At last a testpattern is selected that distinguishes g from f. For small circuits it has been shown that taking the logic dual of small operators (like AND) in the representation of g is a good choice for varying implementations of the module. However, this approach cannot provide 100% fault coverage, either.

In [1] [5] *any* change in the truth table of g is considered. Therefore each possible input vector must be applied at the inputs of the module. This approach guarantees 100% fault coverage of the module independently of the implementation. However, this approach can be used only for modules with a low number of inputs due to its exponential complexity (2^n testpatterns for a function depending on n variables).

2.2 Fault Model for FPGAs

The fault model introduced for FPGAs is based on three assumptions:

1. the unprogrammed LM is fault free
2. the LM has n inputs and one output
3. the number of feasible modules is far less than 2^{2^n}

An example for a FPGA containing this type of LM is the ACT1/2-architecture made by Actel [6].

Let a feasible function g be the completely specified Boolean function associated with a feasible module. The support $sup(g)$ of g is the set of all variables g depends on. The set of all feasible functions is denoted by F. Then P_g is the set of all *personalization faults* p with respect to g:

$$P_g = \{p | p \in F \backslash \{g\} \wedge sup(p) \subseteq sup(g)\} \tag{1}$$

By the restriction of the support of p we do not allow a functional dependancy of p on arbitrary signals of the circuit.

For the rest of the paper we will restrict ourselves to the *single personalization fault model*. This model assumes that only one module of the circuit is affected by a personalization fault. A module with an associated function g is affected by a personalization fault, if it implements a function $p \in P_g$ instead of g.

The application of the personalization fault model requires the knowledge of P_g for every module. According to (1) this requires the knowledge of F which depends only on the LMs contained in a FPGA and the personalization facilities provided by the FPGA. The exhaustive enumeration of all personalizations of a LM delivers the complete set F for the LM. This computation has to be done only once for a given LM.

As an example Figure 1 shows the LM used in the ACT1-FPGAs by Actel. This LM can be personalized by connecting an arbitrary signal (vertical lines in Figure 1) including constant 0 (L) or 1 (H) to any input (horizontal lines). The LM in Figure 1 is personalized to implement the Boolean function $g = xy$. This can be done by establishing a connection at the highlighted crosspoints.

One of the most important characteristics of P_g is its size. The third column of Table 1 shows the size of P_g for this LM. The size of P_g depends only on the

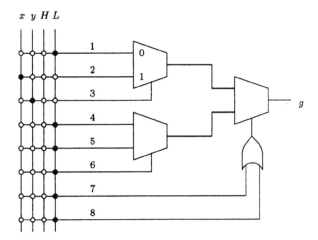

Fig. 1. personalized ACT1 logic module

$|sup(g)|$	worst case	$|P_g|$	$|\overline{P_g^2}|$
1	3	3	2
2	15	15	11
3	255	212	28
4	65535	4,501	39
5	$4 \cdot 10^9$	45,481	43
6	$1 \cdot 10^{19}$	268,443	47
7	$3 \cdot 10^{38}$	1,132,370	51
8	$1 \cdot 10^{72}$	3,806,057	57

Table 1. Personalization fault set sizes

number of elements in $sup(g)$ but not on the truth table of g. The second column shows the size of P_g not taking into account the structure of the LM.
As can be seen, the size of P_g does not explode exponentially even though there is no loss of information compared to a general functional fault model [1]. However, for more than five inputs the size of P_g is still too large for practical purposes. To reduce the size of P_g we now introduce the *programming fault*.
A programming fault reflects a failure of the basic *configuration* element of the FPGA. Therefore the programming fault model requires knowledge about the technology of the FPGA. For the Actel-FPGAs the basic configuration element is the antifuse (AF). For an AF, a programming fault occurs if:

1. an AF should connect two wires but does not
2. an AF falsely connects two wires

A programming fault does not necessarily cause a personalization fault. Furthermore, some $p \in P_g$ may require a multiple programming fault to occur. Therefore a reasonable restriction on the number of programming faults that may occur simultaneously during the configuration of a LM may lower the size of P_g considerably.

Let a denote a specific configuration of a LM. Let A denote the set of all possible configurations a. In case of the Actel- FPGAs a is a set of AFs that are programmed. Then $B \subseteq F \times A$ contains all pairs (f, a) where a feasible function f is realized by the configuration a. Let $X \subseteq A \times A$ contain all pairs of configurations (a_1, a_2) that differ by at most a single programming fault. Then

$$P_g^n = \{f | (f, g) \in B X^n B^{-1} \wedge f \in P_g\} \tag{2}$$

is the set of all personalization faults with respect to g if at most n programming faults can occur simultaneously. This definition allows us to order the personalization fault sets according to n.

$$P_g^1 \subseteq P_g^2 \subseteq P_g^3 \ldots \subseteq P_g \tag{3}$$

The last column of Table 1 shows the average size of P_g^2 for the ACT1-architecture. The average has been calculated for all feasible functions. P_g^2 has been choosen, because it contains the case that AF_1 has been programmed *instead of* AF_2. This is the most likely fault that remains undetected by the programming equipment. As bridges between signal lines can be detected [6], these cases have been eliminated from P_g^2.

In Figure 1 the set a_1 might be denoted by

$$a_1 = \{(L, 1), (x, 2), (y, 3), (L, 4), (L, 5), (L, 6), (L, 7), (L, 8)\}$$

Then the set

$$a_2 = \{(H, 1), (x, 2), (y, 3), (L, 4), (L, 5), (L, 6), (L, 7), (L, 8)\}$$

is transformed into a $p \in P_g^2$ with the LM now implementing the Boolean function $p = x + \overline{y}$. However, it is not possible to find any set a_i such that the LM implements the function $g = \overline{x} + \overline{y}$ regardless of the initial personalization for $g = xy$ if at most two programming faults are allowed.

3   Test Generation

In our approach, test generation is divided into two tasks: first, generation of testvectors using function identification, and secondly, testvector justification and fault propagation by testpatterns. The terms *testpattern* and *testvector* both denote an assignment of logic values to certain signals. A testpattern $\hat{t}$ assigns signal values to all primary inputs t of the circuit, a testvector $\hat{v}_m$ to all inputs v_m of the module m.

3.1 Testvectors

Let $g(\underline{v})$ be a Boolean function and $P = \{\ldots, p_i(\underline{v}), \ldots\}$ a set of Boolean functions $p(\underline{v})$ ($g \notin P$). Then, *function identification* of g with respect to P is the process of calculating a set V of testvectors $\hat{\underline{v}} \in V$ that uniquely distinguishes g from all $p \in P$.

We will now use function identification for test generation. For each module m, we calculate V_m by performing function identification of the associated function g_m with respect to P_g.

In order to obtain a small set of testpatterns, we try to minimize $|V_m|$. For that purpose, we transform the problem of function identification to the well known problem of optimal matrix covering by constructing a coverage matrix $C_m = (c_{ij})$. A row i of C_m corresponds to the i-th personalization fault p_i, a column j to the j-th testvector $\hat{\underline{v}}_j$. An element c_{ij} is set to 1, if personalization fault p_i can be distinguished from g_m by applying testvector $\hat{\underline{v}}_j$ to module m:

$$c_{ij} = \begin{cases} 1 \text{ if } & p_i(\hat{\underline{v}}_j) \neq g_m(\hat{\underline{v}}_j) \\ 0 \text{ else} \end{cases} \tag{4}$$

A row i of the matrix is said to be covered by column j if $c_{ij} = 1$. Then V_m can be directly derived from a minimal set of columns that covers all rows. Most often suboptimal heuristics are used to solve this problem, still guaranteeing a complete coverage of all rows, but possibly using slightly more columns than necessary.

3.2 Testpatterns

In most cases, the modules' inputs and outputs cannot be directly accessed. Therefore, testpatterns have to be found that justify the testvectors, and propagate fault effects from the module's outputs to primary outputs. Some FPGAs provide features that allow direct observation of LM-outputs. However, this is not considered here to avoid loss of generality.

A testpattern $\hat{\underline{t}}$ *applies* a testvector $\hat{\underline{v}}_m$ on module m, if $\hat{\underline{t}}$ justifies the signal lines $\underline{v}_m$ according to $\hat{\underline{v}}_m$, and, at the same time, propagates a fault effect from the output of module m to one or more primary outputs. A testvector $\hat{\underline{v}}_m$ is *available*, if there exists at least one $\hat{\underline{t}}$ that applies $\hat{\underline{v}}_m$ on m.

Sophisticated methods for justifying lines and propagating fault effects are known from test generation for stuck-at faults on gate-level (e.g. [7]). Since all parts of the circuit except the considered module m are assumed to be fault-free, any gate-level representation of the circuit can be used to generate a testpattern $\hat{\underline{t}}$ that applies a testvector $\hat{\underline{v}}_m$ on m. Thus, traditional test generators, like [7], can be easily adapted for this purpose.

3.3 Description of the Algorithm

Figure 2 shows the basic steps of how we calculate a set T of testpatterns $\hat{\underline{t}}$ which applies appropriate testvectors on the modules, such that function identification is performed for each module of the circuit.

preprocessing
input circuit and P_g
sort modules and initialize $T = \{\}$
for each module m: initialize $V_m = \{\}$

■■■■ functional testpattern generation ■■■■
for each module m

	initialize $T_m = \{\}$
	derive C_m from g_m and P_g
	for each $\hat{\underline{v}}_m \in V_m$
		remove all p covered by $\hat{\underline{v}}_m$ from C_m

■■■■ function identification ■■■■
while C_m not empty and still $\hat{\underline{v}}_m$ unaborted

	calculate (sub-)optimal $\hat{\underline{v}}_m$
	try to generate $\hat{\underline{t}}$ that applies $\hat{\underline{v}}_m$

$\hat{\underline{v}}_m$ available ?		
yes	no	unknown
remove all rows covered by $\hat{\underline{v}}_m$ from C_m $V_m = V_m \cup \{\hat{\underline{v}}_m\}$ $T_m = T_m \cup \{\hat{\underline{t}}\}$	remove all rows only covered by $\hat{\underline{v}}_m$ from C_m	mark $\hat{\underline{v}}_m$ as aborted

C_m empty?	
yes	no
	mark p corresponding to remaining rows as aborted faults

■■■■ fault simulation ■■■■
for each module $n \neq m$

	for each $\hat{\underline{t}} \in T_m$

	does $\hat{\underline{t}}$ apply $\hat{\underline{v}}_n$?	
	yes	no
	$V_n = V_n \cup \{\hat{\underline{v}}_n\}$	

$T = T \cup T_m$

■■■■ postprocessing ■■■■
do final test set compaction
output T

Fig. 2. steps of test generation

Special care has to be taken if a testvector is found to be unavailable. Such a $\hat{\underline{v}}_m$ must not be used for function identification. Therefore, approaches like [4], which use a precalculated, fixed V_m to determine T, cannot guarantee a complete coverage of all personalization faults.

Instead, the availability of $\hat{\underline{v}}_m$ must be considered already when calculating V_m such that V_m contains only available testvectors. Since checks for availability are expensive in terms of CPU-time, we only check on demand. For this, we closely combine function identification and testvector application.

The center part of Figure 2 illustrates our suboptimal heuristic for function identification considering availability of testvectors. We choose one (sub-)optimal covering column of C_m and immediately after that check the availability of its corresponding $\hat{\underline{v}}_m$. If $\hat{\underline{v}}_m$ is available, it is added to V_m, all rows covered by the chosen column are removed from C_m, and the testpattern $\hat{\underline{t}}$ applying $\hat{\underline{v}}_m$ is added to T. Testpattern $\hat{\underline{t}}$ was already calculated when checking the availability of $\hat{\underline{v}}_m$.

If however $\hat{\underline{v}}_m$ was found to be unavailable, $\hat{\underline{v}}_m$ must not be added to V_m, and the column corresponding to $\hat{\underline{v}}_m$ is removed from C_m. In addition, all rows in C_m that now contain only zeroes are removed. These rows correspond to *redundant* personalization faults, which can be distinguished from g_m only by unavailable $\hat{\underline{v}}_m$ and thus do not affect the behaviour of the otherwise fault-free circuit.

This process of choosing a $\hat{\underline{v}}_m$, checking its availability and adjusting C_m is repeated until C_m is empty or contains only columns corresponding to aborted $\hat{\underline{v}}_m$. Hereby, a $\hat{\underline{v}}_m$ is called *aborted* if its availability check could not be performed completely, e.g. due to limited computing resources.

To keep $|T|$ small, we take advantage of the fact that a single $\hat{\underline{t}}$ justifies a testvector on *every* module of the circuit. Moreover, some $\hat{\underline{v}}_n$, including $\hat{\underline{v}}_m$ at the module m under test, are not only justified, but also applied by $\hat{\underline{t}}$. For this, $\hat{\underline{t}}$ must also propagate a fault-effect from the output of module n to one or more primary outputs. Then $\hat{\underline{v}}_n$ can be added to V_n of module n at no expense in terms of additional testpatterns.

When processing module n, the $\hat{\underline{v}}_n \in V_n$ applied in advance are considered before performing function identification by removing all rows from C_n that are already covered by the columns corresponding to $\hat{\underline{v}}_n \in V_n$. This most often reduces the size of C_n significantly, requiring less additional $\hat{\underline{v}}_n$ to completely cover C_n. Additionally, the effort to calculate these $\hat{\underline{v}}_n$ is strongly reduced, especially when processing modules with many inputs.

Since function identification is performed on a local basis, considering only one module at a time, and due to the global effect described above, the order in which the modules are processed influences $|T|$ as well as the effort to calculate T. One possible criterion to sort the modules is their number of inputs $|\underline{v}_m|$. The later a module m is processed, the more testvectors $\hat{\underline{v}}_m$ are applied on m before actually starting function identification of g_m. Thus, modules with a high number of inputs should be processed last.

After generation of testpatterns is completed, final test set compaction strategies, which are well known from traditional test generation, can be applied. For example, a testpattern simulation in reverse order of generation ([7]) often provides test set compaction up to 30%.

4 Implementation and Results

The previously described algorithm has been implemented in C. We have used [7] as the required gate-level ATPG system. Table 3 shows some results on several feasible netlists for Actel FPGAs. For each circuit the number of LM, the number of inputs (PI), the number of outputs (PO) as well as the average number of inputs per LM ($\overline{|v|}$) are given in Table 2.

circuit	LM	$\overline{|v|}$	PI	PO
1	51	3.02	8	4
2	166	2.83	22	29
3	295	3.19	135	107
4	440	3.32	38	3

Table 2. Circuit statistics

Table 3 shows the results of test set generation for several approaches. *Enumeration* refers to [1] where all possible testvectors for a given module are enumerated. *Stuckat* assumes an arbitrary two-level representation for each module and performs testpattern generation according to the single stuck-at fault model. For this approach 100% fault coverage (FC) according to the underlying model was achieved for each circuit.

circuit	enumeration			stuck-at				P_g^2			P_g		
1	70	9.4	-[1] -	40	4.9	91.96	99.41	60	7.4	- 99.92	63	8.7 - -
2	156	51.1	- -	88	24.7	93.49	99.80	119	39.0	- 99.93	139	45.1 - -
3	328	238.9	- -	140	91.6	96.64	99.81	221	152.6	- 99.99	229	161.0 - -
4	921	1169.0	- -	366	434.9	93.47	99.80	699	843.4	- 99.98	738	913.0 - -

Table 3. Fault coverages

[1]indicates 100 by definition

P_g^2 and P_g use the single personalization fault model (SPFM) according to section 2. For each approach the four numbers given for each circuit are: the size of the complete test set, the CPU-time used, FC with respect to P_g^2 and FC with respect to P_g.

The enumeration approach achieves 100% FC independently of the fault model. However, it requires significantly more CPU-time than the other approaches and it also generates the largest test sets. The stuck-at model, on the other hand, produces small test sets and uses the least amount of CPU-time. However, the overall FC according to the SPFM is quite low.

The results for P_g^2 show, that the restriction to at most two programming faults still yields FC exceeding 99.9% with respect to the full SPFM.

5 Conclusion

We have developed a new fault model that is suitable for most one-time programmable FPGAs. It is possible to achieve 100% fault coverage without the knowlegde of the actual implementation of the circuit. We have also developed a refinement of this fault model if some knowledge about the FPGA technology is available. An algorithm has been outlined to perform testpattern generation according to this model. The results show, that this approach is feasible for current FPGAs. It is also shown, that previous fault models either require a considerably larger test set even for modules with few inputs or are unable to achieve 100% fault coverage.

References

1. Thirumalai Sridhar and John P. Hayes: "A Functional Approach to Testing Bit-Sliced Microprocessors". - In: *IEEE Transactions on Computers, Vol. 30, No. 8.* (1981) pp. 563-572
2. Utpal J. Dave and Janak H. Patel: "A Functional-Level Test Generation Methodology using Two-Level Representations". - In: *26th ACM/IEEE Design Automation Conference DAC.* (1989) pp. 722-725
3. Chien-Hung Chao, F. Gail Gray: "Micro-Operation Perturbations in Chip Level Fault Modeling". - In: *25th ACM/IEEE Design Automation Conference DAC.* (1988) pp. 579-582
4. Brian T. Murray and John P. Hayes: "Hierarchical Test Generation Using Precomputed Tests for Modules". - In: *IEEE International Test Conference.* (1988) pp. 221-229
5. A. Zemva and F. Brglez and K. Kozminski: "Functionality Test and Don't Care Synthesis in FPGA ICs". - In: *MCNC, Research Triangle Park, NC, Technical Report TR93-04.* (1993)
6. Khalet A. El-Ayat, Abbas El Gamal, Richard Guo et al: "A CMOS Electrically Configurable Gate Array". - In: *IEEE Journal of Solid State Circuits, Vol. 24, No. 3.* (1989) pp. 752-761
7. M. H. Schulz, E. Trischler and T. M. Sarfert: "SOCRATES: A Highly Efficient ATPG System". - In: *IEEE Transactions on Computer-Aided Design Vol. 7, No. 1.* (1988) pp. 126-137

A Test Methodology Applied to Cellular Logic Programmable Gate Arrays

Ricardo de O. Duarte[1] and Mihaïl Nicolaidis

Reliable Integrated Systems Group IMAG/TIMA
46 Av. Felix Viallet, 38031 - GRENOBLE Cedex France
Phone : (+33) 76 57 46 19, Fax : (+33) 76 47 38 14
e-mail: duarte@verdon.imag.fr

Abstract. This paper describes an approach for testing a class of programmable logic devices called Cellular Programmable Gate Arrays. The flexibility in the selection of logic functions and the high number of inter-connections in this class of devices turns test a complex task. It has led to the proposition of an efficient test procedure based on some function properties. The regularity of the procedure permits that all logic cells in the device can be tested completly for functional faults at the same time, whenever is possible. It provides a reduced number of reprogramming times during test mode and a possibility of testing more devices in a defined period of time.

1 Introduction

Nowadays there are a lot of FPGAs architectures moving the market of PLDs, some of them are cellular logic based [1,2]. This means that the architecture is formed by a matrix of basic logic cells, able to be programmed to perform a combinatorial or sequencial logic function of few inputs and outputs and at the same time transmit data to neighboors cells. Such kind of devices are suited for implementation of data paths circuits.

The main concern of FPGAs companies is produce a flexible device sufficient to attend the costumer necessities. Most of these devices present a high degree of testing complexity, because of the large number of connections that they present and the large number of pattern configurations that they can be programmed. When the flexibility increases, the device testing complexity rises in the same direction. By the other side, the users want to program and test their circuit and want to be sure that no faults will come from the programmable device. The guarantee that the device is fault free requires high costs in the final product and high time dedicated to test, if a suited test methodology is not applied.

The reliability of programmable devices is an important subject that interest industries, university researchers and consumers. Although the problem has

---

[1] *Under grant supported by RHAE-CNPq, Brazil*

importance and difficulty to be treated, there are few works approaching test of programmable devices [5,6,7].

The proposal described in this work is to present an efficient test methodology generalized to all cellular logic based PGAs. The method consists in the minimization of the reprogramming times during off line test mode. It generates a set of testing configurations that when programmed, permit to test all logic cells in the matrix at the same time minimizing the final test duration by device. This approach proporcionate that a large number of devices can be tested in short period of time.

2 Problem Illustration

Cellular logic programmable gate arrays are basically composed by a flexible logic cell matrix, where each cell can be programmed with different configurations to perform small logic functions, they are also known as fine-grain architectures [3,10,12]. There are some companies that produce these devices. In order to turn clear and illustrate the problem description, we will fix our attention to one of these devices.

The device comercialized by Algotronix Co. [1], is formed by logic cells arranged in form of matrix, programmable I/O circuits to perform the connection between logic cells and pins, decoders used to do the device address programming. The figure 1 represents well the array architecture at this hierarquical level.

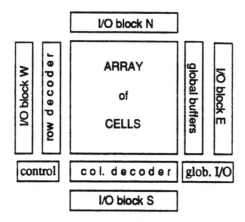

Fig. 1. Logic Programming Device Architecture

Each logic cell in this device, is formed by multiplexers and static RAMs. The multiplexers are mainly responsable by the data selection among the routing signals available to each logic cell during function execution. More specifically, they have three different tasks in the device: they are used to select which logic function the cell will perform (OR, AND, XOR, etc...), the inputs that will take part in the function execution ($N_{in}, S_{in}, E_{in}, W_{in}$, etc...) and the data that will be tranfered or not to a nearest neighboor cell ($N_{out}, S_{out}, E_{out}$ and W_{out}). The two first tasks presented is refered as intra-cell routing (fig. 2a) and the last one as inter-cell routing (fig. 2b) [4,11]. The static RAMs are addressed by decoders and are used to program the multiplexers [11].

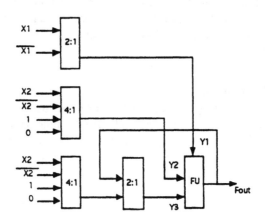

Fig. 2a. Intra-Cell Routing

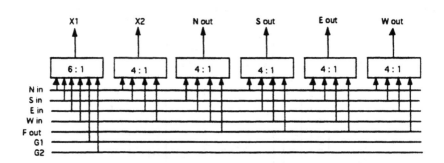

Fig. 2b. Inter-Cell Routing

Each cell can be programmed to implement one of the logic functions described in table 1, the function unit - FU responsable for this task is the Universal Logic Module (ULM) [11] - represented in figure 2a . Each one of these logic functions can be configured in many different ways depending on the combination of pattern inputs (X1 and X2). Such inputs come from the selection of inter and intra-routing multiplexers ($N, \overline{N}, S, \overline{S}, E, ...$).

Table 1. FU Programming Table

n°	Func	Y1	Y2	Y3
0	zero	x1	0	0
1	one	x1	1	1
2	x1	x1	1	0
3	$\overline{x1}$	x1	0	1
4	x2	x1	x2	x2
5	$\overline{x2}$	x1	$\overline{x2}$	$\overline{x2}$
6	x1 · x2	x1	x2	0
7	x1 · $\overline{x2}$	x1	$\overline{x2}$	0
8	$\overline{x1}$ · x2	x1	0	x2
9	$\overline{x1}$ · $\overline{x2}$	x1	0	$\overline{x2}$
10	x1+x2	x1	1	x2
11	x1+ $\overline{x2}$	x1	1	$\overline{x2}$
12	$\overline{x1}$+ x2	x1	x2	1
13	$\overline{x1}$+ $\overline{x2}$	x1	$\overline{x2}$	1
14	x1⊕x2	x1	$\overline{x2}$	x2
15	$\overline{x1⊕x2}$	x1	x2	$\overline{x2}$
16	D clk1	clk1	D	Fout
17	$\overline{D}$ clk1	clk1	$\overline{D}$	Fout
18	D clk2	clk2	D	Fout
19	$\overline{D}$ clk2	clk2	$\overline{D}$	Fout

At this point, we can express the large number of pattern configurations available by a basic unit in a cellular logic based structure in form of a summing. In a general manner the following expression can represent the set of pattern combinations (Z) in terms of available inputs and number of logic functions - f(i) that the function unit can perform.

$$Z = \sum_{i=1}^{nf} f(i) \text{ and } f(i) = C_{n_{npi(i)}}^{n_{nai(i)}}$$

where: $n_{npi(i)}$ is the number of inputs of function f(i);

$n_{nai(i)}$ is the number of available inputs to f(i).

nf is the total number of functions that the FU can perform.

For the case of CAL1024[2] - Algotronix Co. [1], we have this value :

$$Z = 784$$

Concerning the test of cellular logic structures, we have to assure that every logic cell in the array is fault-free, in other words, that every cell can be programmed with whatever function input configuration C, (C ∈ Z) accepted by the programming table, without presenting functional faults. Test the internal connections need programming to be applied, because the lack of testing points

---

[2] *CAL1024 is trademark of Algotronix Ltd.*

within the logic cell causing a low degree of controlability and observability in array level.

Find a group of test vectors that test the array completly could be easily made if the logic cells would be programmed as just one type of logic gate. We could use an ATPG program to do this efficiently. The reason that turns this procedure unacceptable for using in programmable devices is the large number of different logic gates that it can simulate when programmed.

The problem of finding an optimized set of configurations that allows to test the whole array in a minimum time is a NP-complete problem [9]. The solution of problems of this class is heuristic. There are no classical algorithm that can solve or give an answer to the problem in a reasonable CPU time without applying some restrictions that simplify the problem.

While architecture and synthesis in logic programming devices are subjects frequently in discussion, the programmable device testability are rarely rose. The research works come from people that take knowledge from the device architecture data sheets or come from people that work directly in companies that produce them. The approach presented in the next section, determine a regular methodology that can be applied in any cellular logic based architecture, without needing to spend time in finding programs or strategies to test completly the device in a minimum period of time.

Besides the test all logic cells in the array at the same time, the method proporcionate the functional fault coverage in the I/O circuits that perform the connections between the pins and the logic cells. Once the group of configurations is found, the total test time per device can vary proporcionally direct with the number of logic cell that the array content.

3 Description of the Method

In order to become clear the solution proposed, the test of cellular logic architectures will be divided in two hierarquical levels: the logic cell and the array level.

3.1   Logic Cell Level Considerations:

The test in a logic cell level could be performed, through the programming of all pattern configurations (one after the other) and checking the cell outputs (by one primary output) for the application of appropriate test input vectors, when the configuration would be programmed. Although this kind of procedure are very efficient in terms of guarantee of device reliability, this is not feasible in practice due to high consuming time taken to program and test completly each device. The solution proposed is, select determined configurations satisfying certain conditions that will introduce considered reductions in the final off-line test per chip.

The objective in the logic cell level is to find this group of configurations that test all physical connections within a cell for functional faults. The advantage of this logical level fault model, is that it is not dependent on the implementation details. The logic cell can be illustrated by a general model that contains two blocks. One block consists of the functional unit - FU, which implements the expected function, while the second block (SB for Selector Block), selects the physical inputs to be applied to the functional unit. In some cases [1,4], a reconvergent path that brings back the output of the functional unit to the SB input, is included in the model for the representation of the sequential functions.

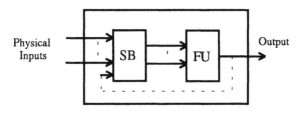

Fig. 3. Logic Cell Test Model

First, all the paths that link the elements contained in SB, have to be covered by the test in order to check any functional fault in this block. This means that any fault p/a, where p/a denotes a single fault on a path p with a logic level a, must be tested by a vector that produces the complement of a on p, and sensitizes p to a SB path. Then, we must ensure that the eventual errors are really propagated by the FU block, through the appropriate programming mode. The objective for reducing the test length, is to test the largest number of paths at the same time. In particular, it must be avoid the worst case occuring when a vector covers a single path. On the other hand, the FU block, has a reduced number of inputs and can be therefore, exhaustively exercised (i.e.application of the 2^n combinations of its n inputs). This will lead to the detection of all the targeted functional faults at the inputs or outputs of the FU block and also of all the faults which are located inside the logic cell. We also note that the interconnections between cells, which constitute in fact the extension of the SB most external paths, are automatically tested when all the paths of the SB are covered. Thus, the only requirement is to verify that all the SB paths are covered.

The representation of Algotronix logic cell structure as the logic cell test model adopted (SB plus FU) is shown in figure 4.

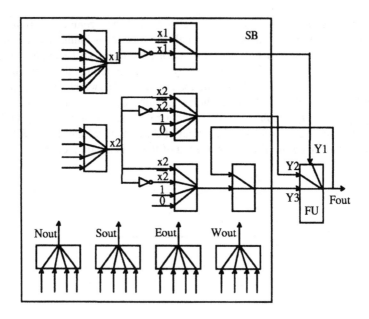

Fig. 4. Algotronix: Logic Cell Test Model

3.2   Array Level Considerations:

At the array level, the objective is find a way of applying the test to all cells at the same time minimizing the number of reprogramming times in the device test procedure.

This can be done by choosing, whenever is possible, cells configurations that permit to test the logic cell and propagate the error through neighboor logic cells until an observable output of the array. Some functions make possible the test of several cells in a row or a column at the same time. Other functions allow only the test a single cell in a row or a column by time, since the test of another would stop the propagation of the eventual errors. In this case, the application of a test vector depends on the number of cells in the row and column (i.e. the number of columns or rows in the array).

This situation introduces the requirement of classication functions that the logic cell can be programmed according with the propagation property:

COMBREP1: combinational functions of one input and one output, that has as characteristic the good propagation considering the array level. Ex.: functions n° 0 to 5 in table I - easy to be tested.

COMBREP2: combinational functions, XOR based. In the specific case, functions n° 14 and 15 of table I - easy to propagate and to be tested too.

NONCOMBREP: functions like ANDs, ORs, NANDs, NORs, etc... Difficult to be tested or propagate errors, once arranged in array form.

SEQUENTIAL: latches and flip-flops. Easy to be tested (good propagation). The cells are arranged as big shift-registers.

In figure 5a and 5b, we can see that every logic cell in the array is programmed with a COMBREP1 or COMBREP2 function respectively. In figure 5a, the application of the input combination {(1),(0)} to all the cells of a row (or column) will not stop the propagation of an eventual fault 1/0 in any cell, where the notation a/b means that under the presence of a fault the correct value a is transformed in a faulty value b. The same association can be done in 5b, with the possible set of inputs {(0,0),(0,1),(1,0),(1,1)} and the respectives output/fault {(0/1),(1/0),(1/0),(0/1)}. These functions present the property of good transmission of possible faults (in function of any inputs configuration) when arranged in arrays.

Configuration: X1 = West   Function: X1 $\oplus$ X2 => Configur.: East $\oplus$ South

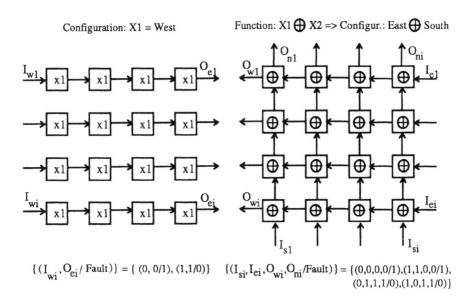

$\{(I_{wi}, O_{ei}/\text{Fault})\} = \{(0, 0/1), (1,1/0)\}$ $\{(I_{si}, I_{ei}, O_{wi}, O_{ni}/\text{Fault})\} = \{(0,0,0,0/1),(1,1,0,0/1),$
$(0,1,1,1/0),(1,0,1,1/0)\}$

Fig. 5a. Function Fout = X1   **Fig. 5b.** Function Fout = X1$\oplus$X2

On the contrary, in the logic array of figure 6, the application of the input combination {(0,0),(0,1),(1,0),(1,1)} to all cells of the row (or a column), will hind the propagation of fault 0/1. They are the NONCOMBREP functions. To test a cell with no observable outputs and controlable inputs using one of this functions, would be necessary to program logic cells to propagate inputs to the target(s) cell(s), program the target(s) cell(s) with the adequated configuration (NONCOMBREP function) and program the remaining cells to propagate the output to a primary output (output pin) of the array.

Target Configuration => North ^ West
Configuration for Transmission => X1 = West

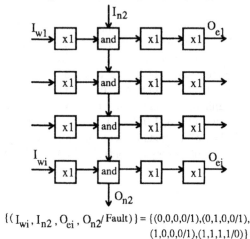

$$\{(I_{wi}, I_{n2}, O_{ei}, O_{n2}/\text{Fault})\} = \{(0,0,0,0/1),(0,1,0,0/1),$$
$$(1,0,0,0/1),(1,1,1,1/0)\}$$

Fig. 6. Function Fout = X1^X2

For the paths which are shared by the SEQUENTIAL and COMBREP functions, is preferable to perform the test by programming COMBREP functions first. Indeed, programming the cells to execute SEQUENTIAL functions results in the construction of shift registers, which require test length proportional to the number of rows or columns of the array like in the case of the NONCOMBREP functions. For that reason, the programmation of SEQUENTIAL functions during the test is done only if necessary, that is, if a path is only dedicated to a SEQUENTIAL function or if the COMBREP functions do not verify the conditions of propagating errors. By the way, there is a main advantage in testing a SEQUENTIAL function configuration (forming a big shift register) beyond testing specific paths, is the test of high working frequency operation in the device.

3.3   The Suggested Proposition:

For our luck, some considerations can be assumed turning the problem easier to be solved. The first consideration take into account is that some function configurations, are easier to be tested in array level (COMBREP1 and 2) than others due to the property of transmission faults when they occur [8]. The second consideration is that some configurations test a path or a set of paths that are not tested by the most remaining configurations. This lead us to conclude that they will certainly appear in the final solution of the problem. A way to distingt one configuration to another is to attribute ponderated weights, according to the paths set description that each one owns when executed. In the beginning an analyse of occurrence of each path in the logic cell model (SB + FU) is performed, after is calculated and attributed a weight w_j for each path j, considering the total number of configurations $C \in Z$. Then the general expression can be formulated as:

$$w_j = \frac{n_C}{n_T};$$ where n_C is the number of occurence in configurations $(C \in Z)$ of path j,

and n_T is the total number of configurations $C \in Z$

The biggest weight would be: $w_j = 1$ in the case of all configurations having this path j in its set of paths. And the smallest would be:

$$w_j = 1/n,$$ where n is the total number of configurations C in Z

meaning that just one configuration C has the occurence of this path j in its paths set description. After calculating the path weigths, every configuration weight w_C can be calculated:

$$w_C = \prod_{l=1}^{m} w_l;$$ where are all weights of paths covered by each configuration C;

m is the number of paths in the set path description of C

Once these simplifications are assumed, we reduce our problem of searching a minimum set of configurations that covers all paths in the logic cell, to a searching in specific sets of configurations (COMBREP1, COMBREP2, etc...). The problem is divided in small problems, where at the end, the solutions of each problem are took into account to find the optimized one. Therefore a greedy algorithm is well suited to be applied to produce the partial results of searching in each set of configuration. In the context, we can summarize in a few lines below, the procedure adopted.

Condition 1 to be satisfied: fill-up all paths in the logic cell (SB + FU).
Condition 2: find the minimum group of configurations.
Divide all configurations in specific groups (Combrep1, Combrep2, etc...)
Calculate weights to the paths to be tested - w_j.
Calculate weights to all configurations - w_C (weights calculed as a product of all paths)
Start selection by Combrep1 group {
　　if(satisfy condition 1) {
　　　　.Find all the set(s) of configurations that satisfy the condition 1
　　　　according with the following selection way (Greedy Algorithm):
　　　　.Choose the configuration $C \in$ Z that has the greatest number of
　　　　paths still not verified, in case of equality among many
　　　　candidates, choose the one(s) with smallest(s) weights w_C.
　　}
　　Select and store which set(s) has the smallest number of configurations to
　　satisfy Condition 2 to this group of functions.
}

Do the same with Combrep2 group {
> *if(satisfy condition 1) {*
>> *.Find all the set(s) of configurations that satisfy the condition 1 according with the following selection way (Greedy Algorithm): .Choose the configuration $C \in$ Z that has the greatest number of paths still not verified, in case of equality among many candidates, choose the one(s) with smallest(s) weights w_C.*
> *}*
>
> *Select and store which set(s) has the smallest number of configurations to satisfy Condition 2 to this group of functions.*

}
Compare and choose the best set taking into account all previous configurations sets coming from the last group analysis, if they exists - Condition 2.

Do the same with Combrep1 + Combrep2
Do the same with Combrep1 + Combrep2 + Sequencial
Compare and choose the best set taking into account all previous configurations sets coming from the last group analysis, if they exists - Condition 2.

if(still not satisfy condition 1)
> *- Choose between the noncomrep configurations, the one(s) that permit to satisfy the conditions and produce the results.*

4 Conclusion and Future Work

At the present time, is being implemented the procedure described. It was tried an implementation of a greedy algorithm without previous selection or dealing with data configurations patterns and it led in a high consumming CPU time. Report of results of this work applied to comercial devices will be published soon.

The method presented in this paper open a wide area of industrial interesting and turn the attention of people involved with programmable devices to test and the reliability.

As was explained, the method enable that cellular logic architectures can be tested in a regular manner, providing an optimized testing time to be applied in such devices. The final aim is to extend and try the method to other programmable devices.

References

1. Algotronix Ltd.: "CAL1024 - Preliminary Data Sheet". Algotronix Ltd. - Edinburgh, U.K., 1988.

2. Concurrent Logic Inc.: "CLI6000 Series Field Programmable Gate Arrays". Preliminary Information, Dec. 1991 - rev.:1.3.

3. S. Hauck, G. Boriello, S. Burns and C. Ebeling: "Montage: An FPGA for Synchronous and Asynchronous Circuits". 2nd. International Workshop on Field Programmable Logic and Applications, Vienna - Austria. Springer pp. 44 - 51, 1992.

4. J.P. Gray and T.A. Kean: "Configurable Hardware: A New Paradigm for Computation". Proceedings of Decennial Caltech Conference on VLSI, Pasadena, CA. March 1989.

5. C. Jordan and W.P. Marnane: "Incoming Inspection of FPGA's". in European Test Conference. pp. 371 - 376, 1993.

6. W.P. Marnane and W.R. Moore: "Testing Regular Arrays: The Boundary Problem". *European Test Conference*, pp. 304 - 311, 1989.

7. Michael Demjanenko and Shambhu J. Upadhyaya: "Dynamic Techniques for Yield Enhancement of Field Programmable Logic Arrays". IEEE International Test Conference. pp. 485 - 491, 1988.

8. Hideo Fujiwara: "Logic Testing and Design for Testability". Computer System series, the MIT Press, 1986.

9. Robert Sedgwick: "Algorithms". Addison-Wesley Publishing Company Inc. 1988.

10. Ricardo O. Duarte, Edil S. T. Fernades, A. C. Mesquita, A.L.V. Azevedo: "Configurable Cells: Towards Dynamics Architectures". Microprocessing and Microprogramming, The EUROMICRO Journal - North Holland Editor, vol. 38 N° 1 - 5 pp. 221 - 224, February 1993.

11. Kean, T.: "Configurable Logic: A Dynamically Programmable Cellular Architecture and its VLSI Implementation". Ph.D. Thesis, University of Edinburgh, Dept. of Computer Science, 1989.

12. J.Rose, A. El Gamal, Sangiovanni-Vicentelli: "Architecture of Field-Programmable Gate Arrays". Proceedings of IEEE,vol. 81, n° 7 - July 1993.

Integrated Layout Synthesis for FPGA's

Michal Z. SERVÍT and Zdeněk MUZIKÁŘ

Czech Technical University, Dept. of Computers
Karlovo nám. 13, CZ - 121 35 Praha 2, Czech Republic

Abstract. A new approach to the layout of FPGA's is presented. This approach integrates replacement and global routing into a compound task. An iterative algorithm solving the compound task is proposed. This algorithm takes into account restrictions imposed by the rigid carrier structure of FPGA's as well as the timing requirements dictated by the clocking scheme. A complex cost function is employed to control the iterative process in order to satisfy all restrictions and to optimize the circuit layout efficiency and performance.

1 Introduction

Field programmable gate arrays (FPGA's) provide a flexible and efficient way of synthesizing complex logic in a regular structure consisting of predefined programmable blocks and predefined programmable routing resources (path patterns). Due to the easy access of CAD tools that provide a fast turnaround time, the popularity of FPGA's is increasing. In comparison to the other implementation styles (e.g. full custom, standard cells), the FPGA layout design is simpler because of a rigid architecture. In spite of this the FPGA layout problem as a whole is intractable[1] [2], [6], [7], and is usually decomposed into subproblems. The typical sequence of subproblems is "initial placement - replacement - global routing - detailed routing" [1], [2], [6], [7].

The aim of the *initial placement* phase is to determine the non-overlapping (legal) locations of all blocks on the chip area.

The *replacement* is the optimization of an initial placement using traditional iterative techniques (displacement of a single block, pairwise interchanges, etc.) with respect to the cost function based on an estimation of the total wire length and/or local density of wires.

The *global routing* is a preliminary planning stage for the detailed routing in channels. The aim of this subtask is to determine a macropath for each net so that detailed routing can be accomplished efficiently. Global routers operate on interchannel connections and determine which segments of channels are traversed by a given net. The primary objective is to avoid channel overflow.

The *detailed routing* phase specifies the detailed routes (physical connections of pins) for all nets. Channel routers or maze routers are often used for this purpose [1], [2], [7].

---

[1] NP-hard.

Traditionally, all these subproblems are solved *independently* using rather poor optimizing criteria based mostly on an approximation of the total length of wires [2], [3], [7]. Moreover, the rigid structure of routing resources is often not taken into account in the placement and global routing phases. This is why the results achieved often do not meet timing requirements and/or are far from the global optimum in spite of the fact that many efficient techniques have been developed that provide near optimum solutions to all the above subproblems.

The rest of the paper is organized as follows: Section 2 outlines the main idea of our approach; in Section 3 there is a discussion of a routing model employed. Section 4 describes an integrated cost function. Sections 5 – 7 describe the basic steps of our approach, i.e. initial routing, replacement and rerouting. In Section 8 we draw conclusions.

2 Main idea of integrated approach

In this article, we propose a new layout technique based on the overlapping of the replacement and global routing subtasks. The main idea of our approach can be briefly stated as follows. First, the initial placement is constructed independently using traditional techniques and a traditional cost function. Then, an *iterative replacement combined with global routing simulation* is performed. A precise estimation of a net topology respecting a carrier structure is employed to provide a sound basis for the subsequent detailed routing.

We decided to integrate replacement and global routing into one compound task because of two reasons:

- The results of global routing provide a good approximation of the final layout.
- Computational time can be held within reasonable limits because models employed (global graphs) are relatively small [5].

The task of replacement is naturally suitable for the application of *iterative algorithms*. The quality of a layout during the iterative process is measured by a cost function which should take into account two (often contradictory) objectives:

- To satisfy restrictions imposed by the predefined routing resources.
- To optimize a layout in terms of performance, i.e. to maximize the applicable clock frequency.

Thus, we proposed an *integrated* cost function. In contrast to traditionally used cost functions, it originates from the estimation of the nets *topology*. The estimated topology of a net has to take into account the available routing resources (predefined paths on a carrier) which are expressed in terms of a *global routing graph*. According to our experience [5] a Steiner tree constructed in a global routing graph provides a good approximation of the final (detailed) topology of a net. Consequently, the delay of a signal can be estimated adequately. The proposed integrated cost function consists of two constituents. The first one is a *nonlinear* function measuring local congestion of wires with respect to the channel capacity constraints. The second constituent is a *nonlinear* function based on an estimate of wiring delay with respect to the delay constraints.

The idea of the integrated layout synthesis for FPGA's can be described as follows:

INTEGRATED LAYOUT SYNTHESIS

1. [*Preliminary timing analysis*]   Generate the bounds on the delays which the placement/global routing algorithms have to satisfy for all signals.
2. [*Initial placement*]   Determine the non-overlapping (legal) locations of all blocks on the chip area minimizing an estimation of the total routing length.
3. [*Initial routing*]   Build a global routing sketch for the initial placement and calculate its cost Q.
4. [*Replacement*]   Synthesize a legal placement permutation.
5. [*Rerouting*]   Modify the sketch and calculate a new cost Q'.
6. [*Selection*]   If $Q' < Q$, accept the permutation (fix the current placement and the routing sketch, set $Q := Q'$). Otherwise restore the previous situation.
7. [*Stopping rules*]   If all placement permutations have been explored (or limits on number of explored permutations and/or calculation time have been exceeded) then continue. Otherwise repeat from Step 4.
8. [*Global routing*]   Detail the current routing sketch up to the necessary level.
9. [*Detailed routing*]   Construct detailed routes through the available paths that comply with the global routes and that obey detailed routing constraints.

Obviously, more complicated rules controlling the iterative process (see Step 6) can be employed, e.g. *simulated annealing*.

Further, we will focus our attention on the core of our approach that is represented by Steps 3 – 5.

3   Global routing graph

Over the last years, several companies have introduced a number of different types of FPGA's [1]. To simplify the argument, we will further consider the so called "symmetrical array" architecture [1], [9]. The central part consists of an array of identical basic cells surrounded by vertical and horizontal routing channels. The I/O cells are located around the central part.

The aim of *global routing* is to determine how wires maneuver around cells. We make this determination by finding paths in an appropriate *global routing graph* $G = (V, E)$ which is a natural abstraction of the chip architecture. The vertices represent cells and channel intersections. Edges represent channel segments. Each edge has two labels: one is called *capacity*, the other *delay*.

The capacity $C : E \to \Re_+$ is an estimate of the actual number of wires that can be placed in the corresponding routing channel segment. Normally, the edge capacity is declared equal to the number of tracks. Notice that may depend on a complex block placement because internal block layout may utilize some tracks.

The delay $L : E \to \Re_+$ is an estimate of the actual delay of wires in the corresponding routing channel segment. Notice that we consider delays instead of traditionally used channel segment lengths. This is more appropriate because the length of a channel segment need not be proportional to its delay [9].

The task of global routing can be described more precisely when using the notion of the global graph. A set of *nets* $N = \{N_1, N_2, ..., N_k\}$ is given, where each net is a subset $V_i \subseteq V$ of vertices of G. A *global routing sketch*, or simply a *global routing*, is to be constructed. The global routing R is a collection of macropaths $R = \{S_1, S_2, ..., S_k\}$. A *macropath* for a net N_i is a connected subgraph S_i of G containing all vertices from V_i, i.e. the corresponding Steiner tree.

The amount of calculations required to construct a global routing sketch (and to modify it during the trial placement permutations) depends on the complexity of the global routing graph, i.e. on the level of abstraction. Anyway, the construction of a macropath for each net is expected to require much more computations[2] than the evaluation of a traditional placement cost function. Therefore, it is necessary to use a fairly simplified chip model to keep the amount of calculations in reasonable bounds.

According to our experience, it is possible to construct a "precise" global routing graph having a reasonable number of vertices and edges. The XILINX XC3000 family [9] can serve as an example. Fig. 1 shows a portion of XC3000 architecture and the corresponding portion of the global routing graph.

4 Integrated cost function

The primary objective of the placement and global routing in FPGAs is to guarantee *routability*. The secondary objective is to guarantee (and/or maximize) the required *performance* in terms of clock frequency.

Let us assume that a global routing sketch R has been constructed. Then, it is easily possible to compute a *demand* C_i^* on edge $e_i \in E$ as the number of macropaths $S_1, S_2, ..., S_k$ containing the edge e_i. The given routing sketch R is often considered as routable (admissible) if $C_i^* \leq C_i$ for any edge e_i [2], [7]. However, according to our observations [8], this condition need not be sufficient because C_i^* represents the global constraint but it may fail to capture local congestions inside the channel segment. Thus it is desirable to have some reserve of extra tracks to increase the probability of successful routing.

Let us assume that a preliminary timing analysis generates the bounds on the delays[3] which the placement/global routing algorithms have to satisfy for all *links*, i.e. source-target connections. Each x-terminal net is decomposed into

---

[2] The minimum Steiner tree problem is NP-hard. However, a couple of efficient heuristics is available [2], [7], [8].

[3] Further we will discuss the maximum delay constraints only because the majority of FPGA architectures support synchronized circuits and have dedicated routing resources intended to implement clocks and other signals that must have minimum skew among multiple targets.

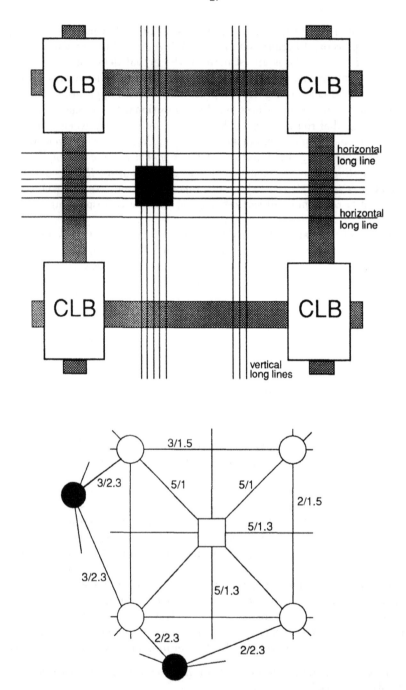

Fig. 1. A portion of XC3000 architecture and its graph model: circle vertices correspond to CLBs, square vertices correspond to switch matrices, full-circle vertices correspond to vertical and horizontal long lines, and edges are labeled with capacity/delay values.

$x - 1$ links and the maximum interconnection delay $D : \mathcal{L} \to \Re_+$ is calculated for each link $l_j \in \mathcal{L}$. Notice that this calculation can be ambiguous because only the maximum delay between a clocked source and a clocked target can be derived from the required synchronization frequency. Let us consider the situation described in Fig. 2 as an example. In this case only the sum $(D_1 + D_2)$ can be calculated. We propose to distribute the available delay uniformly (i.e. $D_1 = D_2$ in our case) provided that there are no hints supporting a different distribution.

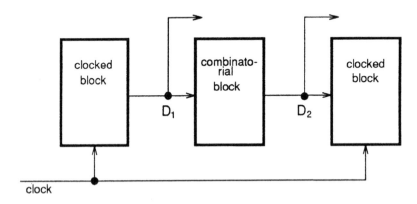

Fig. 2. Maximum delay estimation.

The actual delay D_j^* of a link $l_j \in \mathcal{L}$ can be estimated as the sum of delays of edges forming the corresponding path in R. The given routing sketch R is considered as admissible from the performance point of view if $D_j^* \leq D_j$ for any link l_j.

Let us return to our example (see **Fig. 2**) and assume that the value of D_1^* has been estimated as described above. This allows us to recalculate the value of D_2 as follows: $D_2 = (D_1 + D_2) - D_1^*$.

We propose to relax both the routability and delay constraints, i.e. to deal with the *unconstrained global routing problem* [2], [8] minimizing an appropriate cost function $Q(R)$. Obviously, the cost function $Q(R)$ has to minimize overloads $max[0, (C_i^* - C_i)]$ and overdelays $max[0, (D_j^* - D_j)]$. If there are no overloaded edges or overdelayed links, the performance expressed in terms of clock frequency should be maximized.

Taking into account the above observations, the cost function consisting of two nonlinear constituents seems to be appropriate:

$$Q(R) = \sum_{e_i \in E} F_1(e_i) + \alpha \sum_{l_j \in \mathcal{L}} F_2(l_j) \tag{1}$$

where α is a constant and nonlinear functions F_1, F_2 are defined as shown in Fig. 3.

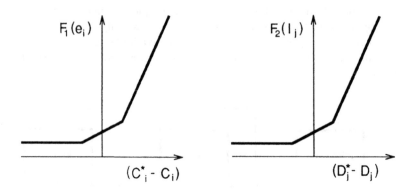

Fig. 3. Nonlinear functions F_1 and F_2.

5 Initial global routing

All versions of the global routing problem are NP–hard [2], [7]. Therefore, only approximate algorithms are of practical interest. A wide variety of such techniques has been developed. The *iterative approach* [5] is adequate to our purposes because it combines initial routing (to construct an initial variant) and rerouting (to optimize the initial variant). Obviously, it allows us to associate replacement and global routing in a natural way.

The proposed initial global routing algorithm consists of two phases denoted here IR1 and IR2.

The IR1 phase starts from the empty solution $R = \emptyset$ which has zero demands $D_n = 0$ on edge capacities C_n. For each net N_i, a macropath S_i is derived. An algorithm constructing suboptimal Steiner tree is employed for this purpose. The cost of a Steiner tree is defined as the sum of delays over its edges.

IR1 PHASE

INPUT: Graph $G = (V, E)$ labelled with edge capacities C_i and delays L_i.
           Set of nets $N = \{N_1, N_2, ..., N_k\}$.
           Set of links $\mathcal{L}$ with associated maximum delays D_j.
OUTPUT: Initial solution R and its cost $Q(R)$.
           Used routing resources (demands) C_i^* and actual delays D_j^*.

1. Initialize: $R := \emptyset$.
2. For each net N_i, $1 \le i \le k$:
   (a) Construct a Steiner tree $S_i = (V_i, E_i)$ interconnecting N_i in G and minimizing the estimated total delay $\sum_{e_j \in E_i} L(e_j)$.

(b) Include S_i into R.

3. Compute the used routing resources (demands) C_i^* for all edges and actual delays D_j^* for all links. Calculate the cost $Q(R)$ according to formula (1).

The IR2 phase aims at improving the initial solution. The attempts to optimize subgraphs S_i for each net N_i in a new environment are made, i.e. the costs of Steiner trees are calculated taking into account the used routing resources and the actual link delays.

IR2 PHASE

INPUT: Graph $G = (V, E)$ labelled with edge capacities C_i and delays L_i.
          Set of nets $N = \{N_1, N_2, ..., N_k\}$.
          Set of links $\mathcal{L}$ with associated maximum delays D_j.
          Initial solution R and its cost $Q(R)$.
          Used routing resources (demands) C_i^* and actual delays D_j^*.

OUTPUT: Improved solution R.
          Updated demands C_i^* and delays D_j^*.

1. For each net $N_i \in N$ calculate the cost of a current Steiner tree S_i according to the formula (1). Sort all nets in the ascending order of their costs[4].
2. [*Iteration*]   Consecutively for all nets do:
   (a) Remove S_i from the current solution and update demands.
   (b) Construct a new Steiner tree S_i' minimizing the cost function (1). Update demands.
3. Compute the cost $Q(R')$ of the new solution. If it holds $Q(R') < Q(R)$, set $R := R'$, $Q(R) := Q(R')$ and repeat the iteration. Otherwise stop.

6   Replacement

Let us assume that a legal (non-overlapping) placement has been constructed (and optimized) using traditional techniques. We are looking for a replacement which:

- transforms legal placement into another legal placement, and
- decreases the cost of the initial variant.

The most popular elementary transformations used in the process of replacement are[5]:

- displacement of a single block,
- pairwise interchange, and
- rotation and/or mirroring of a block.

---

[4] This order has been experimentally justified in [5]
[5] For designs incorporating blocks of different sizes and shapes, more complex transformations are used.

The evaluation of the integrated cost function $Q(R)$ requires significantly larger amount of computations in comparison to the evaluation of traditional length or cut-based functions. Let t_p be the time required to synthesize a replacement and t_c the time required to compute the change in cost. For traditional cost functions, t_c is usually proportional to $|N'|$, where N' is a subset of nets with pins that changed their locations, and $t_p \approx t_c$. In our case, $|N'|$ nets are to be rerouted. The complexity of rerouting a single net is proportional to $|E| \cdot log|V|$ [8], hence the evaluation of the new integrated cost takes $\mathcal{O}(|N'||E| \cdot log|V|)$ elementary operations and $t_p \ll t_c$.

To reduce the amount of computations, it is necessary to use an effective technique which generates transformations with a high probability of a success only. Unfortunately, the cost function (1) does not involve any particular features which could be used as a background in choice of successful transformations, thus we must rely on some domain knowledge to maintain a reasonable probability of a success. Two analogies can be used here: force-directed relaxation and cutline-based interchanges [8].

According to our experience [3], [8], single block displacements and pairwise interchanges are easily and successfully applicable to minimize the integrated cost function (1). Both transformations mentioned above can be treated in an uniform way: a candidate block is selected first and than several alternatives of repositioning are explored. The knowledge of overloaded edges and overdelayed links is useful when selecting a candidate. The knowledge of so called ε-neighborhood of the median of the selected block [8] may help to identify promising alternatives of repositioning.

7   Rerouting

Let us assume that a block B changed its location as a result of replacement. The following modifications concerning global routing (both the graph G and the solution R) must be introduced:

1. [*Correction of edge capacities C_i*]   The routing resources occupied by internal block layout (if any) are set free (the corresponding edge capacities are increased) from the old position and are utilized (edge capacities are decreased) in the new position.
2. [*Partial rerouting*]   Let $N^B \subset N$ denote the subset of nets interconnecting the block B. Obviously, corresponding macropaths (Steiner trees) must be reconstructed because the current global routing variant is no longer valid.

The partial rerouting can be accomplished by a modification of the IR2 algorithm:

PARTIAL REROUTING

INPUT: Graph $G = (V, E)$ labelled with edge capacities C_i and delays L_i.
Set of nets $N^B \subset N$.
Set of links $\mathcal{L}$ with associated maximum delays D_j.
Initial solution R and its cost $Q(R)$.
Used routing resources (demands) C_i^* and actual delays D_j^*.

OUTPUT: Updated solution R.
Updated demands C_i^* and delays D_j^*.

1. For each net $N_i \in N^B$ calculate the cost of its current Steiner tree S_i according to the formula (1). Sort all nets from N^B in the ascending order of their costs.

2. [*Iteration*]   Consecutively for all nets from N^B do:

   (a) Remove S_i from the current solution and update demands.

   (b) Construct a new Steiner tree S_i minimizing the cost function (1). Update demands.

3. Compute the cost $Q(R)$ of the new solution.

8 Conclusion

Decomposition is a principal way to cope with problem complexity, but it has some disadvantages. Generally, decomposition of a problem assumes a decomposition of an available information as well. In layout design, this can be described in terms of inadequacy of cost functions for steps preceding the detailed routing [6].

In our work, we propose a novel approach to cope with the negative impact of decomposition. The approach is based on partial overlapping of placement and global routing subtasks which are usually treated as independent optimization problems.

The proposed method is capable of providing a good delay estimation so that the timing specifications dictated by the clocking scheme can be checked and the iterative process of replacement and global routing can be timing-driven. This appears to be an important feature of this technique because the delay due to the interconnecting wire plays a major role in determining and optimizing the performance of the FPGA chip [1].

In comparison with traditional cost functions, the integrated cost function is more difficult to calculate. We use the most natural approach: an iterative replacement combined with rerouting in a global routing graph.

Limited experimental investigation of our method does not enable us to draw any statistically convincing conclusions. Nevertheless, the results obtained are promising [8] and cannot be explained in terms of undirected random search in a configuration space.

References

1. Brown, S.D., Francis, R.J., Rose, J., Vranesic, Z.G.: Field-Programmable Gate Arrays. Kluwer, Boston, 1992.
2. Lengauer, T.: Combinatorial Algorithms for Integrated Circuit Layout. Wiley-Teubner, Stuttgart-New York, 1990.
3. Muzikář, Z., Schmidt, J.: Experiments with Placement Algorithms on Gate Arrays. APK'92: Proc. of Design Automation Conference, Kaunas, 1992, pp. 86-91.
4. Sapatnekar, S.S., Kang, S.M.: Design Automation for Timing-Driven Layout Synthesis. Kluwer, Boston, 1992.
5. Servít, M.: Iterative Approach to Global Routing. J. Semicustom ICs, Vol.8, No.3, 1991, pp. 18-24.
6. Servít, M.: Algorithmic Problems of VLSI Layout. CTU Workshop'92, Praha 1992, pp. 91-92.
7. Sherwani, N.A.: Algorithms for VLSI Physical Design Automation. Kluwer, Boston, 1993.
8. Tomkevičius, A., Muzikář, Z., Servít, M.: Integrated Approach to Placement and Global Routing in Gate Arrays. Research Report DC-94-04, Czech Technical University, Dept. of Computers, Prague, 1994.
9. XILINX – The Programmable Gate Array Data Book. San Jose, 1992.

This research was supported by the Czech Technical University under grant no. 8095 and by the Czech Grant Agency under grant no. 102/93/0916.

Influence of Logic Block Layout Architecture on FPGA Performance

M. ROBERT, L .TORRES, F. MORAES and D. AUVERGNE

UNIVERSITE MONTPELLIER II, Sciences–LIRMM, UMR 9928 CNRS/UMII.
161 rue Ada (case 477)
34392 Montpellier cedex 5, FRANCE
e–mail : robert@lirmm.lirmm.fr

ABSTRACT : Among the several FPGA technologgies available today, the comparison of tiiming performances is always device dependent.In order to compare accurately the performances of the logic block architectures used in FPGA's families, we have implemented different functions.Using a layout synthesizer we evaluate post layout performances of these functions.A methodology to optimize the size of transistor gates in Look−up Tables is proposed

1− INTRODUCTION

FPGA architectures often make use of complex cells to efficiently implement circuitry with the help of logic synthesis tools. Among the several FPGA technologies available today *(Xilinx, Actel, Altera, etc...)*, the comparison of timing performances is always device *(XC3000, XC4000, ACT1,...)* and technology (CMOS process and programming technique) dependent. For a given FPGA device, the delay of a path depends on the logic block architecture, the number of logic levels crossed, the routing delays associated and the I/O cells. The resulting delay information is directly given by the manufacturers without any technological information on the process parameters used. Consequently, the effect of the logic block structures on the speed of FPGA is quite hard to be analyzed.

In order to focus on the effect of logic block architecture on FPGA timing performances, it is necessary to select a common CMOS process, and to implement a set of logic functions using the different logic blocks alternatives. A first complete study of logic blocks has been proposed in [1][2] : an experimental approach is taken, in which a set of benchmark logic circuits is synthesized into different FPGA's using a same CMOS process. It is shown that the fine grain blocks (Nand2,...) are slow, because they require many levels of logic and consequently require a large routing delay. Five and six look up tables and Actel muxes give the best performances because of the balance obtained between routing and logic delays.This study covers a large selection of combinational circuits, and show the effect of granularity on performance taking into account an approximation of the routing delays. However the limitations come from the evaluation performed at the transistor level without any consideration on the layout and the transistor sizes (always minimum). Moreover the routing delay is taken as a constant similar for each connection.

The effect of logic block layout on the speed of FPGA's is studied in this paper. Using a layout generator, we evaluate the post−layout performances of the basic cells taking into account the load capacitances and the transistor widths. To generate the different CMOS logic functions, we use an efficient layout methodology, where the cells are implemented as an array of rows with different heights [4]. Optimized cells are generated in a two pass procedure: first the automatic generator transform the electrical netlist in a symbolic description, then this description is translated into a layout after technology interpretation and compaction procedures.The main differences with the traditional cell based approach are : no explicit routing channel between rows, variable row heights, all cells are vertically transparent to the second metal layer used to vertically cross the cells, technology independence, and optimization of the transistor widths depending on the user constraints.

This paper is divided into 5 sections. Section 2 presents the logic block investigated, and the layout synthesis approach used. In section 3 we describe the experimental procedure used to evaluate the post layout performances of the different logic blocks, and the results obtained. In section 4, we discuss on the improvement of the look up table logic blocks using a methodology to size the transmission gates. Finally in section 5 we compare our results with the Xilinx devices.

2 - DESCRIPTION OF LOGIC BLOCKS

The FPGA architectures consists in an array of identical cells separated by wiring channels. The Actel architectures implement 4−to−1 muxes in a single level (mux A, *figure 1*) [3]. The logic block of Xilinx XC series uses a programmed look up table (LUT) to implement boolean functions : 4 input look up table for the XC2000 series LCA (*figure 2*), 5 input look up table for the XC3000 and XC 4000 series.

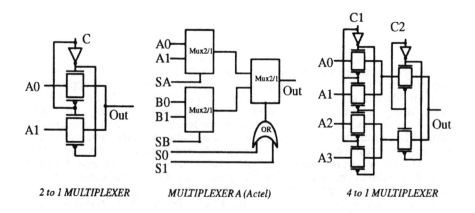

| 2 to 1 MULTIPLEXER | MULTIPLEXER A (Actel) | 4 to 1 MULTIPLEXER |

Figure 1 : Multiplexer basic modules

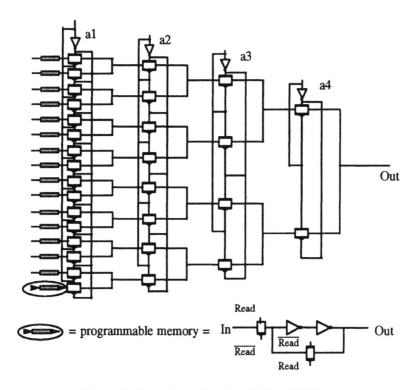

Figure 2 : Four–input Lookup Table (LTK4)

2.1 Layout methodology

To generate the different CMOS logic functions, we use an efficient layout methodology, where the cells are implemented as an array of rows with different heights *(Tropic* generator) [4] [5]. Optimized cells are generated in a two pass procedure: first the automatic generator transform the electrical netlist in a symbolic description , then this description is translated into a layout after technology interpretation and compaction procedures. Interconnection is realized with two metal layers for intra−cell and inter−row routing.

The proposed layout style [4] is characterized by the placement of cells in horizontal rows, where the even rows are horizontally mirrored *(figure 3a)*. The main differences with the traditional cell based approach are : no explicit routing channel between rows, variable row heights, all cells are vertically transparent to the second metal layer used to vertically cross the cells, technology independence, and optimization of the transistor widths depending on the user constraints.

Figure 3a illustrates the organization of the layout, as a direct abutment of cells. As shown in *figure 3b*, each cell is divided into 5 parts : two parts dedicated to the diffusion rows for the transistor implementation, and three parts devoted to the routing regions.

Figure 4 illustrates the automatic implementation of the logic block structures obtained with the *Tropic* generator from an electrical netlist (SPICE FORMAT).

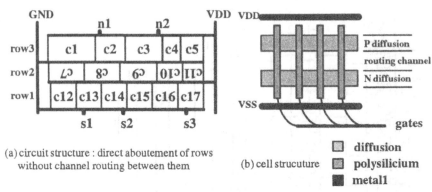

(a) circuit structure : direct aboutement of rows without channel routing between them

(b) cell strucuture

diffusion

polysilicium

metal1

Figure 3 : Tropic layout style

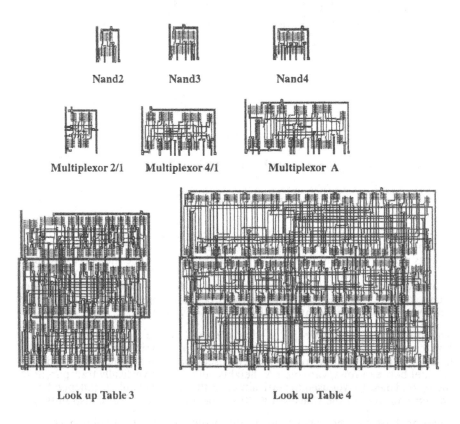

Nand2

Nand3

Nand4

Multiplexor 2/1

Multiplexor 4/1

Multiplexor A

Look up Table 3

Look up Table 4

Figure 4 : Layouts of logic blocks obtained with the tropic generator

3 – EXPERIMENTAL PROCEDURE

The experimental procedure is described in *figure 5*. Using the transistor level schematics (Spice netlist) of the logic blocks presented previously (*figures 1,2,4*) we have implemented two logic functions :

$$f1 = abc + ab\overline{d} + ac\overline{d}, \text{ proposed in } [1]$$
$$f2 = ab + fc\overline{d} + \overline{f}be + \overline{b}dc$$

The choice of simple functions avoid a shift in the results due to the quality of the logic synthesis tools. Our main objective in this first study is to analyzes the influence of the layout level. For more complex functions the results obtained at the electrical level in [1] can be used as a reference.

As described in *figure 5*, different types of transistor widths have been analyzed (W minimum, and W = 16 μm) and the output load has been fixed to Cin and 10 Cin, where Cin represent the input capacitance of an inverter). After layout generation, the extracted netlist contains all the parasitic capacitances (diffusion, routing,...).

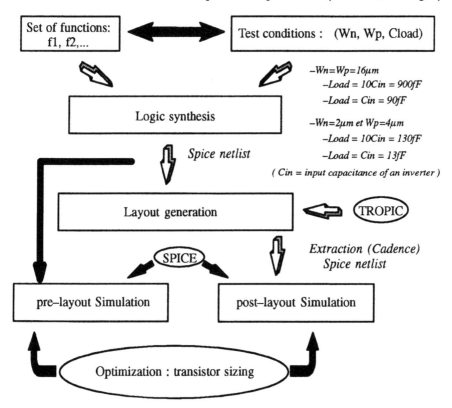

Figure 5 : Summary of the experimental procedure

3.1 Results

In order to don't overload this section, we present here some of the main results obtained with these experiments. As shown in the results given in *figures 6 and 7*, the Mux 2/1 architecture exhibit the lowest delay before and after layout. The effect of the layout capacitances reverse the ranking between Nand gates and Mux A, Mux 4/1 gates. The timing performances of the look up tables implementation is not as good as reported in [1]. One of the reason is that these modules are not optimally used in this test configuration. For more complex functions, the full capacity of the look up table will be used. However, at his level it is interesting to analyze the effect of the transistor sizing on the timing performances of the look up table modules.

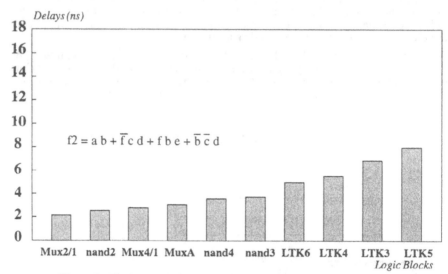

Figure 6 : pre–layout evaluation of the delays for the logic function f2.
(Wn=2μm and Wp=4μm and CL=130fF= 10 Cin)

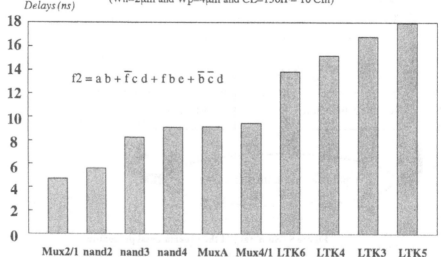

Figure 7 : Post–layout evaluation of the delays for the logic function f2.
(Wn=2μm and Wp=4μm and CL=130fF= 10 Cin)

4 – LOOK UP TABLES OPTIMIZATION

Transistor sizing at layout level is necessary to improve the overall performance of integrated circuits. Based on a local optimization defined through an explicit formulation of delays [6], a sizing methodology has been applied to the look up table cells. A Typical critical path is represented in *figure 8* where we separate a control block (Structure I) and a data block (Structure II). The optimization strategy begins by the sizing of the structure II from the output (where the load capacitance C_L is fixed) to the data inputs. The local sizing rules used for this "ANDORI" structure are given as follow [6] :

$$\begin{cases} XN = \dfrac{\sqrt{Y}\ (1 + 12K'(n_2-1))}{[1 + \frac{\mu n}{\mu p} + 12K'[(n_2-1) + \frac{\mu n}{\mu p}(n_1-1)]]^{\frac{1}{2}}} \\[4mm] XP = \dfrac{\mu n}{\mu p}\dfrac{\sqrt{Y}\ (1 + 12K'(n_1-1))}{[1 + \frac{\mu n}{\mu p} + 12K'[(n_2-1) + \frac{\mu n}{\mu p}(n_1-1)]]^{\frac{1}{2}}} \end{cases}$$

Where :

* $Y = \dfrac{C_L}{C_{ref}}$ *represents the output load, with respect to a reference capacitance*

$$C_{ref} = C_{ox}L_{min}W_{min}$$

* $X_N = \dfrac{W_N}{W_{ref}}$, $X_P = \dfrac{W_P}{W_{ref}}$, *represents the transistor size, with respect to a refernece width (Wmin)*

* n_1 and n_2 *respectively represents the numbers of parallel transistors in P and N array,*

* $K' = \dfrac{1}{24K}\dfrac{\frac{Vcc}{2} - V_T}{\frac{Vcc}{2} - V_T^*}$ *with* $K = \dfrac{8Vcc(Vcc - V_T)}{7Vcc^2 + 4V_T^2 - 12VccV_T}$

where K is a slowly varying technological coefficient.

For the 1.5µm CMOS process under consideration the value of these coefficients are : C_{ref}=4.4 fF, K= 1.274 and K' = 0.049 .

Depending on the logic level to be transmitted the structure of the look up table has been replaced by an equivalent "ANDORI" configuration allowing full application of the sizing procedure, as follows : low or high levels to be passed transform the structure in equivalent N or P serial arrays (*figure 9*), directly sized as for general "ANDORI" configuration [6]. Then the P or N companion transistor is sized in the ratio of mobility of its partner (μ_n/μ_p = 2.4).

For example, as shown in *figure 8*, with a given load of 130 fF the transistor width obtained for the transmission gates TG5, TG6, TG7, TG8, of the structure II are : W_N = 5 µm and W_P = 13µm.

After optimization of the structure II the input load of structure I is known (this is the input of *inv3* – node "6" – in *figure 8*). The same backward process is applied to the array of structure I from the output (here node "6") to the input (node "IN"). Figures of the resulting transistor sizes are given in the corresponding buble of *figure 8*.

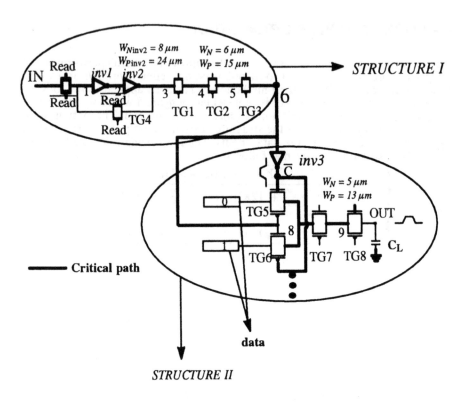

Figure 8 : Critical path for the implementation of the function 1 in a Look up Table 3

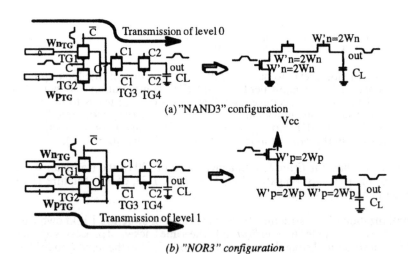

(a) "NAND3" configuration

(b) "NOR3" configuration

Figure 9 : Optimization procedure

As shown in *Table 1,* where we compared Look up Table implementations of the function f1 and f2 with different transistor sizes alternative, it appears clearly that sized solutions result in the fastest implementations of the look up table cells (up to 50% reduction of the delay). The comparison with the other logic block structures (*figures 10 and 11*) show that after optimization of the transistor widths the look up tables exhibit very good speed performances.

		function1								function2	
		LTK3				LTK4				LTK6	
		C_L=130fF		C_L=900fF		C_L=130fF		C_L=900fF		C_L=130fF	
		t_{HL}	t_{LH}	t_{HL}	t_{LH}	t_{HL}	t_{LH}	t_{HL}	t_{LH}	t_{HL}	t_{LH}
Fixed width	Wn=16μm Wp=16μm			5.40	2.20			3.96	2.25		
	Wn=2 μm Wp=4 μm	5.33	3.6			3.53	2.87			5.06	4.34
Optimization of transistor width		3.27	3.07	3.51	3.44	1.77	1.64	2.78	2.61	2.24	2.21
Benefits		39%		35%		50%		30%		50%	

Table 1 : Optimization results for the functions f1 and f2 implemented with differents look up tables (delays in ns) and transistor sizing alternative (minimum and constant sizes, optimized sizes for a given load).

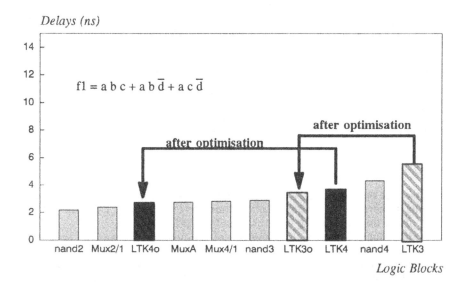

Figure 10 : delays of the function f1 (with Wn=2μm and Wp=4μm and CL=130fF) and performance improvements of the look up table 3/4 implementation after transistor sizing

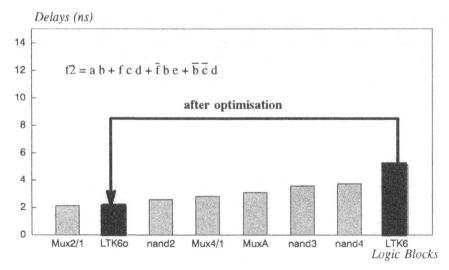

Figure 11 : delays of the function f2 (with Wn=2μm and Wp=4μm and CL=130fF) and performance improvements of the look up table 6 implementation after transistor sizing

5 – FPGA IMPLEMENTATION

We present in this section the implementation of f1 and f2 functions with Xilinx devices : XC3020 and XC4002.The results obtained with XACT tools are summarized in table 2.

	XC 4002					XC 3020				
	PAD to PAD	IOB to IOB	Blk–LUT	T_{rout}	Nbr CLB	PAD to PAD	IOB to IOB	Blk–LUT	T_{rout}	Nbr CLB
Function 1 (ns)	17.2	7.2	4.5	2.7	1	14.7	7.3	4.6	2.7	1
Function 2 (ns)	2.03	10.3	7	3.3	1	17.8	10.5	2*4.6	1.3	2

"PAD to PAD" : delay between input and output PADs,"IOB to IOB" : delay between input and output IOB,"Blk–LUT" : delay through the CLB,"Trout" : routing delay,"Nbr CLB":Number of CLB used to implement f1 and f2.

Table 2 : Delays obtained with xilinx devices.

To compare the performances, we consider the delays between IOB to IOB, which represents the sum of the routing delay and the delay through the CLBs (the "PAD to PAD" delay depends on the buffer size options).

The results obtained for the implementation of functions f1 and f2 with the XC4002 and XC3020 families are the same. As we can see in Table2, the function f2 needs two CLBs with the XC3000 and only one with the XC4000. In othe side, the routing delays are differents.

If we compare the results given by the layout synthsizer (in Table 1 for Wn=Wp=16 μm and C_L=900fF, we have a critical delay of 3.96ns) and xilinx implementation, we observe approximatively 10% difference between then. This prove that our layout approach allows to characterize the performances of the different logic block architectures.

6 – CONCLUSION

We explored in this work the comparative performances of the different logic blocks taking into account the layout evaluation. The main conclusion is the confirmation of the good performances of the look up table and multiplexer modules. A significant reduction in delay is obtained by optimizing the transistor widths of the look up tables. The use of a layout generator is of great help to evaluate architectural alternatives as well as the fast migration of circuits in different processes. The high regularity of the layout style used, gives the possibility to parametrize the layout capacitances, allowing an accurate prediction of performances for logic synthesis tools.

Selection of FPGA structures as a technology management alternative for performance driven design is a widely opened problem. Using automatic layout generator combined to transistor sizing and performance evaluation we were able to compare different logic blocks unit used to implement FPGAs. Examples are obtained in different logic paths and compared in terms of speed and area. As a surprising result evidence is given of LUT based high performance implementation, through the definition of optimal sizing for logic units.

REFERENCES

[1] S. Singh, J. Rose, P. Chow, D. Lewis "The effect of logic Block architecture on FPGA performance"IEEE Journal of Solid State Circuits, Vol. 27 , N°3, March 1992.

[2] S. Singh, "The effect of logic Block architecture on the speed of field programmable gate array",M.A. Sc. thesis, Dept. Elect. Eng., Univ. of Toronto, Ont. Canada, Aug. 1991

[3] A. E. Gamal "An architecture for electrically configurable gate arrays" IEEE Journal of Solid State Circuits, Vol. 24 , N°2, April 1989.

[4] F. Moraes, N.Azemard, M.Robert, D. Auvergne "Flexible macrocell layout generator",Proc. of 4th ACM/SIGDA physical Design Workshop, Los Angeles, 1993, p. 105–116

[5] F. Moraes, N.Azemard, M.Robert, D. Auvergne "Tool box for performance driven macrocell layout ggenerator",Fourth Eurochip Workshop on VLSI design training,Toledo,September 1993.

[6] D.Auvergne, N.Azemard, V. Bonzom, D. Deschacht, M.Robert "Formal sizing rules of CMOS circuits",The European Design Automation Conference, Amsterdam, February 1991.

A Global Routing Heuristic for FPGAs Based on Mean Field Annealing

Ismail Haritaoğlu and Cevdet Aykanat

Dept. of Computer. Eng & Information. Sci. Bilkent University 06533 Bilkent,
Ankara, TURKEY
hismail@bilkent.edu.tr

Abstract. In this paper, we propose an order-independent global routing algorithm for SRAM type FPGAs based on Mean Field Annealing. The performance of the proposed global routing algorithm is evaluated in comparison with LocusRoute global router on *ACM/SIGDA Design Automation* benchmarks. Experimental results indicate that the proposed MFA heuristic performs better than the LocusRoute in terms of the distribution of the channel densities.

1 Introduction

This paper investigates the routing problem in Static RAM (SRAM) based Field Programmable Gate Arrays (FPGAs) [7]. As the routing in FPGAs is a very complex combinatorial optimization problem, routing process can be carried out in two phases: *global routing* followed by *detailed routing* [5]. Global routing determines the course of wires through sequences of channel segments. Detailed routing determines the wire segment allocation for the channel segment routes found in the first phase which enables feasible switch box interconnection configurations [5, 9, 10].

Global routing in FPGA can be done by using global routing algorithms proposed for standard cells [5]. *LocusRoute* global router is one of this type of router used for global routing in FPGAs [4] which divides the multi-pin nets into two-pin nets and considers only two or less bend, minimum distance routes for these two-pin nets. The objective in LocusRoute is to distribute the connections among channels so that channel densities are balanced. In this work, we propose a new approach for the solution of global routing problem in FPGAs by using *Mean Field Annealing* (MFA) technique.

MFA merges collective computation and annealing properties of Hopfield neural networks [2] and simulated annealing [3], respectively, to obtain a general algorithm for solving combinatorial optimization problems [1]. MFA can be used for solving a combinatorial optimization problem by choosing a representation scheme in which the final states of the spins can be decoded as a solution to the target problem. Then, an energy function is constructed whose global minimum value corresponds to the *best solution* of the target problem. MFA is expected to compute the best solution to the target problem, starting from a randomly chosen initial state, by minimizing this energy function. Steps of applying MFA technique to a problem can be summarized as follows.

1) Choose a representation scheme which encodes the configuration space of the target optimization problem using spins. In order to get a good performance, number of possible configurations in the problem domain and the spin domain must be equal, i.e., there must be a one-to-one mapping between the configurations of spins and the problem.

2) Formulate the cost function of the problem in terms of spins, i.e., derive the energy function of the system. Global minimum of the energy function should correspond to the global minimum of the cost function.

3) Derive the mean field theory equations using this energy function, i.e., derive equations for updating averages (expected values) of spins.

4) Select the energy function and the cooling schedule parameters.

The FPGA model used in this paper are given in Section 2. The proposed formulation of the MFA algorithm for the global routing problem following these steps is presented in Section 3. The performance of the proposed MFA algorithm is evaluated in comparison with LocusRoute algorithm. Section 4 summarizes the implementation details of these two-algorithms. Finally, experimental results are presented in Section 5.

2   Global Routing Problem in FPGAs

The form of commercial FPGA consists of a two dimensional regular array of programmable logic blocks (LB's), a programmable routing network and switch boxes (SB's) [6, 13, 14]. Logic blocks are used to provide the functionality of a circuit. Routing network makes connections between LB's and input/output pads. Routing network of FPGA consists of wiring segments and connection blocks. Wiring segments have three type of routing resources in the commercial SRAM based FPGA [13]: channel segments, long lines and direct-interconnections. A horizontal (vertical) channel segment consists of a number of parallel wire segments connecting two successive SB's in a horizontal (vertical) channel. The SB's allow programmed interconnection between these channel segments. Direct-interconnection provides the connections between neighbor LB's. Long lines cross the routing area of FPGA vertically and horizontally. Connection blocks provide the connectivity from the input/output pins of LB's to the wiring segments of the respective channel segments. Each pin can be connected to a limited number of wiring segments in a channel and this is called as flexibility of connection block [7]. In this paper, it is assumed that each LB pin can be connected to all wiring segments in the respective channels. Therefore, we can omit the connection block in our FPGA model.

Since the *direct-interconnections* are used by neigbor LB's to provide minimum propagation delay and the *long lines* are used by signals which must travel long distances (i.e., global clock), these interconnection resources are not considered in the global routing. Hence, our FPGA model for global routing considers

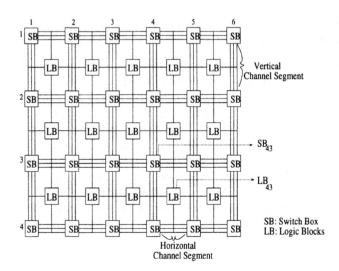

Fig. 1. The FPGA model used for Global Routing

only the LB's, SB's and channel segments. An FPGA can be modeled as a two dimensional array of LB's which are connected to the vertical and horizontal channel segments, and SB's which make connections between the horizontal and vertical channel segments (Fig. 1).

In this work, we divide all multi-pin nets into two-pin nets using minimum spanning tree algorithm [12] as in LocusRoute. Hence, a net refers to a two-pin net here, and hereafter. Consider the possible routings for a two-pin net with a Manhattan distance of $d_h + d_v$ where d_h and d_v denote the horizontal and vertical distances, respectively, between the two pins of the net on the LB grid. The routing area of this net is restricted to a $(d_h+1) \times (d_v+1)$ LB grid as shown in Fig. 2.a. Then, the shortest distance routing of this net can be decomposed into three *independent* routings as follows. Each pin of this net has only one neighbor SB in the optimal routing area. Hence, each pin can be connected to its unique neighbor SB either through a horizontal or a vertical channel segment (Fig. 2). Meanwhile, the optimal routing area for the connection of these two unique SB's is restricted to a $d_h \times d_v$ SB grid embedded in the LB grid (Fig. 2). Hence, by exploiting this fact, we further subdivide each net into three two-pin subnets referred here as *LS, SS* and *SL* subnets (Fig. 2.b). Here, *LS* and *SL* subnets represent the LB-to-SB and SB-to-LB connections, respectively, and *SS* subnets represent the SB-to-SB connection for a particular net. Therefore, we consider only two possible routings for both *LS* and *SL* subnets and $d_h + d_v - 2$ possible one or two bend routings for *SS* subnets for routing the original net.

We define an FPGA graph $\mathbf{F}(L, S, C)$ for modeling the global routing problem in FPGAs. This graph is a $P \times Q$ two-dimensional mesh where L, S and C denote the set of LB's, SB's and channel segments, respectively. Here, P and Q

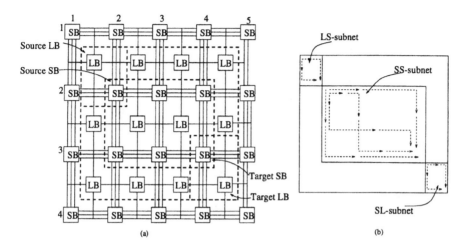

Fig. 2. (a) The routing area of the two-pin net and its subnets, (b) The possible routes for each subnets

is the number of horizontal and vertical channels in the FPGA. Each grid point (vertex) s_{pq} of the mesh represents the SB at horizontal channel p and vertical channel q. Each cell L_{pq} of the mesh represents the LB which is adjacent to four SB's s_{pq}, $s_{p,q+1}$, $s_{p+1,q+1}$ and $s_{p+1,q}$. Edges are labeled such that the horizontal (vertical) edge c_{pq}^h (c_{pq}^v) corresponds to the channel segment between the two consecutive SB's s_{pq} and $s_{p,q+1}$ ($s_{p+1,q}$) on the horizontal (vertical) channel p (q), respectively. Figure 3 displays a 8×6 sample FPGA graph. Then, the pins of the LS/SL and SS type subnets are assigned to the respective cell-vertex and vertex-vertex pairs of the graph as is in mentioned earlier.

The global routing problem reduces to searching for most uniform possible distribution of the routes for these subnets. The uniform distribution of the routes is expected to increase the likelihood of finding a feasible routing in the following detailed routing phase. Hence, we need to define an objective function which rewards *balanced* routings. We associate weights with the edges of FPGA graph in order to simplify the computation of the balance quality of a given routing. The weight w_{pq}^h (w_{pq}^v) of a horizontal (vertical) edge c_{pq}^h (c_{pq}^v) denotes the density of the respective channel segment. Here, the density of a channel segment denotes the total number of nets passing through that segment for a given routing. Using this model, we can express the balance quality B of a given routing $\mathbf{R}$ as

$$B(\mathbf{R}) = \sum_{p=1}^{P} \sum_{q=1}^{Q-1} (w_{pq}^h(\mathbf{R}))^2 + \sum_{q=1}^{Q} \sum_{p=1}^{P-1} (w_{pq}^v(\mathbf{R}))^2 \tag{1}$$

As is seen in Eq. (1), each channel segment contributes the square of its density to the objective function thus penalizing imbalanced routing distributions. Hence,

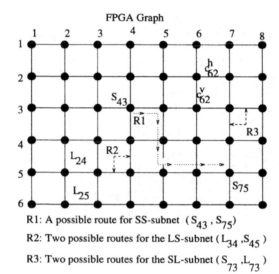

R1: A possible route for SS-subnet (S_{43}, S_{75})

R2: Two possible routes for the LS-subnet (L_{34}, S_{45})

R3: Two possible routes for the SL-subnet (S_{73}, L_{73})

Fig. 3. The Cost Graph for FPGA model

the global routing problem reduces to the minimization of the objective function given in Eq. (1).

3   MFA Formulation

The MFA algorithm is derived by analogy to *Ising* and *Potts* models which are used to estimate the state of a system of particles, called spins, in thermal equilibrium. In Ising model, spins can be in one of the two states represented by 0 and 1, whereas in Potts model they can be in one of the K states. All LS/SL subnets are represented by Ising spins since they have only two possible routes. In Ising spin encoding of each LS/SL subnet m, $u_m = 1$ (0) denotes that the LB-to-SB or SB-to-LB routing is achieved through a single horizontal (vertical) channel segment. Each SS subnet n having $K_n \geq 2$ possible routes is represented by a K_n-state Potts spin. The states of a K_n-state Potts spin is represented using a K_n dimensional vector

$$\mathbf{v}_n = [v_{n1}, \ldots, v_{nr}, \ldots, v_{n,K_n}]^t \tag{2}$$

where "t" denotes the vector transpose operation. Each Potts spin $\mathbf{v}_n$ is allowed to be equal to one of the principal unit vectors $\mathbf{e}_1, \ldots, \mathbf{e}_r, \ldots, \mathbf{e}_{K_n}$, and can not take any other value. Principal unit vector $\mathbf{e}_r$ is defined to be a vector which has all its components equal to 0 except its r'th component which is equal to 1. Potts spin $\mathbf{v}_n$ is said to be in state r if $\mathbf{v}_n = \mathbf{e}_r$. Hence, a K_n-state Potts spin $\mathbf{v}_n$

is composed of K_n two state variables $v_{n1}, \ldots, v_{nr}, \ldots, v_{nK_n}$, where $v_{nr} \in \{0, 1\}$, with the following constraint

$$\sum_{r=1}^{K_n} v_{nr} = 1 \tag{3}$$

If Potts spin n is in state r (i.e., $v_{nr} = 1$ for $1 \leq r \leq K_n$) we say that the corresponding net n is routed by using the route r.

In the MFA algorithm, the aim is to find the spin values minimizing the energy function of the system. In order to achieve this goal, the average (expected) values $\langle u_m \rangle$ and $\langle \mathbf{v}_n \rangle = [\langle v_{n1} \rangle, \ldots, \langle v_{nr} \rangle, \ldots, \langle v_{nK_n} \rangle]^t$ of all Ising and Potts spins, respectively, are computed and iteratively updated until the system stabilizes at some fixed point. Note that for each Ising spin m, $u_m \in \{0, 1\}$, i.e., u_m can take only two values 0 and 1, whereas $\langle u_m \rangle \in [0, 1]$, i.e., $\langle u_m \rangle$ can take any real value between 0 and 1. Similarly, for each Potts spin n, $v_{nr} \in \{0, 1\}$ whereas $\langle v_{nr} \rangle \in [0, 1]$. When the system is stabilized, $\langle u_m \rangle$ and $\langle v_{nr} \rangle$ values are expected to converge to either 0 or 1 with the constraints $\sum_{r=1}^{K_n} \langle v_{nr} \rangle = 1$ for the Potts spins.

In order to construct an energy function it is helpful to associate the following meaning to the values $\langle u_m \rangle$ for LS/SL subnets.

$\langle u_m \rangle = \mathcal{P}(\text{subnet } m \text{ is routed by using the horizontal channel segment})$

$1 - \langle u_m \rangle = \mathcal{P}(\text{subnet } m \text{ is routed by using the vertical channel segment})$

That is, $\langle u_m \rangle$ and $1-\langle u_m \rangle$ denote the probabilities of finding Ising spin m at states 1 and 0, respectively. In other words, $\langle u_m \rangle$ and $1-\langle u_m \rangle$ denote the probabilities of routing subnet m through a single horizontal and vertical channel segment, respectively. Similarly, for SS subnets represented with Potts spins

$$\langle v_{nr} \rangle = \mathcal{P}(\text{subnet } n \text{ is routed through route r}) \quad \text{for} \quad 1 \leq r \leq K_n \tag{4}$$

That is, $\langle v_{nr} \rangle$ denotes the probability of finding Potts spin at state r for $1 \leq r \leq K_n$. In other words, $\langle v_{nr} \rangle$ denotes the probability of routing net n through route r. Here and hereafter, u_m and v_{nr} will be used to denote the respective expected values ($\langle u_m \rangle$ and $\langle v_{nr} \rangle$, respectively) for the sake of simplicity. Now, we formulate the total density cost of global routing problem as an energy term

$$E_B(\mathbf{U}, \mathbf{V}) = \sum_{p=1}^{P} \sum_{q=1}^{Q-1} [w_{pq}^h(\mathbf{U}) + w_{pq}^h(\mathbf{V})]^2 + \sum_{q=1}^{Q} \sum_{p=1}^{P-1} [w_{pq}^v(\mathbf{U}) + w_{pq}^v(\mathbf{V})]^2 \tag{5}$$

where $w_{pq}^h(\mathbf{U}) = \sum_{m \ni c_{pq}^h} u_m$ and $w_{pq}^h(\mathbf{V}) = \sum_{n \ni c_{pq}^h} \sum_{r \in R_n, r \ni c_{pq}^h} v_{nr}$

$w_{pq}^v(\mathbf{U}) = \sum_{m \ni c_{pq}^v} (1 - u_m)$ and $w_{pq}^v(\mathbf{V}) = \sum_{n \ni c_{pq}^v} \sum_{r \in R_n, r \ni c_{pq}^v} v_{nr}$

where $\mathbf{U} = \{u_1, u_2, \ldots\}$ and $\mathbf{V} = \{v_1, v_2, \ldots\}$ represent the sets of Ising and Potts spins corresponding to the LS/SL and SS subnets, respectively. For

LS/SL subnets, "$m \ni c_{pq}$" denotes "for each LS/SL subnet m whose pair of pins share the horizontal or vertical channel segment c_{pq}". For SS subnets "$n \ni c_{pq}$" denotes "for each SS subnet n whose routing area contains the horizontal and vertical channel c_{pq}". Furthermore, "$r \in R_n, r \ni c_{pq}$" denotes "for each possible route r of SS subnet n which passes through the horizontal or vertical channel segment c_{pq}". Here, $w_{pq}(\mathbf{U})$ and $w_{pq}(\mathbf{V})$ represent the *probabilistic densities* of the horizontal or vertical channel segment c_{pq} for the current routing states of LS/SL and SS subnets, respectively. Hence, $w_{pq}(\mathbf{U}, \mathbf{V}) = w_{pq}(\mathbf{U}) + w_{pq}(\mathbf{V})$ represents the total probabilistic density of horizontal or vertical channel segment c_{pq} for the overall current routing state.

Mean field theory equations, needed to minimize the energy function E_B, can be derived as

$$\phi_m(\mathbf{U}, \mathbf{V}) = E_B(\mathbf{U}, \mathbf{V})|_{u_m=0} - E_B(\mathbf{U}, \mathbf{V})|_{u_m=1}$$
$$= -2\big[w_{pq}^h(\mathbf{U}, \mathbf{V}) - w_{pq}^v(\mathbf{U}, \mathbf{V}) - 2(u_m - 0.5)\big] \qquad (6)$$
$$\text{where} \quad c_{pq}^h, c_{pq}^v \in m$$

for an Ising spin m and

$$\phi_{nr}(\mathbf{U}, \mathbf{V}) = E_B(\mathbf{U}, \mathbf{V})|_{\mathbf{v}_n=0} - E_B(\mathbf{U}, \mathbf{V})|_{\mathbf{v}_n=e_r} \qquad (7)$$
$$= -2\big[\sum_{c_{pq}^h \in r} (w_{pq}^h(\mathbf{U}, \mathbf{V}) - v_{nr}) + \sum_{c_{pq}^v \in r} (w_{pq}^v(\mathbf{U}, \mathbf{V}) - v_{nr}) \big]$$
$$\text{for} \quad 1 \le r \le K_n$$

for a Potts spin n, respectively. Mean field values ϕ_m and ϕ_{nr} can be interpreted as the increases in the energy function $E_B(\mathbf{U}, \mathbf{V})$ when Ising and Potts spins m and n are assigned to states 1 and r, respectively. Hence, $-\phi_m$ and $-\phi_{nr}$ may be interpreted as the decreases in the overall solution qualities by routing LS/SL and SS subnets m and n through the horizontal channel and route r, respectively. Then, u_m and v_{nr} values are updated such that probabilities of routing subnets m and n through horizontal channel and route r increase with increasing mean field values ϕ_m and ϕ_{nr} as follows:

$$u_m = \frac{e^{\phi_m/T}}{1 + e^{\phi_m/T}} \qquad (8)$$

$$v_{nr} = \frac{e^{\phi_{nr}/T}}{\sum_{k=1}^{K_n} e^{\phi_{nk}/T}} \quad \text{for} \quad r = 1, 2, \ldots, K_n \qquad (9)$$

respectively. After the mean field equations (Eqs. (6-7)) are derived, the MFA algorithm can be summarized as follows. First, an initial high temperature spin average is assigned to each spin, and an initial temperature T is chosen. Each u_m value is initialized to $0.5 \pm \delta_m$ and each v_{nr} value is assigned to $1/K_n \pm \delta_{nr}$ where δ_m and δ_{nr} denote randomly selected small disturbance values. Note that $\lim_{T \to \infty} u_m = 0.5$ and $\lim_{T \to \infty} v_{nr} = 1/K_n$. In each MFA iteration, the mean field effecting a randomly selected spin is computed using either Eq. (6) or Eq. (7). Then, the average of this spin is updated using either Eq. (8) or Eq. (9). This process is repeated for a random sequence of spins until the system is stabilized for the current temperature. The system is observed after each spin update in order to detect the convergence to an equilibrium state for a given temperature. If energy function E_B does not decrease in most of the successive spin updates, this means that the system is stabilized for that temperature. Then, T is decreased according to a cooling schedule, and iterative process is re-initialized. At the end of this cooling schedule, each Ising spin m is set to state 1 if $u_m \geq 0.5$ or to state 0, otherwise. Similarly, maximum element in each Potts spin vector is set to 1 and all other element are set to 0. Then, the resulting global routing is decoded as mentioned earlier.

4 Implementation

The performance of the proposed MFA algorithm for the global routing problem is evaluated in comparison with the well-known LocusRoute algorithm [4].

The MFA global router is implemented efficiently as described in Section 3. Average of each Ising spin m is initialized by randomly selecting u_m^{init} in the range $0.45 \leq u_m \leq 0.55$. Similarly, average of each Potts spin n is initialized by randomly selecting $K_n v_{nr}$ values in the range $0.9/K_n \leq v_{nr} \leq 1.1/K_n$ and normalizing $v_{nr}^{init} = v_{nr}/\sum_{k=1}^{K_n} v_{nk}$ for $r = 1, 2, \ldots, K_n$. Note that random selections are achieved by using uniform distribution in the given ranges.

The initial temperature parameter used in mean field computation is estimated using the initial spin averages values. Selection of initial temperature parameters T_0 is crucial to obtain good routing. In previous applications of MFA, it is experimentally observed that spin averages tend to converge at a critical temperature. Although there are some methods proposed for the estimation of critical temperature, we prefer an experimental way for computing T_0 which is easy to implement and successful as the results of experiments indicate. We compute the initial average mean field as

$$\phi_{avg}^{init} = \Big(\sum_{m=1}^{N_m} \phi_m^{init} + \sum_{n=1}^{N_n}\sum_{k=1}^{K_n} \phi_{nr}^{init}\Big)/(N_m + \sum_{n=1}^{N_n} K_n)$$

Note that initial mean field values ϕ_m^{init} and ϕ_{nr}^{init} are computed according to Eqs. (6) and (7) using initial spin values u_m^{init} and v_{nr}^{init}. Here, N_m and N_n denote the total number of Ising and Potts spins, respectively, where $N = N_m + N_n$ denotes the total number of spins (subnets). Then, initial temperature is computed as $T_0 = C\phi_{avg}^{init}$ where constant C is chosen as 540 for all experiments.

The cooling schedule is an important factor in the performance of MFA global router. For a particular temperature, MFA proceeds for randomly selected unconverged net spin updates until $\Delta E < \epsilon$ for M consecutive iterations respectively where $M = N$ initially and $\epsilon = 0.05$. Average spin values are tested for convergence after each update. For an Ising spin m, if either $u_m \leq 0.05$ or $u_m \geq 0.95$ is detected, then spin m is assumed to converge to state 0 or state 1, respectively. For a Potts spin n, if $v_{nr} \geq 0.95$ is detected for a particular $r = 1, 2, \ldots, K_n$, then spin n is assumed to converge to state r. The cooling process is realized in two phases, slow cooling followed by fast cooling, similar to the cooling schedules used for Simulated annealing. In the slow cooling phase, temperature is decreased by $T = \alpha \times T$ where $\alpha = 0.9$ until $T < T_0/1.5$. Then, in the fast cooling phase, M is set to $M/2$, α is set to 0.8. Cooling schedule continues until 90% of the spins converge. At the end of this cooling process, each unconverged Ising spin m is assumed to converge to state 0 or state 1 if $u_m < 0.5$ or $u_m \geq 0.5$, respectively. Similarly, each unconverged Potts spin n is assumed to converge to state r where $v_{nr} = \max\{v_{nk} : k = 1, 2, \ldots, K_n\}$. Then, the result is decoded as described in Section 3, and the resulting global routing is found.

The LocusRoute algorithm is implemented as in [4]. As the LocusRoute depends on rip-up and reroute method, LocusRoute is allowed to reroute the circuits 5 times. No bend reduction has been done as in [6]. Both algorithms are implemented in the C programming language.

5 Experimental Results

This section presents experimental performance evaluation of the proposed MFA algorithm in comparison with LocusRoute algorithm. Both algorithms are tested for the global routing of thirteen *ACM SIGDA Design Automation* benchmarks (MCNC) on SUN SPARC 10 . The first 4 columns of Table 1 illustrate the properties of these benchmark circuits.

These two algorithms yield the same total wiring length for global routing since two or less bend routing scheme is adopted in both of them. Last six columns of Table 1 illustrate the performance results of these two algorithms for the benchmark circuits. The MFA algorithm is executed 10 times for each circuit starting from different, randomly chosen initial configurations. The results given for the MFA algorithm in Table 1 illustrate the average of these executions. Global routing cost values of the solutions found by both algorithms are computed using Eq. (1) and then normalized with respect to those of MFA. In Table 1, maximum channel density denotes the number of routes assigned to the maximally loaded channels. That is, it denotes the minimum number of tracks required in a channel for 100% routability.

As is seen in Table 1, global routing costs of the solutions found by MFA are 3.1%-10.5% better than those of LocusRoute. As is also seen in this table, maximum channel density requirements of the solutions found by MFA are less than those of LocusRoute in almost all circuits except *alu2* and *term1*. Both algorithms obtain the same maximum channel density for these two circuit.

Figures 4 and 5 contain visual illustrations as pictures (left) and histograms (right) for the channel density distributions of the solutions found by MFA and LocusRoute, respectively, for the circuit *C1355*. The pictures are painted such that the darkness of each channel increases with increasing channel density. Global routing solutions found by these two algorithms are tested by using SEGA [5] detailed router for FPGA. Figure 6 illustrates the results of the SEGA detailed router for the circuit *C1355*

Table 1. The performance results of the MFA and LocusRoute algorithms for the global routing of MCNC benchmark circuits

Benchmarks				Performance Results					
Circuits				MFA			LocusRoute		
name	number of nets	number of 2-pin nets	FPGA size	global routing cost	max. channel density	exec. time (sec)	global routing cost	max. channel density	exec. time (sec)
9symml	71	259	10x9	1.000	12.0	0.36	1.032	14	0.28
too−large	177	519	14x13	1.000	16.0	0.88	1.071	17	0.64
apex7	124	300	11x9	1.000	14.0	0.42	1.073	16	0.29
example2	197	444	13x11	1.000	15.0	0.64	1.097	16	0.72
vda	216	722	16x15	1.000	17.0	0.42	1.055	18	0.10
alu2	137	511	14x12	1.000	17.0	0.30	1.080	17	0.32
alu4	236	851	18x16	1.000	17.0	0.68	1.073	19	0.50
term1	87	202	9x8	1.000	14.0	0.34	1.093	14	0.27
C1355	142	360	12x11	1.000	13.0	0.56	1.119	15	0.43
C499	142	360	12x11	1.000	15.0	0.48	1.075	16	0.36
C880	173	427	13x11	1.000	15.4	0.68	1.065	17	0.38
K2	388	1256	21x19	1.000	20.2	0.94	1.038	22	0.60
Z03D4	575	2135	26x25	1.000	17.0	2.34	1.117	18	1.84

6 Conclusion

In this paper, we have proposed an order-independent global routing algorithm for FPGA, based on Mean Field Annealing. The performance of the proposed global routing algorithm is evaluated in comparison with the LocusRoute global router for 13 MCNC benchmark circuits. Experimental results indicate that the proposed MFA heuristic performs better than the LocusRoute.

7 Acknowledgments

The authors would like to thank Jonathan Rose for providing the benchmarks and necessary tools for FPGA.

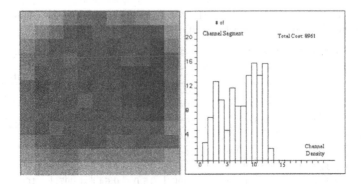

Fig. 4. Channel density distribution obtained by MFA for the circuit *C1355*

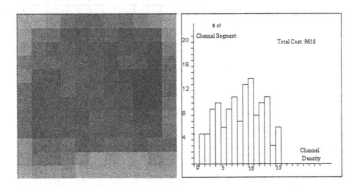

Fig. 5. Channel density distribution obtained by LocusRoute for the circuit *C1355*

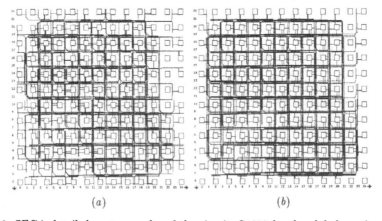

Fig. 6. SEGA detailed router results of the circuit *C1355* for the global routing solutions obtained by (a) MFA (b) LocusRoute

References

1. T. Bultan and C. Aykanat, " A new mapping heuristic based on mean field annealing, " *Journal of Parallel and Distributed Computing*, 16 (1992) 292-305.
2. J.J.Hopfield, , and D.W.Tank, "Neural Computation of Decisions in Optimization Problems", *Biolog. Cybern.*, vol. 52, pp. 141-152, 1985.
3. S.Kirkpatrick, C.D.Gelatt, and M.P.Vecchi. " Optimization by simulated annealing", *Science*, vol. 220, pp. 671-680, 1983.
4. J.Rose, "Parallel Global Routing for Standart Cells" *IEEE Transactions on Computer-Aided Design* Vol. 9 No. 10 pp:1085-1095 October 1990.
5. S.Brown, J.Rose, Z.Vranesic, "A Detailed Router for Field Programmable Gate Arrays" *Proc. International Conference on Computer Aided Desing 1990.*
6. J.Rose and B.Fallah, " Timing-Driven Routing Segment Assignment in FPGAs " *Proc. Canadian Conference on VLSI*
7. J.Rose, A. El Gamal, A. Sangiovanni-vincentalli, " Architecture of Filed-Programmable Gate Arrays " *Proceedings of the IEEE* pp:1013-1029 .Vol:81, No:7, July 1993.
8. C.Sechen, "VLSI Placement and Global Routing Using Simulated Annealing", *Kluwer Academic Publishers.* 1988
9. J.Greene, V.Roychowdhury, S.Kaptanoglu, A.E.Gamal, "Segmented Channel Routing", *27th ACM/IEEE Design Automation Conference* pp:567-572 1990.
10. S.Burman, C.Kamalanathan, N.Sherwani, "New Channel Segmentation Model and Routing for High Performance FPGAs", *Proc. International Conference on Computer Aided Desing, pp:22-25 1992*
11. N.Sherwani, "Algorithms for VLSI Physical Design Automation", *Kluwer Academic Publishers.* 1993
12. T.Lengauer, "Combinatorial Algorithms for Integrated Circuit Layout" 1990 *Wiley-Teubner Series.*
13. "Fundamentals of Placement and Routing", *Xilinx Co. 1990*
14. "The Programmable Gate Array Data Book", *Xilinx Co. 1992.*

Power Dissipation Driven FPGA Place and Route under Delay Constraints

Kaushik Roy[1] and Sharat Prasad[2]

[1] Electrical Engineering, Purdue University, West Lafayette, IN, USA
[2] Integrated Systems Lab., Texas Instruments, Dallas, TX, USA

Abstract. In this paper we address the problem of FPGA place and route for low power dissipation with critical path delay constraints. The presence of a large number of unprogrammed antifuses in the routing architecture adds to the capacitive loading of each net. Hence, a considerable amount of power is dissipated in the routing architecture due to signal transitions occurring at the output of logic modules. Based on primary input signal distributions, signal activities at the internal nodes of a circuit are estimated. Placement and routing are then carried out based on the signal activity measure so as to achieve routability with low power dissipation and required timing. Results show that more than 40% reduction in power dissipation due to routing capacitances can be achieved compared to layout based only on area and timing.

1 Introduction

The Field Programmable Gate Arrays (FPGA's) combine the flexibility of mask programmable gate arrays with the convenience of field programmability. The FPGA's which were once used only for prototyping has found application in larger volume productions too. Hence, it is extremely important to achieve high performance and lower power dissipation out of these devices. Depending on the technology, the FPGA based designs can have large interconnect capacitances. In this paper we will consider minimization of power dissipation due to wiring capacitances for the row-based FPGA's under delay constraint.

With the widespread use of portable systems, power dissipation of circuits have become a very important design consideration for longer battery life and enhanced reliability. And if power dissipation is low enough, expensive ceramic packages can be replaced by plastic ones which cost about 25% less. There are various ways in which power dissipation can be minimized. One of the conventional ways of minimizing power comes from scaling down the supply voltage at the cost of larger circuit delays. Considerable improvement in power dissipation can be achieved at the cost of higher circuit delays. Hence, lowering supply voltage make delay constrained power optimization even more desirable in view of the longer delay times. Recently there has been a lot of research effort to minimize power dissipation during different phases of a design such as high level synthesis, logic synthesis, and circuit synthesis [1, 2]. These synthesis procedures are based on average number of signal transitions on circuit nodes. Due to presence of a large number of antifuses in the routing architecture, the FPGA-based designs can have large wiring capacitance. In this paper we will model the power dissipation due to interconnects using signal transitions and minimize power dissipation during placement and routing for FPGA's.

Figure 1 shows a row-based FPGA architecture [4]. There are rows of Logic Modules (LM's) each of which can implement a large number of logic functions. The routing channels are in between the rows of logic modules and are used for routing of the nets. The routing tracks are laid out. The tracks are segmented and the adjacent segments are separated by horizontal antifuses (*hfuses*). In the unprogrammed state, the antifuses have a very large resistance and a small capacitance. However, programming these antifuses produces a low-resistance bidirectional connection between adjacent segments. The pins of the logic mod-

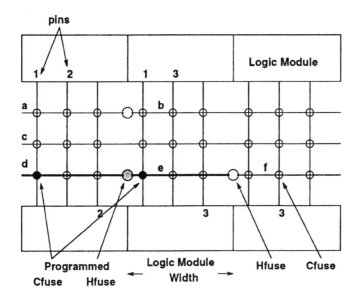

Fig. 1. Row-based FPGA Architecture

ules can be connected to the routing tracks using the vertical lines as shown in Figure 1. There is a cross antifuse (*cfuse*) present at the crossing of each horizontal and vertical line. The *cfuse* have the same electrical characteristic as the *hfuse*. In the programmed state a low-resistance connection is obtained between a pin of the LM and a routing track. Figure 1 shows the routing of net 1 using segments *d* and *e* by programming two *cfuses* and an *hfuse*. If there are T tracks in the channel, and M number of vertical lines, then at most one *cfuse* per vertical line gets programmed, and hence, at least $M(T-1)$ *cfuses* remain unprogrammed in each channel. Each of the unprogrammed antifuses contribute a small capacitance, and due to the presence of large number of such antifuses on each segment, the capacitive loading can be significant. Power dissipation in CMOS circuits is associated with charging and discharging of LM load capacitances, and hence, consideration of power dissipation during layout is very important for FPGA's.

The paper is organized as follows. Section 2 on preliminaries and definitions introduces the reader to signal activity estimation at the output of CMOS logic gates. Section 3 considers estimation of power dissipation due to layout capacitances. The details of the power dissipation driven FPGA layout algorithms are given in Section 4. Results of our analysis is given in Section 5, and finally conclusions are drawn in Section 6.

2 Preliminaries and Definitions

2.1 Multilevel logic representation

Multilevel logic can be described by a set $\mathcal{F}$ of completely specified Boolean functions. Each Boolean function $f \in \mathcal{F}$, maps one or more input and intermediate signals to an output or a new intermediate signal. A circuit is represented as a *Boolean network*. Each *node* has a Boolean *variable* and a Boolean *expression* associated with it. There is a directed edge to a node g from a node f, if the expression associated with node g contains in it the variable associated with f in either true or complemented form. A circuit is also viewed as a *set of gates*. Each gate has one or more input pins and (generally) one output pin. Several pins are electrically tied together by a signal. Each signal connects to the output pin of exactly one gate, called the driver gate.

2.2 Signal Probability and Signal Activity

Power dissipation can be estimated if signal transitions are accurately estimated for all the nets. Research on estimation of the average number of signal transitions has been reported in [3, 1]. Digital circuit signals can be represented as a steady state stationary stochastic process [3], each signal being associated with a *signal probability* and an *activity*. *Signal probability* is defined as the probability that a particular signal has a logic value of ONE, and *signal activity* is defined as the average number of signal transitions at the nets. We assume that signal probability and activity of the primary input signals are known, and can be obtained from system level simulations of the design with real life inputs. The signal activities at the internal nodes of a circuit can be efficiently and accurately estimated using the methods described in [1].

Let us consider a multi-input, multi-output logic module M which implements a Boolean function. M can be a single logic gate or a higher level circuit block. We assume that the inputs to M, $g_1, g_2, ..g_n$ are mutually independent processes each having a signal probability of $P(g_i)$, and a signal activity of $A(g_i)$, $i \leq n$. The signal probability at the output can be easily computed using one of the methods described in [10]. For example, if P_1, P_2, and P_3 are the input signal probabilities to a three input AND gate, the output signal probability is given by $P_1 P_2 P_3$, whereas, for an OR gate the output signal probability is $1 - (1 - P_1)(1 - P_2)(1 - P_3)$. For an inverter, the ouput signal probability is simply $(1 - P_1)$, where P_1 is the input signal probability. The signal activity at any output h_j, of M is given by

$$A(h_j) = \sum_{i=1}^{n} P\left(\frac{\partial h_j}{\partial g_i}\right) D(x_i) \qquad (1)$$

Here $x_i, i = 1, .., n$ are the module inputs and $\partial h / \partial g$ is the boolean difference of function g with respect to h and is defined by

$$\frac{\partial h}{\partial g} = h \mid_{g=1} \oplus h \mid_{g=0} = h_g \oplus h_{\overline{g}} \qquad (2)$$

Figure 2 shows the propagation of signal activity through AND, OR, and NOT gates. The signal probabilities and the circuit activities at the primary inputs to a circuit are assumed to be available.

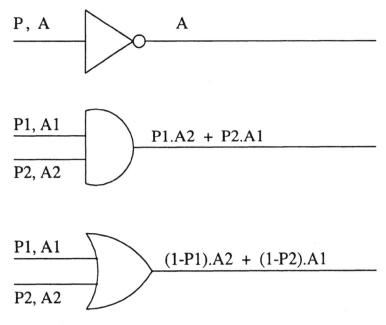

Fig. 2. Propagation of circuit activities through basic gates

3 Calculation of Power Dissipation

The three different sources of power dissipation in CMOS circuits are – leakage current, short circuit current, and signal transitions to charge or discharge load capacitances. Of these three, the last is the most dominant one and will only be considered in the following discussions. The overall load capacitance that each logic module or logic gate experiences is due to the routing capacitances, the number of fanouts of the LM, and the transistor gate capacitance per fanout connection.

For the FPGA's, the segmented routing tracks are laid out, and each segment has a number of antifuses on it. Programming these antifuses produces a bidirectional low-resistance connection between the segment and the corresponding vertical line connected to a logic module pin (in case of *cfuses*) or between two segments (in case of *hfuses*). However, in the unprogrammed state each antifuse is associated with a small capacitance, and due to the presence of a large number of such antifuses on each segment the capacitive loading can be significant. If the signal activity associated with a net having large wiring capacitance is high, larger power will be dissipated due to the interconnect. The average power dissipation can be given by:

$$Power_{avg} = \sum_{i \in all \ LM's} V_{DD}^2 . A_i . C_i \qquad (3)$$

where V_{DD} is the supply voltage, C_i is the capacitive load, and A_i represents the signal activity associated with each LM output. The power dissipation internal to an LM has not been considered in the above equation. The capacitive load C_i that each logic module i experiences can be approximated by

$$C_i = C_{ri} + \sum_{f=1}^{fanout_i} C_{gf} \qquad (4)$$

where the C_{ri} represents the total wiring capacitance due to the metal line forming the track segment(s) and the unprogrammed *cfuses* on it, $fanout_i$ represents the number of fanout for logic module i, and C_{gf} represents the transistor gate capacitances associated with each fanout f. If the wiring capacitance is comparable to the the fanout gate capacitances considerable improvement in total power dissipation can be achieved if a signal activity based layout algorithm can be used.

4 FPGA Layout

From the previous section it can be noted that power dissipation due to wiring capacitances can be minimized if it is possible to assign the nets with higher activity to routing tracks associated with lower capacitance. Logic synthesis to achieve low power dissipation has been considered in [1], where multilevel logic was synthesized based on signal activity measure. Previous research on FPGA layout mainly concentrated on routability and critical path delays [4, 5, 7]. In this research power minimization has been considered along with routability and performance optimization during placement and channel routing. The signal activities on different nets are determined from the signal probabilities and activities of the primary input signal using Equation 1. We have developed techniques with efficient datastructures to accurately determine signal activities even in the presence of signal correlations, the details of the procedure is given in [1].

4.1 Placement

The placement algorithm is based on simulated annealing [9]. The cost function not only considers timing and wire length penalty, but it also considers the activity measures of the signals. The FPGA architecture details and constraints are incorporated in the array template. Unlike the gate arrays, the feedthrough cells cannot be inserted for vertical routing. Whenever a pin cannot be reached by a module output through its dedicated vertical tracks, uncommitted feedthrough segments have to be used. This appears as an extra cost in the cost function for placement optimization.

The cost function ($\mathcal{C}$) consists of total wire length (W), power dissipation penalty ($W * A$), timing path penalty (P), and extra cost (F) for using uncommitted feedthrough. The activity associated with the net under consideration is represented as A. The complete expression for the cost function is given by

$$\mathcal{C_P} = W + a(W * A) + bP + cF \tag{5}$$

where a, b, and c are the relative weights of the three terms in the cost function. The wire length of a net is estimated as half the perimeter of the minimum rectangle, a bounding box, that encompasses the net. The wire length W is proportional to the FPGA routing capacitances, and hence, $W * A$ is proportional to the power dissipated due to the routing capacitances. For each critical timing path, an upper bound is put on the wire length of all the nets in the path. A penalty is assigned for a path that has the wire length beyond this upper bound.

In order to provide a homogeneous cost factor, the extra cost (F) of using uncommitted feedthrough segment is represented by the vertical distances between module output vertical span and the bounding box of the net. To facilitate calculation, a concept of *module driver* is introduced. A *module driver* is a pin that drives the rest of the pins in a net. With the location of the *module driver* specified, F is readily available. For our timing and power dissipation driven placement, the identification of the driver for a net will help the accuracy of delay and power estimation. Global routing follows placement. It efficiently uses the scarce vertical routing resources or feedthroughs to connect same nets in different channels.

4.2 Detailed Routing

A routing channel contains segmented routing tracks as shown in Figure 1. The channel routing problem is formulated as an assignment problem where each net within a channel is assigned to one or more unassigned segments. Each net within a channel is allowed to use at most one track due to a technology constraint which does not allow programming of antifuses connected in an L-shaped fashion [4].

The programming of such antifuses can lead to programming two antifuses at the same time which can degrade the performance of the programmed antifuses.

The cost of routing is determined by the number of segments used by the critical nets in a channel, and the length of the segments assigned to different nets. The number (H_x) of *hfuses* to be programmed is equal to the number of segments assigned to the corresponding net (x) minus one. Depending on the technology, the resistance associated with programmed antifuses can be detrimentally high. If a net x of length L_x is routed with p ($1 \leq p \leq K$, maximum of K segments allowed) segments, each of length L_j ($j = 1, ..., p$), then $\{(\sum_{j=1}^{p} L_j) - L_x\}$ gives a measure of the wasted (or excess) length of segment(s). It can be observed that the wasted segment is associated with unprogrammed *cfuses* which increases the capacitive loading on the net. This has two detrimental effects: a large segment wastage means a larger delay on that net, and if the net is critical, timing of the circuit might get affected. Secondly, if the net is associated with large activity, higher power will be dissipated due to higher capacitance. Besides, these two performance effects, the segment wastage is also associated with routability of the channel. For K-segment routing, in which a maximum of K segments can be used by each net, we define the cost C_x of routing a net x as

$$C_x = w_1.\alpha + w_2.\beta$$

where

$$\alpha = \{(\sum_{j=1}^{p} L_j) - L_x\}A_x$$

$$\beta = H_x$$

The factor α is a product of segment wastage due to the assignment of a net to a segment(s) and the activity of that net. The factor β is associated with routing performance, because the programmed *hfuses* add to delays of a net. The weights w_1, w_2 assigned to the wastage factor, and the horizontal antifuse usage factor respectively, are determined by the technology under consideration. For example, the metal-metal horizontal antifuse has a much lower programmed resistance than a programmed ONO [4] antifuse, and hence, w_2 for the latter technology should be higher than the metal-metal antifuse technology. The total cost of routing all the nets in a channel is $\sum_p C_p$, $0 < p \leq V$, where V is the total number of nets in a channel.

Green et. al. [8] have shown that K-segment ($K > 1$) channel routing problem is NP-complete. We use a routing algorithm based on net ordering. The nets are ordered in terms of their length in a channel times the activity associated with that net ($L_x A_x$), and the nets with higher length-activity measure is routed first. Hence, the nets which are routed first can be assigned to the best possible segment with least segment wastage. All timing critical net can be routed first for timing critical designs. If a net is unroutable, we resort to backtracking to determine if undoing a previous net-segment assignment can achieve routability [5]. If exhaustive backtracking is unable to route the nets in the channel, then the channel is not routable. Average power dissipation due to channel routing is measured using Equation 3.

5 Implementation and Results

The power dissipation and performance driven FPGA layout algorithms have been implemented in C on SPARC 10 workstation. Table 1 shows the results of our analysis on two MCNC benchmarks (*bw* and *duke2*) from Microelectronic Center of North Carolina and some industrial designs. The primary input signals were assigned signal probabilities of 0.5. Signal activities of primary inputs were randomly assigned a number between 1 and 7. The logic was synthesized and internal node activities were calculated using our algorithm of [1]. The FPGA's used for experimentation had twenty-five routing tracks per channel with channel segmentation of *TPC1010*. There were forty-four logic modules per row. For larger designs we increased the number of rows of logic module to fit the design. The second column of Table 1 shows the average percentage segment wastage over all nets of the design. The results have been compared with traditional layout for FPGA's based on timing and routability (R), and our power dissipation driven layout (P). The number of *hfuses* to be programmed is shown in the next column. The last column shows the percentage change in power dissipation due to routing capacitances. Considerable improvement in power dissipation was obtained for all the examples satisfying all the critical path constraints. The nets were modeled as RC trees [7] and some of the nets were analyzed using SPICE. SPICE results indeed show a large improvement in average power dissipation.

Table 1. Layout results for some examples

Design	Wastage (%)		Hfuses		% Change
	R	P	R	P	Power
bw	40.9	41.4	0	0	34.3
duke2	45.8	46.3	3	7	29.7
f104667	51.6	51.8	0	0	32.6
f104243	44.9	45.2	0	0	23.1
f104780	47.9	48.1	8	9	37.2
f103918	43.8	44.3	3	3	42.1
cf92382a	52.0	52.2	0	0	22.0

6 Conclusions

Power dissipation and performance driven layout for row-based FPGA's has been considered in this paper. Power dissipation in CMOS circuits is dependent on the nature of primary inputs. Hence, probabilistic measures were used to determine the signal activities. Experimental analysis shows the feasibility of achieving considerable improvement in power dissipation due to routing capacitances for FPGA's.

7   Acknowledgment

The research was sponsored in part by IBM Corporation under the SUR program.

References

1. K. Roy and S. Prasad, "Circuit Activity Based Logic Synthesis for Low Power Reliable Operations," *IEEE Trans. on VLSI Systems* Dec. 1993, pp. 503-513.
2. A. Chandrakashan, S. Sheng, and R. Brodersen, "Low Power CMOS Digital Design," *IEEE Journal on Solid-State Circuits*, Apr. 1992, pp. 473-484.
3. F.N. Najm, "Transition Density, A Stochastic Measure of Activity in Digital Circuits," *Design Automation Conf.*, 1991, pp. 644-649.
4. A. El Gammal, J. Greene, J. Reyneri, E. Rogoyski, and A. Mohsen, "An Architecture for Electrically Configurable Gate Array," *IEEE Journal of Solid State Circuits*, vol. 24, Apr. 1989, pp. 394-397.
5. K. Roy, "A Bounded Search Algorithm for Segmented Channel Routing for FPGA's and Associated Channel Architecture Issues," *IEEE Trans. on Computer-Aided-Design*, Nov. 1993, pp. 1695-1705.
6. C. Shaw, M. Mehendale, D. Edmondson, K. Roy, M. Raghu, D. Wilmoth, M. Harward, and A. Shah, "An FPGA Architecture Evaluation Framework," *FPGA-92 workshop*, Berkeley, Feb. 1992.
7. S. Nag and K. Roy, "Iterative Wirability and Performance Improvement for FPGA's," *ACM/IEEE Design Automation Conf.*, 1993, pp. 321-325.
8. J. Greene, V. Roychowdhury, S. Kaptanaglu, and A. El Gammal, "Segmented Channel Routing," *Design Automation Conf.*, pp. 567-572, 1990.
9. C. Sechen and K. Lee, "An Improved Simulated Annealing Algorithm for Row-Based Placement," *Intl. Conf. on Computer-Aided-Design*, 1987. pp. 942-995.
10. S. Ercolani, M. Favalli, M. Damiani, P. Olivio, and B. Ricco, "Estimation of signal Probability in Combinational Logic Network," *1989 European Test Conference*, pp. 132-138.

FPGA Technology Mapping for Power Minimization [1]

Amir H. Farrahi   and   Majid Sarrafzadeh

Department of Electrical Engineering and Computer Science
Northwestern University
Evanston, IL 60208

Abstract. The technology mapping problem for lookup table-based FPGAs is studied in this paper. The problem is formulated as assigning LUTs to nodes of a circuit so as to minimize the total estimated power consumption. We show that the decision version of this problem is NP-complete, even for simple classes of inputs such as 3-level circuits. The same proof is extended to conclude that the general library-based technology mapping for power minimization is NP-complete. A heuristic algorithm for mapping the network onto K-input LUTs in polynomial time, aimed at minimizing the power consumption is presented. Despite the fact that the Boolean properties of the network are not exploited in the mapping procedure, the experimental results show %14.8 improvement on the average power consumption compared to the results obtained from a mapping algorithm aimed at minimizing the number of LUTs. On the average, the number of LUTs is increased by %7.1.

1 Introduction

With the rapid development and advances in VLSI technology, the average transistor count in a chip is increased enormously, allowing more sophisticated functionality. Moreover, the advent of personal communication and computing services has stirred a great deal of interest in both the commercial and research areas. The minimization of power consumption in modern circuits, is therefore of great importance. In particular, battery operated products such as portable computers and cellular phones, have come to a point in which minimization of the power consumption is among the most crucial issues. Due to the importance of power consumption issue, there has been a great shift of attention in the logic and layout synthesis areas from the delay and area minimization issues towards this issue [18, 17, 9, 14].

An FPGA is an array of programmable logic blocks (PLBs) that can be interconnected in a fairly general way. The interconnection between these blocks are also user programmable. In a Lookup-Table (LUT) based FPGA, the PLB is a K-input LUT (K-LUT) that can implement any Boolean function of up to K

---

[1]This work has been supported in part by the National Science Foundation under grant MIP 9207267.

variables. The technology mapping problem for LUT-based FPGAs is to generate a mapping of a set of Boolean functions onto K-LUTs. Previous mapping
algorithms have focused on three main issues: minimization of the number of
levels of LUTs in the mapped network [4, 7, 11, 15] , minimization of the number of LUTs used in the mapping solution [6, 10, 5], routability of the mapping
solution [16, 3], or combinations of these.

In this paper we study the technology mapping problem for LUT-based FPGAs
for power minimization. We formulate the problem as assigning LUTs to vertices in the network so as to minimize the total estimated power consumption
of the mapped circuit. We show that the decision version of this problem is
NP-complete, even for simple classes of inputs. We use the transition density
metric for power estimation, and derive formulas for propagation of the transition densities from primary inputs to all the nodes in the network for a general
AND/OR/INVERT circuit. The rest of this paper is organized as follows. Section 2 sets up the notation. Section 3 describes the formulation of the power
estimation for a circuit being mapped onto LUTs. Section 4 summarizes our
NP-completeness results for the formulation presented in Section 3. Section 5
presents modeling of a general AND/OR/INVERT combinational circuit and describes the transition density propagation formulas derived for this model. A
polynomial time heuristic algorithm to solve the problem is presented in Section
6. The experimental results are presented in Section 7, and Section 8 summarizes
the key features of this paper and provides directions for further research in this
area.

2 Problem Formulation and Notation

Consider the representation of a combinational logic circuit as a directed acyclic
graph (DAG) $G(V, E)$ where each vertex v_i in V represents a boolean function,
and each directed edge (v_i, v_j) represents a connection between the output of v_i
and the input of v_j. A *primary input* (PI) vertex has no in-coming edge and a
primary output (PO) has no out-going edge. Given a vertex $v \in V$, by $input(v)$
we mean the set of vertices that supply inputs to vertex v i.e. $input(v) = \{u \in
V | (u, v) \in E\}$. In addition, given a subset $V1$ of V, not including any PIs or POs,
by $input(V1)$ we mean the set of vertices in V-$V1$ that supply inputs to vertices
in $V1$ i.e. $input(V1) = \{u \in (V - V1) | \exists v \in V1 : (u, v) \in E\}$, and by $output(V1)$
we mean the set of vertices in $V1$ that supply inputs to vertices in V-$V1$ i.e.
$output(V1) = \{u \in V1 | \exists v \in (V - V1) : (u, v) \in E\}$. In case $V1$ contains PIs
or POs, each PI in $V1$ will also be included in $input(V1)$, and each PO in $V1$
will also be included in $output(V1)$. For a set S of elements, the notation $|S|$ is
used to denote the number of elements in S (i.e. *cardinality* of S). Vertex u is a
predecessor of vertex v if there is a directed path from u to v in the network. A
K-feasible cone at v, denoted by C_v, is defined as a subgraph consisting of v and
its predecessors such that any path connecting a vertex in C_v and v lies entirely
in C_v, and $|input(C_v)| \leq K$.

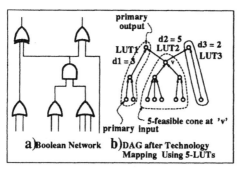

Figure 1. DAG Representation of a Boolean Network and Mapping it onto 5-LUTs

A *K-input lookup-table* (K-LUT) is a programmable logic block capable of implementing any K-input (and single output) boolean function. Therefore, each K-LUT can implement a K-feasible cone in the boolean network. The technology mapping problem for K-LUT based FPGAs is to map a boolean network using K-LUTs, which can be viewed as covering the DAG representation of the boolean network with K-feasible cones. Figure 1 shows a boolean network, its corresponding DAG and its mapping onto 5-LUTs.

Now, consider a DAG $G(V, E)$ and a subset L of V. Assume that corresponding to each vertex l in L, there is a LUT "placed at" this vertex in the graph [2], that is, the output of vertex l will be supplied by this LUT. To simplify the notation, the corresponding LUT will also be denoted by l. The dependency y_l of LUT l is defined as the number of PIs or LUTs that feed vertex l. That is, y_l represents the number of inputs to the LUT l if the assignment of LUTs to vertices of G are as specified in L. This means that the LUT l corresponds to a K-feasible cone at vertex l (a K-LUT) if and only if $y_l \leq K$. For a node v, the *contribution* quantity, denoted by Z_v represents the contribution of node v to the dependency of its fanout nodes. That is, Z_v is equal to 1 or y_v f v is or is not assigned a LUT in the mapping, respectively.

Intuitively, in technology mapping for LUT-based FPGAs for power minimization, we desire to map the circuit onto K-LUTs such that the activity at the LUT outputs, and hence the power consumption due to these activities are minimized. That is, a mapping would have low power consumption if the highly active signals (edges) are hidden inside the LUTs in this mapping. As in [12], let us model a logic signal by a function $x(t)$, $t \in (-\infty, +\infty)$, which only takes values 0 or 1. Note that such a model ignores waveform details such as over/under-shoots and rise/fall times. The *equilibrium probability* (EP) and the *transition density* (TD)

---

[2]Note that the terms network and graph are used interchangeably in this paper.

of a logic signal $x(t)$, denoted by $p(x)$ and $d(x)$ respectively, are defined as:

$$p(x) = \lim_{T \to \infty} 1/T \int_{-T/2}^{+T/2} x(t)dt \tag{1}$$

$$d(x) = \lim_{T \to \infty} \frac{n_x(T)}{T} \tag{2}$$

Where $n_x(T)$ represents the number of transitions of x(t) in the time interval $(-T/2, +T/2]$. It is shown in [12] that under a reasonable model, the limit in (2) always exists. Consider a Boolean function $y = f(x_1, x_2, ..., x_n)$. Then the *Boolean difference* of y with respect to x_i, denoted by $\frac{\partial y}{\partial x_i}$, is defined as:

$$\frac{\partial y}{\partial x_i} = y|_{x_i=0} \oplus y|_{x_i=1} \tag{3}$$

Consider a logic module $\mathcal{M}$ with inputs $x_1, ..., x_n$ and outputs $y_1, ..., y_m$. If there is no propagation delay associated with module $\mathcal{M}$, the module is known as a *zero-delay* logic module. The following theorem quoted from [12] relates the TDs $d(y_j)$ to TDs $d(x_i)$:

Theorem 1: If the inputs $x_i(t)$, $i = 1, 2, ..., n$ of a zero-delay logic module $\mathcal{M}$ are pairwise independent signals with TDs $d(x_i)$, then the TDs $d(y_j)$, $j = 1, ..., m$ are given by:

$$d(y_j) = \sum_{i=1}^{n} p(\frac{\partial y_j}{\partial x_i})d(x_i) \tag{4}$$

This theorem provides a tool to propagate the TDs at the PIs into the network to compute the TDs at any point in the network. The assumption that the inputs to a node are independent, however, may not be true for every node. Even though this independence holds for the PIs, the existence of reconvergent paths may cause correlation between the values of the inputs to a node in the circuit. It has been mentioned in [12], however, that if the modules are large enough so that tightly coupled nodes are kept inside the same module, then the coupling effect, outside the modules are sufficiently low to justify the independence assumption. In this paper we assume such an independence to simplify the propagation of TDs. The importance of TDs at different sites in a network is due to the fact that the average dynamic power consumption in a CMOS gate with output signal $x(t)$ is given by:

$$P_{av}(x) = \frac{1}{2}CV_{dd}^2\{\lim_{T \to \infty} \frac{n_x(T)}{T}\} = \frac{1}{2}CV_{dd}^2 d(x) \tag{5}$$

Where C and V_{dd} represent the load capacity at the output of the gate and the power supply voltage, respectively. Note that for a CMOS gate, the static power consumption is negligible, so, to minimize the total power consumption in a network, the summation of the dynamic power consumption over all the nodes in a network should be minimized. Note that at this stage of the design, the routing information is not available. A rough estimate for the length of a net can be the number of destinations of that net, which is taken into account in our power estimation model when computing capacity of each net. In a LUT-based FPGA environment, since the only modules are the LUTs, the summation should be minimized over all the LUTs used to map the Boolean network. This idea is also in accordance with technical data sheet information, e.g., [1] confirms that the power consumption is proportional to the summation of the average activities at the LUT outputs in a mapped network.

In this paper, we address the LUT-based technology mapping problem for minimizing the total power consumption of the mapping result, to be referred to as K-input LUT power minimization problem (*K-PMP*). This can be stated as mapping a Boolean network onto K-LUTs such that the average power consumption over the entire mapped circuit is minimized. To simplify the problem, we assume that $|input(v)|$ for each vertex v in the boolean network is less than or equal to K (any circuit can be transformed to attain such property). We also assume that the covering procedure is not allowed to decompose a node into its fanin [3] nodes. This restricted version of the problem will be denoted as *K-RPMP*, which can be viewed as assigning LUTs to the nodes of a network such that each LUT corresponds to a K-feasible cone, and that the power consumption of the mapping result is minimized. Let *PMP* denote the library-based technology mapping targeted at minimization of the power consumption of the mapping result. This problem is studied in [17]. Under the simplified assumptions of the constant load model, this problem is NP-hard [2] [4]. Note that *PMP*, *K-PMP* and *K-RPMP* can each be formulated as an optimization or a decision problem. Unless otherwise stated, in the rest of this paper, *PMP*, *K-PMP* and *K-RPMP* refer to the decision versions of the problems.

In our formulation of *K-RPMP*, we use the zero-delay model and TD for estimating the power consumption. Not only does this allow the application of Theorem 1 to compute the TDs at all nodes in the circuit, but it also simplifies the problem and allows us to focus on the mapping algorithm rather than the transition density propagation.

3   Power Estimation Model

As mentioned earlier, the major energy consumption term in CMOS circuits is due to dynamic power dissipation, which happens at transition times. In LUT-based FPGAs, the same is true with the difference that the transitions only take place at the input/output of LUTs. Therefore, the average power consumption

---

[3]For each node v, *fanin(v)* and *input(v)* represent the same set.

[4]The decision version of this problem under the constant load model is NP-complete.

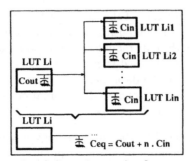

Figure 2. Load Capacity at the Output of a LUT

in a network $\mathcal{N}$ mapped onto LUTs can be approximated by the following:

$$P_{av}(\mathcal{N}) = \sum_{PI \; p_i} (n_i \frac{1}{2} C_{in} V_{dd}^2 \; d(p_i) + \sum_{LUT \; L_i} \{\frac{1}{2}[C_{out} + fanout(L_i)C_{in}]V_{dd}^2 \; d(L_i)\}$$

(6)

In this formula, the term $[C_{out} + fanout(L_i)C_{in}]$ accounts for the equivalent load capacity at the output of LUT L_i as shown in Figure 2, and the term n_i represents the number of LUTs receiving input from the primary input p_i. Note that by introducing the equivalent load capacity, we are in essence taking into account the power consumption at the inputs of the fanout LUTs as well as the power consumption at the output of current LUT.

4 Complexity Issues

This section summarizes our results on the complexity of K-$RPMP$ [5]. Consider the 3-Satisfiability problem *(3-SAT)*. It is well-known that *3-SAT* is an NP-complete problem [8]. It is reported in [8] that *3-SAT* remains NP-complete if for each variable x_i, there are at most 5 clauses that contain either of the literals x_i ,$\bar{x}_i$. We will refer to this version of *3-SAT* as *R3-SAT* (restricted *3-SAT*). Lemma 1 forms the foundation of our complexity results for K-$RPMP$.

Lemma 1. *R3-SAT* is polynomial time transformable to K-$RPMP$, for any value $K \geq 5$.

Based on Lemma 1, and the fact that K-$RPMP$ is in class NP we conclude the following:

Theorem 2. K-$RPMP$ is NP-complete for $K \geq 5$.

---

[5]See Northwestern University, EECS Department, Technical Report June 1993 for proofs.

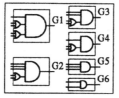

Figure 3. Candidate Gates in a Gate Library

Furthermore, we can conclude the following, simply from the transformation presented in the proof of Lemma 1:

Corollary 1. *K-RPMP* with $K \geq 5$ remains NP-complete for the following ("simple") networks: 3-level networks, networks with bounded fanin and/or fanout, AND/OR networks, combinations of these.

Corollary 2. Let *K-RLMP* denote the technology mapping problem for LUT-based FPGAs targeted at minimization of the number of LUTs, under the same assumptions as *K-RPMP*. Then *K-RLMP* is NP-Complete for all $K \geq 5$ [6].

An interesting observation here, is that the same transformation can be used to show the NP-completeness of both *K-RLMP* and *K-RPMP* for all $K \geq 5$.

Corollary 3. General library-based technology mapping problem for power minimization (*PMP*) is NP-complete, even if our library consists only of the ("simple") gates G1, G2,...,G6 shown in Figure 3.

Note that Corollary 3 solves a previously open problem regarding the NP-completeness of the *PMP* under this model.

We shall conclude this section by pointing out that the same transformation does not work for values $K < 5$, due to technical problems.

5 Circuit Model And Transition Density Propagation

Consider a general Boolean network consisting of AND/OR/INVERTER gates. We can view such a circuit as a network consisting of AND, OR gates with arbitrary input/output polarities for each gate . We shall refer to these gates as *polarized* gates. We will show how we can apply Theorem 1 to propagate the TDs from the PIs into such a network. Since every Boolean network can be transformed into such a representation, we can then use our technique for propagation of TDs into the network. A polarized AND gate can be modeled as an n-input AND gate with programmable *inverter blocks* at each input and output as shown in Figure 4. Define *polarity* function $pol(x)$ for each input/output as

---

[6]This has also been shown in [5].

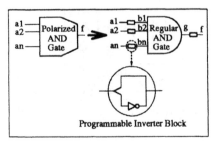

Figure 4. Modeling Polarized AND Gate

0 or 1 if the corresponding programmable inverter block acts as an inverting or non-inverting buffer respectively . Now, consider such an AND gate with inputs $a_1, ..., a_n$ and output f. Associated with each input a_i the *polar input* b_i and associated with the output f, the polar output g are introduced. Then we have:

$$p(b_i) = \begin{cases} p(a_i), & \text{if } pol(a_i) = 1 \\ 1 - p(a_i), & \text{if } pol(a_i) = 0 \end{cases}$$

$$= pol(a_i)p(a_i) + [1 - pol(a_i)][1 - p(a_i)] \tag{7}$$

$$p(f) = pol(f)p(g) + [1 - pol(f)][1 - p(g)] \tag{8}$$

Then by applying Theorem 1, with the assumption that the inputs a_i are independent logic signals, we obtain:

$$d(f) = d(g) = \sum_{i=1}^{n} p(\frac{\partial g}{\partial b_i})d(b_i) = \sum_{i=1}^{n} [(\prod_{k=1, k \neq i}^{n} p(b_k))d(a_i)] \tag{9}$$

$$p(f) = pol(f)\prod_{i=1}^{n} p(b_i) + [1 - pol(f)][1 - \prod_{i=1}^{n} p(b_i)] \tag{10}$$

A polarized OR gate is modeled similarly. By a similar analysis we obtain the following:

$$p(f) = pol(f)p(g) + [1 - pol(f)][1 - p(g)]; \quad p(g) = 1 - \prod_{i=1}^{n}[1 - p(b_i)] \tag{11}$$

$$d(f) = \sum_{i=1}^{n} \{[\prod_{k=1, k \neq i}^{n} (1 - p(b_k))]d(a_i)\} \tag{12}$$

Where $p(b_i)$ is obtained from (7). This means that we can easily propagate the TDs at PIs into the network to compute the TDs at any site in the network.

```
Tech_Mapping_for_Low_Power
Input    : DAG  G(V,E)
Output : A 5-feasible mapping of G with minimal power
Begin
   Perform a topological sort on G(V,E) ;
   Compute TDs for all nodes in the network ;
   Starting from PIs towards POs for each node v do
        Compute dependency  y(v) of v ;
        While (y(v) > K)
             Sv = Set of fanin nodes of v to which no LUT is assigned;
             Mv = Minimal_set (Sv) ;
             Find the element  u  of  Mv  with maximum priority ;
             Assign a LUT  to  u and update y(v);
   End.
```

Figure 5. Technology Mapping Algorithm for Power Minimization

6 A Heuristic Mapping Algorithm

As mentioned in Section 2, our mapping algorithm is restricted only to assigning LUTs to nodes in the network such that the whole network is mapped onto K-feasible cones (K-LUTs). Our heuristic algorithm is shown in Figure 5.The main idea in this algorithm is to scan the network starting from the PIs towards POs and make all LUTs K-feasible. At each node v, as we go along, if the dependency y_v of v is larger than K, a fanin node of v is chosen and a LUT is assigned to it, thus lowering y_v. This procedure gets repeated until $y_v \leq K$. The key point in the algorithm is to consider both the contribution and TD factors in the selection of fanin nodes for LUT assignment. Consider current node v, and assume $S_v = \{u_1, u2, ..., u_m\}$ is the set of fanin nodes of v to which no LUTs are assigned yet. Let the ordered pair (Z_{u_i}, d_{u_i}) represent the contribution Z_{u_i} of u_i to the dependency y_v of node v, and TD d_{u_i} associated with u_i respectively. We say u_i *dominates* u_j if $d_{u_i} \geq d_{u_j}$ and $Z_{u_i} \leq Z_{u_j}$. The *minimal set* of v is defined as the set of members of S_v that are not dominated by any other element in S_v, and is denoted by M_v. Figure 6 shows a set of points in two dimensional space (Z, d) and its corresponding minimal set. The minimal set of m elements in two-dimensional space can be found easily in $O(m \ log \ m)$ time (see [13]). Suppose u_1 and u_2 are members of S_v. Note that if u_1 dominates node u_2, from a local standpoint, we can claim that assignment of a LUT to u_2 is more beneficial for power minimization than assigning a LUT to u_1. That is, for selecting a fanin node from S_v for LUT assignment, we can just focus on elements in M_v. A priority function $priority(u_i) = P_{av}(u_i) \times (F_c.Z(u_i) - 1)$ is used to select the node in M_v for next LUT assignment. The parameter F_c, called *contribution factor* is introduced to allow control over the relative significance of contribution and TD factors. The time complexity of the algorithm can be shown to be $O(nK^2)$ where n is the total number of nodes in the network, and K is the maximum input capacity of the LUTs.

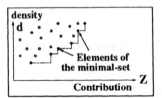

Figure 6. Minimal-set of a Set of Points

Network	Level-Map Results		Power Min. Alg. Results	
name	LUTs	Power(mW)	LUTs	Power(mW)
5xp1	25	187	25	182
9sym	60	374	62	365
9symml	56	385	58	376
C499	83	1410	91	1076
C880	103	1177	111	1060
alu2	127	894	146	836
apex6	235	1413	237	1404
apex7	69	502	77	459
count	31	226	31	227
duke2	175	526	190	478
misex1	18	121	16	106
rd84	23	345	27	344
rot	211	1945	238	1749
vg2	25	159	25	160
z4ml	5	80	5	80
Total	1246	10219	1339	8902
Comparison	1	+%14.8	+%7.1	1

Table 1. Comparison of Power and LUT Minimization Algorithms

7   Experimental Results

The algorithm was implemented in C and tested using a number of MCNC Benchmark circuits. The results were compared to Level-Map [5] algorithm which addresses LUT minimization. Both algorithms were run a number of times with 30 different values of F_c (F_f for Level-Map) in the range $[0, 20]$ and the best result for power consumption was recorded. The input/output capacity for the LUTs were set at $10pF$ and the examples were run with constant EP $p = 0.5$ and TD $d = 10000$ for all PIs. The algorithms were run for 5 input LUTs. On the average, our power minimization algorithm shows %14.8 improvement on the total power consumption. In addition, experiments show an average of %7.1 increase on the number of LUTs, compared to the results obtained from Level-Map. These results are shown in Table 1. The running times of both algorithms on each of the benchmark circuits were less than 10 seconds on a SUN SPARC Station 1. Note that our main focus in this work has been on the mapping algorithm for power minimization, and not on the propagation of the TDs into the network. The propagation of the TDs into the circuit could be computed using other, perhaps more realistic, approaches, e.g., using binary decision diagrams as in [17], or could even be given as part of the input. In order to compare the delay of mapping results of different technology mappers, we have also performed

Network	Level-Map Results		Power Min. Alg. Results		FlowMap Results	
	Depth	Delay Depth	Delay Depth	Delay		
	(ns)	(mW)	(ns)	(mW)	(ns)	(mW)
5xp1	5	70.6	5	73.4	3	64.3
9sym	7	66.4	7	64.0	7	59.5
9symml	7	61.7	6	54.1	5	94.7
count	5	64.3	5	70.1	5	64.3
misex1	5	56.4	6	73.3	3	78.6
rd84	5	47.5	5	61.6	3	38.1
vg2	6	80.2	8	62.9	4	75.5
z4ml	2	39.8	2	24.4	2	24.4
Total	42	487.2	44	483.8	32	449.4
Comparison	+%31	+%0.8	+%37	1	1	+%3.1

Table2. Comparison of LUT, Power, and Depth Minimization Algorithms

placement and routing steps on a number of these benchmarks that would fit into the XILINX 3090 FPGA architecture. The depth of the mapping [7] and the delay of these benchmark circuits after placement and routing by XILINX tools are shown in Table 2. As you can see, minimizing depth does not necessarily correspond to minimizing the delay. This is mainly because of the fact that the applied placement and routing algorithms were not directed to minimize the delay as the primary objective, and that they used simulated annealing technique which is a probabilistic approach.

8   Conclusion

In this work, we studied the technology mapping step for LUT-based FPGAs aimed at minimization of the power consumption of the mapping result. We showed that even restricted versions of the problem are NP-complete for values $K \geq 5$, where K is the input capacity of the LUTs. A polynomial time algorithm was presented to solve the problem using a greedy heuristic. Experimental results show substantial reduction on the total power consumption on a number of MCNC Benchmark examples. This is a promising result as no Boolean properties were exploited in the mapping algorithm. Our future research in this area will be in the following directions: exploiting the Boolean properties of the circuit to achieve better results, intergating the technology mapping and the placement and routing steps to improve routability and taking into account the resistive and capacitive effect of the routing paths on the power consumption, improving the power estimation metric to allow taking into account the power consumption due to glitches, and finally recognizing and effectively exploiting reconvergent paths to improve the quality of the mapping algorithm.

---

[7]The maximum number of LUTs on any PI to PO path in the mapping solution.

References

[1] *"The Programmable Gate Array Data Book"*. XILINX Inc., San Jose, CA, 1992.

[2] R. K. Brayton, G. D. Hachtel, and A L. Sangiovanni-Vincentelli. "Multilevel Logic Synthesis". *Proceedings of the IEEE*, 78(2):264–300, February 1990.

[3] P. K. Chan and J. Y. Zien M. D. F. Schlag. "On Routability Prediction for Field-Programmable Gate Arrays". In *Design Automation Conference*, pages 326–330. ACM/IEEE, 1993.

[4] J. Cong and Y. Ding. "An Optimal Technology Mapping Algorithm For Delay Optimization In Lookup-Table Based FPGA Design". Technical Report CSD-920022, University of California at Los Angeles, May 1992. Also appeared in Proceedings of the ICCAD, 1992.

[5] A.H. Farrahi and M. Sarrafzadeh. "On the Look-up Table Minimization Problem for FPGA Technology Mapping". In *International ACM/SIGDA Workshop on Field Programmable Gate Arrays*. IEEE/ACM, 1994.

[6] R. Francis, J. Rose, and K. Chung. "Chortle: A Technology Mapping Program for Lookup Table-Based Field Programmable Gate Arrays". In *Design Automation Conference*, pages 613–619. IEEE/ACM, 1990.

[7] R. Francis, J. Rose, and Z. Vranesic. "Technology Mapping for Lookup Table-Based FPGAs for Performance". In *International Conference on Computer-Aided Design*, pages 568–571. IEEE, 1991.

[8] M. R. Garey and D. S. Johnson. *Computers and Intractability: A Guide to the Theory of NP-completeness*. Freeman, 1979.

[9] B. Lin and H. DeMan. "Low-Power Driven Technology Mapping Under Timing Constraints". In *International Conference on Computer Design*, pages 421–427. IEEE, 1993.

[10] R. Murgai, Y. Nishizaki, N. Shenoy, R. K. Brayton, and A. Sangiovanni-Vincentelli. "Logic Synthesis Algorithms for Programmable Gate Arrays". In *Design Automation Conference*, pages 620–625. IEEE/ACM, 1990.

[11] R. Murgai, N. Shenoy, R. K. Brayton, and A. Sangiovanni-Vincentelli. "Performance Directed Synthesis for Table Look Up Programmable Gate Arrays". In *International Conference on Computer-Aided Design*, pages 572–575. IEEE, 1991.

[12] F. Najm. "Transition Density: A New Measure of Activity in Digital Circuits". *IEEE Transactions on Computer Aided Design*, 12(2):310–323, 1992.

[13] F. P. Preparata and M. I. Shamos. *Computational Geometry: An Introduction*. Springer-Verlag, 1985.

[14] K. Roy and S. Prasad. "Circuit Activity Based Logic Synthesis for Low Power Reliable Operations". *IEEE Transactions on VLSI Systems*, 1(4):503–513, 1993.

[15] P. Sawkar and D. Thomas. "Performance Directed Technology Mapping for Look-Up Table Based FPGAs". In *Design Automation Conference*, pages 208–212. IEEE/ACM, 1993.

[16] M. Schlag, J. Kong, and K. Chan. "Routability Driven Technology Mapping for Lookup Table-Based FPGAs". In *International Conference on Computer Design*, pages 89–90. IEEE, 1992.

[17] C. Tsui, M. Pedram, and A.M. Despain. "Technology Decomposition and Mapping Targeting Low Power Dissipation". In *Design Automation Conference*, pages 68–73. ACM/IEEE, 1993.

[18] H. Vaishnav and M. Pedram. "A Performance Driven Placement Algorithm for Low Power Designs". In *EURO-DAC*, 1993.

Specification and Synthesis of Complex Arithmetic Operators for FPGAs*

Hans-Juergen Brand   Dietmar Mueller

Technical University Chemnitz-Zwickau, Faculty of Electrical Engineering
PSF 964, 09009 Chemnitz, Germany
Phone: ++ 49 - 371 - 531 - 3158 / Fax: ++ 49 - 371 - 531 - 3186
E-Mail: brand@infotech.tu-chemnitz.de

Wolfgang Rosenstiel

University of Tuebingen, Faculty of Informatics
Sand 13, 72076 Tuebingen, Germany
Phone: ++49 - 7071 - 29 - 5482 / Fax: ++49 - 7071 - 610399
E-Mail: rosenstiel@peanuts.informatik.uni-tuebingen.de

Abstract. This paper describes the application of the experimental system LORTGEN for the technology-specific specification and synthesis of high performance arithmetic operators for FPGAs. Using multiplier and adder designs for the XC3xxx-LCA-family as example we demonstrate that the implemented architecture-specific techniques boost the performance and density of the designs. Consequently, LORTGEN enables the implementation of complex operators (e.g. multipliers) in one FPGA, frees the designers from device-specific implementation details and allows them to focus more on actually designing the application.

1   Introduction

The behavioral description of complex systems using HDLs - e.g. VHDL - often contains arithmetic operators like addition or multiplication. During the synthesis step these operators have to be replaced by an implementation satisfying the design requirements.

Usually commercial synthesis tools like the Design Compiler [1] or AutoLogic [2] synthesise operator implementations with a sufficient performance but they do not apply special methods for data path synthesis. However, if an extensive design space exploration is necessary to adjust the implementation to additional restrictions in time, area and so on, the synthesis process is often computationally infeasible for large datapath-intensive ASIC-designs[1] and involves the following difficulties:

---

* This work is partly supported by the BMFT under the contract 01M3007C

[1] The optimisation of one 32x32 bit multiplier implementation with the Design Compiler takes about 8 hours on a SPARC10.

- Commercial synthesis systems do not support the mathematical decomposition of arithmetic operators into proper subfunctions [3]. Consequently, the designer must specify this decomposition in the specification which requires a great deal of VHDL and design knowledge.
- The quality of the synthesis results depends largely on a suited specification style and optimisation strategy.
- The synthesis of a certain implementation in the design space often requires several synthesis cycles.
- The used synthesis and optimisation (mapping) techniques do not utilise all architecture-specific features of FPGAs.

Consequently,
- commercial synthesis tools produce suboptimum implementations of arithmetic operators for FPGAs,
- the quality of the synthesis result (as well as the design effort) depends on the designer's experience,
- the determination of the implementation best adjusted to the design requirements often results in a high design effort and consequently in a low design efficiency.

To simplify and accelerate the synthesis of arithmetic operators CAD-tool vendors supply special tools often based on a library of technology-independent generic modules or module generators (e.g. Design Ware [4], AutoLogic BLOCKS [2] or X-BLOX [5]). This results in a better performance but usually the number of design alternatives can explode so that the approach of a predefined library storing all possible implementations becomes disadvantageous.

To handle this problem for arithmetic operators, the following goals must be achieved:
(1) we need accurate and fast design space exploration methods to select the implementation best adjusted to the design requirements,
(2) parameterisable and efficient generation algorithms are required instead of predefined libraries.

Our main contributions are:
- Presentation of new specification and synthesis strategies for arithmetic operators especially suited for FPGA design.
- Implementation of the new strategies in a test version of a synthesis tool called LORTGEN.
- Development of fast and accurate evaluation methods not only for design quality metrics like area or delay, but also for fuzzy criteria like regularity or modularity.
- Improvement of the implementation performance and a considerable reduction of the design effort for arithmetic operators.

Our basic ideas are as follows:
- We integrate the evaluation of design alternatives into the specification process.
- We use function- and technology-specific parameterisable generation algorithms which enable the prediction of synthesis results.

- The result of the specification process must be a unique, complete and consistent formal specification description of the design alternative best adjusted to the design requirements.
- The fixed relation between the specification description and the (technology-specific) synthesis result is used to derive the evaluation models.
- We apply fuzzy methods for the evaluation of circuit characteristics important for VLSI design like regularity or modularity.

In the paper we present an approach that supports the interactive specification of arithmetic operators including a real (technology-specific) design space exploration without any synthesis activity as well as the following architecture-specific synthesis of the specified implementation for FPGAs. We ensure that only one specification and one synthesis process have to be carried out independent from the designer's experience and the design requirements.

The paper is organised as follows. In section 2 we give a short description of our experimental system LORTGEN. Section 3 describes the application of LORTGEN for the specification of arithmetic operators using the example of a multiplier. Section 4 summarises the synthesis results for adders and multipliers and gives a comparison with other synthesis tools.

2 LORTGEN

The synthesis tool LORTGEN (*LO*gic and *R*egister-*T*ransfer *GEN*erator) was implemented based on the methods of "constraint-driven" specification and "parameter-driven" module generation [6, 7]. Figure 1 shows the system structure of the experimental system. LORTGEN supports the data path design in two phases of the design process:

- the specification and modelling of datapath components in the system design phase and
- the technology-specific synthesis of data path components in the realisation phase.

It addresses the following aims:

- to support the specification, evaluation and synthesis of arithmetic operators,
- to provide the existing design alternatives,
- to ensure that only one specification and synthesis step has to be carried out independent from the design requirements given in the design project and the designers experience,
- to select the optimal design alternative before starting the synthesis process and without any synthesis activity,
- to utilize the architecture-specific features of FPGAs,
- to ensure that the quality of the final solution is independent from the designers experience.

The consideration of realistically modelled design characteristics during the specification task results in two main advantages which can reduce the design effort drastically:

(1) It avoids that incomplete and abstract models could adversely affect the quality of the final solution or require to go through additional design iteration steps.

(2) The designer has to specify and synthesise one design alternative only independent from the given constraints.

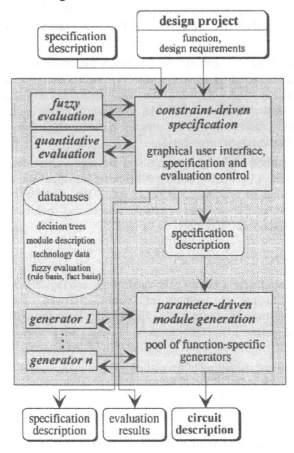

Fig. 1. System structure of LORTGEN

Figure 2 illustrates the achieved design effort reduction and design efficiency improvement.

The design flow using LORTGEN consists of two main steps. In a first step the constraint-driven specification produces a unique, complete and consistent specification description. The designer has to fix the function and the design requirements as input to LORTGEN. Using quantitative and fuzzy evaluation the suitable design alternative will be specified. The interactive specification process is controlled by the function-specific decision tree visualised as browser in the graphical user interface of LORTGEN. The nodes of the used decision trees do not represent the existing design alternatives but the functional and algorithmic parameters to describe the different implementations. Thus we avoid the exponential growth of the specification effort with a growing number of design alternatives[2].

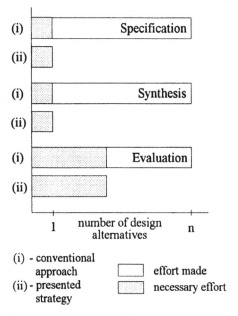

Fig. 2. Comparison between different synthesis approaches

In the second step the parameter-driven module generation transforms the specification description into a functional model (for the system verification) and/or into a technology-specific netlist (e.g. as structural VHDL-description or MAP-file[3]).

The integration of evaluation activities into the specification task allows the efficient and reliable selection of the design alternative best adjusted to the design requirements without any synthesis activity. Thus expensive design iteration steps can be avoided.

3 Operator Specification with LORTGEN

3.1 Design example - Fuzzy-Pattern Classification Accelerator

In recent years fuzzy technology has become an interesting alternative issue to cope with analysis and control of complex systems. It is especially used if the system model is described incompletely, inaccurately or there is only empirical knowledge about the system behavior. The fuzzy pattern classification (FPC) [9] - a non rule-

---

[2] Although the number of design alternatives significantly increases from about 100 for a 16x16 bit multiplier to about 480 for a 22x9 bit multiplier the node number in the corresponding decision tree is constant.

[3] The MAP-file generation process includes transformation steps using tools described in [8].

based fuzzy approach - is being successfully applied for quality inspection, recognition of standardised objects or for process control. Up to now only software FPC systems were available which are too slow for many practical applications especially in the real-time domain.

Therefore a special hardware FPC accelerator system was developed. Starting from the idea of rapid prototyping FPGAs (XC3xxx family from XILINX) were chosen for a first system implementation [10].

A complex multiplier contained in the accelerator determines the delay and area characteristics of the whole system. The specification of such a multiplier subcircuit is used to demonstrate the design process with LORTGEN. The multiplier design has to satisfy the following requirements:

- operand word length: 22 and 9 bit
- result word length: 31 bit
- data format: unsigned (integer)
- target technology: LCA
- master: XC3090PG175-100
- critical path length: 12 stages (CLBs)
- maximum area: 280 CLBs (active area)
- optimisation goals: fastest implementation, area efficient, regular and
             local connections.

LORTGEN provides about 480 design alternatives[4] for such a 22x9 bit multiplier. The selection of the optimal design alternative using commercial synthesis systems would require to specify these alternatives using VHDL and to synthesise an implementation. The design process with LORTGEN can be carried out in one specification process (selection of the appropriate design alternative) and one synthesis process (generation of the netlist).

3.2 Specification with LORTGEN

LORTGEN uses hierarchical decision trees for the control of the specification process. Each decision tree node represents one or several specification steps which contribute to the setting of the functional and algorithmic parameters distinguishing between the different design alternatives.

Figure 3 shows as example the two-operand integer multiplier decision tree. The figure includes short descriptions of the activities done in the decision tree nodes containing both the activity type and the activity description in italics (specification of parameters or selection of subtrees in branch nodes). The detailed explanation of these activity types using the multiplier design as example will be a part of the presentation.

---

4) The number of design alternatives corresponds to the number of all possible combinations of three algorithms for generating (Shift and Add, Booth and Baugh-Wooley) and four algorithms for adding partial products (CPA-field, CSA-field, Wallace tree, Dadda algorithm) and 40 different implementations for the vector merging addder (considering different block sizes for carry select adders).

We distinguish between five activity types (examples are given in figure 3):

(1) *interactive setting of given values*:
The designer enters values for functional parameters (e.g. operand word length for multipliers) which are given by the design project.

(2) *fuzzy evaluation*
The designer makes decisions concerning the selection of the optimal design alternative basing on the results of a fuzzy evaluation of non-metric design characteristics (e.g. regularity or modularity) or of metric design characteristics of the existing design alternatives in early specification steps[5] . LORTGEN realises this design space exploration automatically using the according facts and rules from the fuzzy evaluation database considering design requirements given by the design project.

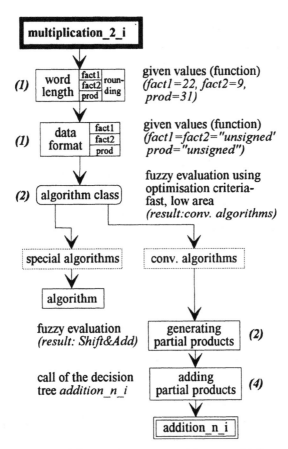

Fig. 3. Decision tree for two-operand integer multipliers

---

[5] The area of an incompletely specified design alternative can not be calculated.

(3) *quantitative evaluation*

The designer makes decisions concerning the selection of the optimal design alternative basing on the results of a quantitative evaluation of metric design characteristics (e.g. delay or area) of the existing design alternatives. LORTGEN realises this design space exploration automatically using the according evaluation models [7, 11].

Both evaluation activity types contribute to the specification of algorithmic parameters. Due to this evaluation support, the designer need not know the features of the different algorithmic parameters for each component type.

(4) *decision tree call*

The designer continues the interactive specification process in the called decision tree. The calling node will be set automatically after the specification in the called tree has been finished.

(5) *automatic setting of parameters*

LORTGEN analyses the already specified parameters and sets a not specified parameter automatically if the value of this parameter can be calculated uniquely from the already specified parameters. This contributes to the consistency of the specification without limiting the design space.

In the following an example for a quantitative evaluation will be explained. After the specification of the algorithms for generating and adding the partial products as subfunctions of a multiplication the algorithm for the vector-merging adder (26 bit carry propagate adder) must be specified. The optimal adder will be selected using the results of a quantitative evaluation. The vector merging adder may occupy 64 CLBs and the critical path delay is limited to 7 CLBs[6] .

Table 1 summarises the evaluation results calculated by LORTGEN. It contains only carry-select adders composed of different sized blocks because all other design alternatives exceed the delay limit and therefore the main optimisation criterion could not be satisfied.

All design alternatives in table 1 occupy more than 64 CLBs. The highlighted implementation with the block sizes 9/7/5/5 represents the smallest design and consequently it will be selected as the optimal one. It gives an example for the fact, that the fastest design can be the smallest as well.

Table 1 simultaneously shows the fine resolution of the design space by the design alternatives available in LORTGEN. This enables a good adaption to the design requirements.

The entire multiplier specification process comprises 19 decision tree nodes. Two nodes are necessary to specify the functional parameters. The evaluation of the existing design alternatives (about 480) and the selection of the appropriate one requires nine nodes (about 50 percent of the entire specification process) - four nodes

---

[6] This values are calculated by subtracting the area and critical path length of the already specified multiplier components for partial product generation and addition from the limits given in the design project.

use fuzzy and five quantitative evaluation methods. Six nodes (about 30 percent) will be set automatically by LORTGEN. The remaining two nodes are call nodes.

carry-select block sizes	delay (CLB)	area (CLB)	carry-select block sizes	delay (CLB)	area (CLB)
12/10/4	7	74	11/8/7	7	67
12/10/3/1	7	79	11/8/6/1	7	77
12/10/2/2	7	79	11/8/5/2	7	74
12/9/5	7	71	11/8/4/3	7	72
12/9/4/1	7	77	11/8/4/2/1	7	77
12/9/3/2	7	76	11/7/5/3	7	70
12/8/6	7	70	11/7/5/2/1	7	75
12/8/5/1	7	77	11/7/4/4	7	69
12/8/4/2	7	76	11/7/4/2/2	7	73
12/8/3/3	7	74	10/8/8	6	66
12/7/7	7	68	10/8/6/2	6	75
12/7/5/2	7	74	10/8/5/3	6	72
12/7/4/3	7	72	10/8/5/2/1	6	74
12/7/4/2/1	7	77	10/8/4/4	6	71
12/6/4/4	7	71	10/8/4/2/2	6	71
12/6/4/2/2	7	75	10/7/5/4	6	69
11/9/6	7	68	10/7/5/3/1	6	74
11/9/5/1	7	75	10/7/5/2/2	6	74
11/9/4/2	7	74	9/7/5/5	6	65
11/9/3/3	7	72	9/7/5/3/2	6	70

Table 1. Feasible design alternatives for 26 bit vector merging adder

4 Synthesis Results

This section summarises the results achieved for the XC3xxx-LCA-family.
In a first experiment we compared 16-bit adders synthesised by AutoLogic [12] and LORTGEN, respectively, with manually optimised designs [13].

Table 2 summerises the results for three different optimisation goals. It shows that AutoLogic allows no adaption to the design requirements [12]. The manually optimised adder with 8 CLBs delay requires an additonal area of two CLBs compared with the adder generated by LORTGEN. As a result of this experiment we found a generation algrorithm producing a 16-bit adder with a delay of 3 CLBs.

Secondly, we present the synthesis results of the multiplier design as described in section 3. In the synthesis step LORTGEN generates a 22x9-bit multiplier occupying 281 CLBs with a critical path length of 11 CLBs (170 ns). The overall design (multiplier and some registers) uses 286 CLBs and was realised in an XC3090. Thus we realised a device utilisation of 88 percent compared with a typical 70 percent device utilisation. This result could be achieved because the Dadda scheme used for adding partial products requires mostly local connections. The accelerator was successfully applied in two real-time applications (clock frequency 5 MHz).

optimisation goal	AutoLogic [12]		LORTGEN		manually optimised [13]	
	delay (CLB)	area (CLB)	delay (CLB)	area (CLB)	delay (CLB)	area (CLB)
small design	16	16	16	16	16	16
area/speed compromise	16	16	8	22	8	24
fast design	16	16	4	42	3	41

Table 2. Synthesis results for 16-bit adders

bit x bit	tool from [14]		LORTGEN		Comparison	
	delay (CLB)	area (CLB)	delay (CLB)	area (CLB)	delay (%)	area (%)
22x7	28	144	11	217	- 61	+ 51
22x9	30	189	11	281	- 63	+ 49
22x16	37	347	13	407	- 65	+ 17

Table 3. Synthesis results for different multipliers

A comparison of LORTGEN with X-BLOX concerning the multiplier design was not possible because X-BLOX does support neither the XC3xxx-family nor the multiplier design. In [12] synthesis results of AutoLogic and AutoLogic BLOCKS for a 22x9-Bit multiplier are presented. However, the designs use at least 450 CLBs, so that they can not be integrated in one XC3xxx-LCA.

Therefore, we compared LORTGEN with a generator presented at the EURO-DAC'93 [14]. Table 3 summarises the synthesis results for three different multipliers. The table shows that LORTGEN achieves a performance improvement ranging between 61 and 65 percent at the expense of up to 51 percent increase in area.

5 Summary

The results achieved with LORTGEN have shown that our approach could improve the performance and minimise the area of arithmetic operators for FPGAs.

The presented techniques enable the integration of complex operators like multipliers in one FPGA. LORTGEN significantly improves the designer's productivity. We free the designers from device-specific implementation details and allow them to focus more on actually designing the application.

Moreover, we avoid expensive design iteration steps by ensuring that only one design alternative has to be specified and synthesised independent from the designer's experience and the given design constraints.

Thus, in contrary to commercial synthesis tools, LORTGEN is especially suited for system designers because of the provided comprehensive design support.

6 References

1. Design Compiler Version 3.0 - Reference Manual, Synopsys, 1992.
2. AutoLogic Family for Top-Down Design - Product Description. Mentor Graphics, 1992.
3. BRAYTON, B. K.; RUDELL, R.; SANGIOVANNI-VINCENTELLI, A.; WANG, A.R.: MIS: A Multi-Level Logic Optimization System. IEEE Transactions on CAD of Integrated Circuits and Systems. November 1987, pp. 1062-1081.
4. Design Ware Version 3.0 - Databook, Synopsys, 1992.
5. KELEM, S. H.; FAWCETT, B. K.: Module Generators for Field Programmable Gate Arrays. In Proceedings of the 2nd International Workshop on Field-Programmable Logic and Applications 1992.
6. BRAND, H.-J.: Specification and Synthesis of Arithmetic Data Path Components (in German). PhD thesis, Technical University Chemnitz-Zwickau, 1993.
7. BRAND, H.-J.; MUELLER, D.; ROSENSTIEL, W.: An Efficient Data Path Synthesis Strategy. In Proceedings of the Synthesis and Simulation Meeting and International Interchange SASIMI'93 in Nara/Japan, October 1993, pp. 155-164.
8. SCHMIDT, J.; MUELLER, D.; SCHUPPAN, H.: Logic Synthesis of Logic Cell Arrays Using Register-Transfer Description. 2nd International Workshop on Field-Programmable Logic and Applications, Wien 1992.
9. BOCKLISCH, S.; ORLOVSKI, S.; PESCHEL, M.; NISHIWAKI, Y.: Fuzzy Sets Application, Methodological Approaches and Results. Akademie-Verlag, Berlin, 1986.
10. SCHLEGEL, P.; EICHHORN, K.; BRAND, H.-J.; MÜLLER, D.: Accelerated Fuzzy Pattern Classification with ASICs. 6. IEEE ASIC Conference and Exhibit in Rochester, 27. September - 1. Oktober 1993. pp. 250-253.
11. BRAND, H.-J.; MUELLER, D.; ROSENSTIEL, W.: Design of High Throughput Data Path Components. In Proceedings of the 4th ACM/SIGDA Physical Design Workshop, Lake Arrowhead, April 1993, pp. 141-151.
12. SCHUBERT, E.; ROSENSTIEL, W.: Synthesis of Register-Transfer Elements for High-Level Synthesis Using VHDL (in German). SMT/ASIC/Hybrid'94, Nuernberg, 1994.
13. The Programmable Logic Data Book. XILINX, 1993, pp. 8/72-8/78.
14. GASTEIER, M.; WEHN, N.; GLESNER, M.: Synthesis of Complex VHDL Operators. In Proceedings of the EURO-DAC'93, Hamburg, September 1993, pp. 566-571.

A Speed-Up Technique for Synchronous Circuits Realized as LUT-Based FPGAs

Toshiaki Miyazaki, Hiroshi Nakada, Akihiro Tsutsui
Kazuhisa Yamada, Naohisa Ohta

NTT Transmission Systems Laboratories
Y-807C, 1-2356 Take, Yokosuka-shi, Kanagawa, 238-03 JAPAN
e-mail: miyazaki@ntttsd.ntt.jp

Abstract. This paper presents a new technique for improving the performance of a synchronous circuit configured as a look-up table based FPGA without changing the initial circuit configuration except for latch location. One of the most significant benefits realized by this approach is that the time-consuming and user-uncontrollable reconfiguration processes, i.e., re-mapping, re-placement and re-routing, are unnecessary to improve circuit performance.

1 Introduction

Field Programmable Gate Arrays (FPGAs) have been widely used because of their usefulness. The circuit programming of FPGAs is often performed with CAD tools[1], and the execution speed of the programmed circuit really depends on the CAD tools. For example, if considering a look-up table (LUT) based FPGA, circuit programming requires the solution of the mapping and placement problems inherent in assigning the input logic onto LUTs, and the routing problem of interconnecting the LUTs. However, because the problems are known to be *NP-hard*, most CAD tools adopt some form of heuristics[2][3][4][5][6]. This means that the CAD tools do not guarantee to produce optimal programming results. Thus, particularly with regard to the propagation delay, some tune-up is often necessary to improve the execution speed of the FPGA-based circuit.

To improve circuit performance, several techniques have been developed. One of the most famous techniques is *retiming*[7][8][9][10]. It minimizes the combinational circuit delay between two latches by only inserting and/or removing the latches without changing the logic of the original circuit design. In another development, single-phase clocked synchronous circuits are often used in communication or digital signal transport systems, and a common design technique is to insert latches into the original circuit at the cost of increasing the input clock frequency[11]. Compared to retiming, this technique does cause some *clock response delay*, but it is often applied as a technique that can reliably improve circuit performance. A similar technique can be found in the pipeline sequence design of CPUs[12][13].

All the above methods, unfortunately, do not pay much attention to the upper limit of the number of latches inserted between two combination logic parts. With

FPGAs, however, it is very important to consider this limitation because the number of usable latches is finite in an FPGA, and the latch-insertion points are restricted given the condition that the initial routing result remains unchanged, which is our major premise as described hereafter. Accordingly, existing design methods cannot directly be applied to the task of improving FPGA-based circuit performance.

One solution to improve the performance of FPGA-based circuits is to optimize the delay using *try-and-error* or *patch* based approaches. This is because if placement and routing tools are invoked again, they often change the circuit configuration drastically and the designer cannot keep track of the delay information easily; after delay optimization, LUT placement and routing results should not be changed. Thus, a natural question arises with regard to the performance improvement, is it possible to speed up the circuit without re-placement and re-routing ?

In this paper, we present a method that answers this question. Here, we consider that the FPGA has latches located at the output of each LUT, and a mechanism is available to control latch usage without changing the initial circuit programming result. The basic approach of the method is to insert as many latches into the initial circuit as possible, under the latch limitation described before, to increase the input clock frequency. One of the most significant benefits of our method is that the initial routing result is not changed because placement and routing are not re-executed.

Kukimoto and Fujita[14] presented a method that rectified a circuit programmed onto an FPGA by changing only the logic configuration of LUTs. This is, the netlist is not changed and the net delay in the original circuit is preserved after modifying the logic of the circuit as in our method. However, there are few cases to which the method is applicable. Especially for logic optimized circuits, it is hard to apply their method because such circuits employ rather complex logic in each LUT and there is little room in which to rectify the logic. In addition, their method does not improve circuit performance itself. Thus, it seems that the method is not so useful in practice. Unlike their method, whether our approach is applicable to a circuit or not, depends on the topology of the nets connecting the LUTs, not the complexity of the configured logic in each LUT. Our technique is very effective for topologically simple circuits such as combination circuits which do not have any feedback loops.

This paper is organized as follows: Section 2 formulates the problem addressed. In Section 3, the speed-up method to solve the problem formulated in Section 2 is described in detail. Section 4 shows the experimental results gained using some benchmark data. Section 5 concludes this paper with a discussion of possible directions for future research.

2 Problem

We consider an LUT-based FPGA containing basic cells (BCs) as shown in Fig. 1. Each basic cell has an LUT and a latch. An n-input LUT can implement 2^{2^n}

different Boolean functions[1], and the output of the LUT can be programmed freely as to whether it is latched or not. In addition, we assume that the FPGA-implemented circuit does not contain any loops that do not include any latches, i.e., the circuit has no asynchronous feedback.

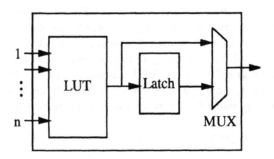

Fig. 1. A basic cell in the FPGA.

Under the above assumptions, we consider latch insertion without changing the initial placement and routing results. If some latches can be inserted in the critical path, it is guaranteed that the performance of the circuit is improved because the net delays are not changed. Here, our problem is defined as follows:

Problem 1. If an FPGA has a mechanism that permits independent control of the latch located at the output of each LUT, is it possible to realize a *q-clock response delay circuit* compared to the original circuit by controlling only the output value of each LUT by latching it or unlatching it ?

Here, *q-clock response delay circuit* is defined as follows[11]. Suppose A is a synchronous circuit. Let $Y_A(X, t_n)$ be the output vector at time $t_n \mid n > 0$ corresponding to the input signal X for all input pins of A. Suppose B is a synchronous circuit which has external pins corresponding to the external pins of A one by one. Here, let t_0 be the initial time. If an integer value $q \mid q > 0$ exists for every X and the following equation is satisfied

$$Y_B(X, t_{n+q}) = Y_A(X, t_n),\qquad(1)$$

we say "B is the *q-clock response delay circuit* of A."

2.1 Formulation

Let $G(V, E, s, t)$ be a circuit graph, where V is a node set representing BCs, i.e., pairs of an LUT and its latch, E is an edge set representing connections among BCs, s is the input node, and t is the output node. Here, each edge is weighted with the propagation delay. In addition, nodes s and t are dummy nodes that are connected to all input pins and from all output pins, respectively, and the weights of all edges connected to s and t are zero.

An example of the circuit graph is shown in Fig. 2(1). In the graph, a solid node indicates that the corresponding BC is latched.

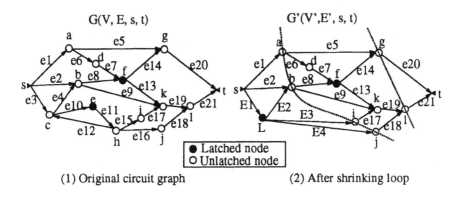

Fig. 2. A circuit graph and a loop-shrunk circuit graph.

It is known that the number of latches in a loop cannot be changed by retiming[7]. Our approach is not to actually perform retiming, but to simplify the latch insertion algorithm as described hereafter, we do not insert any latches into the nodes in the loops. Thus, shrinking the loops in G is used as a pre-process. In Fig. 2(1), nodes c, e and h form a loop so they are shrunk to node L and edges $E1$, $E2$, $E3$ and $E4$ are re-linked to the appropriate nodes as shown in Fig. 2(2). Let the loop-shrunk circuit graph be $G'(V', E', s, t)$. Problem 1 becomes the problem of finding the nodes which should be latched, under the constraint that the number of the nodes on each path from node s to node t must be the same in G'. Finding such latch insertion points is equal to finding the *articulation sets* of graph G' in graph theory. Here, the *articulation set* A is defined as follows:

Definition 1 (Articulation set). *Let $G'(V', E', s, t)$ be a loop-shrunk circuit graph. Articulation set A is defined as satisfying the following conditions: $A \subset V'$ and each path from s to t has only one element in A.* In other words, if all nodes in A with connected edges are removed from G', G' is divided into two sub-graphs; one contains s and the other contains t. In Fig. 2(2), nodes $\{a,b,i,j\}$ and $\{g,l\}$ are the articulation sets.

Changing all nodes in an articulation set into latched nodes means inserting a latch into each path from node s to node t in G'. The addition of the latches causes *1-clock response delay*. Thus, in general, if n articulation sets are found, the circuit becomes an n-clock response delay circuit. Here, consider the insertion of as many latches as possible to improve the circuit performance, ignoring the increase in the clock response delay. This is often important when designing communication or digital signal transport hardware as described in Section 1. With the addition of the above constraint, the problem we should solve is formulated as Problem 2:

Problem 2. Let $G'(V', E', s, t)$ be a loop-shrunk circuit graph. Find as many articulation sets AS as possible, where $\forall A$, $\forall B \in AS$, $\forall a \in A$, $\forall b \in B$, $a \neq b$.

3 Speed-Up Algorithm

3.1 Loop Shrinking

Let $G(V, E, s, t)$ be a circuit graph, where V is a node set representing the BCs, E is an edge set representing connections among BCs, s is the input node, and t is the output node. Here, the nodes s and t are dummy nodes and are connected to all input pins and from all output pins, respectively.

 G is a directed graph, and generally contains loops. As mentioned before, we preserve the initial latch location in loops. However, to simplify the latch insertion algorithm described in 3.2, the loops in G are shrunk. Based on graph theory, the loops can be found as *strongly connected components* (SCCs). Thus, we adopt the Depth First Search (DFS) based algorithm to find SCCs, and shrink each SCC found into a node. Here, the node is always treated as a latched node because we assumed a loop must contain at least one latched node. Let $G'(V', E', s, t)$ be the loop-shrunk graph of G. G' is a directed acyclic graph (DAG).

3.2 Latch Insertion

Our problem is to equally insert as many latches into every path as possible from node s to node t. One straightforward method is to formulate this task as a linear programming problem as in *retiming*[7]. In fact, we can obtain a solution if the problem maximizing the number of latches is solved under the constraint equations that say that the latches should exist equally in *all paths*. However, the shortcoming of the method is the combinatorial explosion in the process of searching all paths from s to t in G'. This is a very serious weakness because of its computational complexity. Thus, we adopt a graph based algorithm to solve the above problem.

 The latch insertion algorithm, i.e., articulation-set finding algorithm, is now presented. The algorithm utilizes heuristics so it does not find all articulation sets. However, it is guaranteed that the found node sets by the algorithm are the articulation sets. The latch insertion algorithm is shown in Fig. 3, and the proof is shown below.

```
Notation
x → y: y is a child of x.
x ⤳ y: y is a descendant of x.
x →?: all children of x.

1:      Procedure find_articulation_sets(G'(V', E', s, t)) {
2:          /* find descendants for each node */
3:          depth_first_search(G'(V', E', s, t));
4:          A := {s};
5:          repeat {
6:              A := {a →? | ∀a ∈ A};
7:              if (t ∈ A) exit;
8:              while (∀u ∈ A | u is a latched node, or u ⤳ v, v ∈ A) {
9:                  A := A + {u →?} − {u};
10:                 if (t ∈ A) exit;
11:             }
12:             output(A); /* a result */
13:         }
14:     }
```

Fig. 3. The latch insertion algorithm.

Proof. It is proved that the algorithm shown in Fig. 3 can find the articulation sets. $G'(V', E', s, t)$ is a DAG, so all paths started from node s must reach node t. In Fig. 3, the process in line 6 guarantees that all paths from s to t contain at least one element in the set A, because A clearly contains the descendants of node s. In addition, the processes in lines 8-11 eliminate the ancestor-descendant relationship between any two elements, or the initially latched nodes in A. This means that a path passing through a node in A never passes any other node in A. Thus, at line 12, A satisfies the conditions in Definition 1, i.e., $A \subset V'$ and each path from s to t must contain only one element in A. □

An algorithm trace of the latch insertion algorithm as applied to the example in Fig. 2(2), is shown in Fig. 4. At first, nodes $\{a, b, L\}$ are defined as set A. However, node L is a latched node, so nodes $\{i, j\}$ which are the children of L are added to A and L is removed from A. No two nodes in $\{a, b, i, j\}$ have the ancestor-descendant relationship, so $\{a, b, i, j\}$ is found as an articulation set (Fig. 4(1)). In the second iteration, $\{d, g, k, l\}$ is re-defined as A. Here, node f is a child of node b, but it is a latched node, so nodes g and k are registered in A (Fig. 4(2)). Next, node d is an ancestor of nodes g, k and l, so d is removed and f is added. However, f is a latched node and also a parent of g and k, so A becomes $\{g, k, l\}$ (Fig. 4(3)). Finally, node k is removed, because k is a parent of node l. Remaining nodes g and l satisfy the conditions of the articulation set. Consequently, two articulation sets indicated by the dotted lines in Fig. 4(4) are obtained by the algorithm in this example.

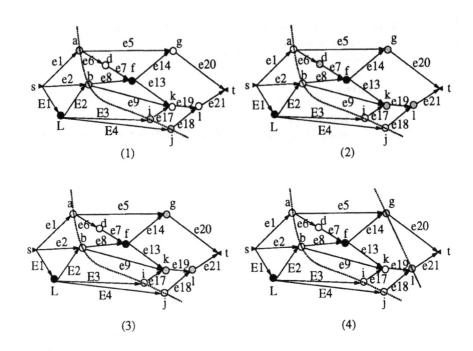

Fig. 4. A trace example of the latch insertion algorithm.

3.3 The Algorithm

The detailed procedure of our method is shown in Fig. 5. In the procedure, the fourth process is a post-process that eliminates the articulation sets that do not contribute to circuit performance.

4 Experimental Results

The presented algorithm was implemented using C language in a SUN SPARC-server 630 with 64 MB of memory. The FPGA used for this experiment was Xilinx LCA, and its placement and routing were performed by Xilinx CAD tools[15]. In Xilinx LCA, the basic block is called the Configurable Logic Block (CLB). Each CLB has two latches, and three modes: "F", "FG" and "FGM", as determined by its configuration. In creating the circuit graph, we had to consider

---

Procedure speed_up () {
    1. Create a circuit graph G from input FPGA configuration data.
    2. Shrink loops in G and make loop-shrunk circuit graph G'.
    3. Apply the latch insertion algorithm to G'. If the algorithm fails, exit.
    4. If any articulation set does not satisfy the delay constraint described below, discard the articulation set.
       Delay constraint: After changing all elements in all found articulation sets to latched nodes, there is at least one element in the currently considered articulation set which increases the maximum delay if it is changed to an unlatched node.
    5. Make new FPGA configuration data so as to insert the latches at the basic cells corresponding to the elements in the found articulation sets.
}

Fig. 5. The speed-up method for the circuit configured in an FPGA

the CLB modes, e.g., we created two separate nodes from one CLB if the CLB was in the "FG" mode which allows one CLB to act as two different logic blocks. We analyzed LCA XC3090PG175 which has 320 CLBs and 144 I/O blocks.

The experimental results determined using partitioning benchmark data (Partitioning93)[16] are shown in Table 1. In the table, "data" are data names, and data starting with character "c" indicate combinational circuits. Data starting with character "s" indicate sequential circuits. "CLB" and "I/O" indicate the numbers of CLBs and I/O blocks used. In addition, columns "A" and "B" show the maximum delay (and the minimum clock frequency, if it is a sequential circuit) of the initial circuit and of the circuit obtained after applying the presented algorithm respectively. Furthermore, "delay" indicates the *clock response delay* caused by the proposed algorithm, while "CPU" shows the execution time of the algorithm.

In the case of the Xilinx LCA, if a latch is inserted, it is necessary to route the clock signal from the clock input pin to the latch. In the above analysis, we invoked the routing tool again after latch insertion, but did not re-route any nets except to add the appropriate clock nets if routing succeeded. The initial routing data was given to the routing tool as a guide. Furthermore, no added clock nets were critical paths. Thus, our assumption for the presented algorithm, i.e., the original placement and routing results are unchanged, was satisfied in practice. For two circuits "c3540xc3" and "s953xc3", unfortunately, the algorithm succeeded but re-routing failed, i.e., some unrouted nets remained. This fact is indicated by the "NG" notation in column "B" in Table 1.

The average performance improvement ratios are 0.27 for the combination circuits, and 0.81 for the sequential circuits. This fact indicates that the proposed algorithm effectively inserts latches into the rather long critical paths common in combination circuits.

Table 1. Experimental results (Used Xilinx 3090PG175-150: 320CLB, 144I/O)

data	CLB	I/O	A: init. [ns] (MHz)	B: applied [ns] (MHz)	B/A	delay	CPU [s]
c1355xc3	70	73	98.6	22.0 (45.5)	0.22	3	2.9
c17xc3	2	7	41.1	22.0 (45.5)	0.54	3	0.0
c1908xc3	116	58	162.4	22.0 (45.5)	0.14	3	2.84
c3540xc3	283	72	265.7	NG	-	3	21.5
c432xc3	50	43	204.7	52.5 (19.0)	0.26	2	1.7
c499xc3	66	73	99.0	22.0 (45.5)	0.22	3	2.4
c880xc3	84	86	141.8	37.0 (27.0)	0.26	2	3.1
s1196xc3	143	30	139.9 (7.1)	102.2 (9.8)	0.73	2	5.0
s1238xc3	158	30	147.8 (6.8)	105.1 (9.5)	0.71	2	6.1
s1423xc3	112	24	336.1 (3.0)	317.4 (3.2)	0.94	2	3.0
s208xc3	25	15	87.4 (11.4)	65.6 (15.2)	0.75	3	0.5
s27xc3	3	7	30.9 (32.4)	22.0 (45.5)	0.71	2	0.0
s298xc3	26	11	68.4 (14.6)	54.9 (18.2)	0.80	2	0.5
s344xc3	20	22	61.8 (16.2)	47.2 (21.2)	0.76	2	0.4
s349xc3	20	22	64.1 (15.6)	49.8 (20.1)	0.78	2	0.4
s382xc3	31	11	67.7 (14.8)	53.6 (18.6)	0.79	2	0.6
s400xc3	32	11	62.3 (16.1)	51.0 (19.6)	0.82	2	0.7
s420xc3	50	23	153.8 (6.5)	132.6 (7.5)	0.86	3	1.2
s444xc3	32	11	67.3 (14.9)	52.9 (18.9)	0.79	2	0.6
s510xc3	68	28	85.2 (11.7)	71.8 (13.9)	0.84	3	1.4
s526nxc3	55	11	68.4 (14.6)	54.1 (18.5)	0.79	2	1.2
s526xc3	55	11	69.6 (14.4)	60.4 (16.5)	0.87	2	1.1
s820xc3	91	39	76.2 (13.1)	63.5 (15.8)	0.83	3	2.5
s832xc3	91	39	82.1 (12.2)	68.5 (14.6)	0.83	3	2.5
s838xc3	102	39	300.0 (3.3)	269.7 (3.7)	0.90	3	4.2
s953xc3	107	41	156.8 (6.4)	NG	-	2	3.6

5 Conclusion

We have presented a new latch insertion algorithm for synchronous circuits realized as LUT-based FPGAs. It offers considerable user support by improving circuit performance without changing the initial placement and routing results. According to experimental results gained with benchmark data, the performance of almost all circuits, including both combination and sequential circuits, is improved by the presented algorithm. For a few circuits, our algorithm succeeded, but actual latch insertion failed. However, that was due to the constraint of the Xilinx FPGA used, not the algorithm itself. In other words, if the FPGA had had a mechanism that controlled the closure of the latch located at the output of each LUT without changing the original configuration, the proposed algorithm would improve the performance of all circuits. The execution speed of the algorithm is fast so it can be applied to large scale FPGAsr, i.e., multi-chip FPGA systems.

If latch insertion fails, it is useful to apply *retiming* before invoking the presented algorithm in order to change the initial latch locations. However, the constraints of the location and number of latches in the FPGA should be considered if *retiming* is applied. This is one of our future tasks.

References

1. Brown S. D., Francis R. J., Rose J. and Vranesic Z. G. : "Field-Programmable Gate Arrays," Readings, Kluwer Academic Publishers, 1992.
2. Francis R. J., Rose J. and Chung K.: "Chortle: a Technology Mapping Program for Lookup Table-Based Field-Programmable Gate Arrays," Proc. 27th DAC, pp. 613-619, June 1990.
3. Francis R. J., Rose J. and Vranesic Z.: "Chortle-crf: Fast Technology Mapping for Lookup Table-Based FPGAs," Proc. 28th DAC, pp. 227-233, June 1991.
4. Karplus K.: "Xmap: a Technology Mapper for Table-lookup Field-Programmable Gate Arrays," Proc. 28th DAC, pp. 240-243, June 1991.
5. Rose J.: "Parallel Global Routing for Standard Cells," IEEE Trans. on CAD, Vol. 9, No. 10, pp. 1085-1095, October 1990.
6. Rose J. and Brown S.: "The Effect of Switch Box Flexibility on Routability of Field Programmable Gate Arrays," Proc. 1990 CICC, pp. 27.5.1-27.5.4, May 1990.
7. Leiserson C. E., Rose F. M. and Saxe J. B.: "Optimizing Synchronous Circuitry by Retiming," Proc. Third Caltech Conf. on VLSI, pp. 87-116, March 1983.
8. Dey S., Potkonjak M. and Rothweiler S. G: "Performance Optimization of Sequential Circuits by Eliminating Retiming Bottlenecks," Proc. ICCAD-92, pp. 504-509, November 1992.
9. Iqbal Z., Potkonjak M., Dey S. and Parker A.: "Critical Path Minimization Using Retiming and Algebraic Speed-Up," Proc. 30th DAC, pp. 573-577, June 1993.
10. Malik S., Singh K. J. and Brayton R. K.: "Performance Optimization of Pipelined Logic Circuits Using Peripheral Retiming and Resynthesis," IEEE Trans. CAD, Vol. 12, No. 5, pp. 568-578, May 1993.
11. Nakada H., Yamada K., Tsutsui A. and Ohota N.: "Latch Insertion Method without Logical Change for Synchronous Circuit Design," IEICE Trans., Vol.J75-A, No.12 pp. 1849-1858, December 1992.
12. Cloutier R. J. and Thomas D. E.: "Synthesis pf Pipeline Instruction Set Processors," Proc. 30th DAC, pp. 583-588, June 1993.
13. Park N. and Parker A. C.: "Sehwa: A Software Package for Synthesis of Pipelines from Behavioral Specifications," IEEE Trans. on CAD, Vol. 7, No. 3, pp. 356-370, March 1988.
14. Kukimoto Y. and Fujita M.: "Rectification Method for Lookup-Table Type FPGA's," Proc. ICCAD-92, pp. 54-60, November 1992.
15. —: "The Field Programmable Gate Array Data Book," Xilinx Inc., 1992.
16. Kuźnar R., Brglez F. and Kozminski K.: "Cost Minimization of Partitions into Multiple Devices," Proc. 30th DAC, pp. 315-320, June 1993.

An Efficient Technique for Mapping RTL Structures onto FPGAs

A R Naseer, M Balakrishnan, Anshul Kumar

Department of Computer Science and Engineering
Indian Institute of Technology, Delhi
New Delhi - 110 001

Abstract. This paper presents an efficient technique for realizing Data Path using FPGAs. The approach is based on exploiting the iterative structure of the datapath modules and identifying 'largest' slices of connected modules that can be mapped onto each CLB. The mapping process employs a fast decomposition algorithm for checking whether a set of slices can be realized by a single CLB. Comparison with manufacturer's proprietary software for a set of High-level synthesis benchmark structures show a significant reduction in CLB count. Another advantage of our technique is that CLB boundaries in the final design are aligned to RTL module boundaries providing modularity and ease in testing. Thus this technique is very suitable for integration as a technology mapping phase with a high-level synthesis package.

1   INTRODUCTION

Currently a number of technology options are available ranging from Fullcustom, semi-custom to the Field Programmable Gate Arrays (FPGAs). FPGAs are rapidly gaining popularity due to the short design cycle time and low manufacturing cost. There are two main classes of FPGAs[1] - Look-Up-Table (LUT) based FPGAs and Multiplexer (MUX) based FPGAs. In this paper, we are presenting a methodology for mapping RTL structures onto Look-up table based FPGAs (as exemplified by XILINX[2]. A LUT based FPGAs typically consists of a 2-dimensional array of CLBs (Configurable Logic Blocks). A variety of devices with similar architecture, but differing in the number of inputs, outputs and flipflops per CLB, are available. Further, the way in which each function generator in a CLB can be configured to implement a logic function also varies. These differences are important for both technology mapping and decomposition. In particular, these parameters define the limits on the portion of a network which can be realized by a CLB.

Recently several different approaches for LUT based technology mapping have been reported, e. g. , Chortle[3], Chortle-crf[4], MIS-pga[1], Hydra[5], X-MAP[6], VIS-MAP[7]. All these algorithms optimize the number of CLBs in the generated solutions. On the other hand some approaches like Chortle-d[8], DAG-MAP[9] and Improved MIS-pga[10], emphasize on minimizing the delay of the mapped solutions. In all these approaches, the input is a boolean network in which each node is a gate or a boolean function.

In the context of High level synthesis, the above approaches are suitable for implementing the control part of the design only. For synthesizing the data path, it is beneficial to take into account the iterative structure of data part modules. Further, there are some advantages in aligning CLB boundaries to RTL module boundaries for improving testability and enhancing modularity.

The main objective of our work is to integrate RTL structure synthesis with mapping onto FPGA technology. In this paper, we present an approach (part of a system called FAST[14, 15]) which directly realizes an RTL structure (of data path) in terms of FPGAs. FAST forms a backend to a Data path Synthesizer[13] and is being integrated to IDEAS[12]. Starting from a VHDL like behavioral description language and a global time constraint RTL data path is generated automatically. As the RTL modules are generic in nature, they cannot be directly mapped onto a single CLB. We use a dynamic slicing technique based on the iterative structure of RTL modules to partition them into component parts. Each RTL module is viewed as consisting of slices of one or more bits. Closely connected slices of different modules are considered together and mapped onto one or more logic blocks. At each stage an attempt is made to maximally utilize the logic block.

The rest of the paper is organized in five sections. Classification of RTL component cells and slicing structures are described in section 2. Section 3 presents the preliminary definitions and terms used in this paper and the expressions for computation of cost of nodes, cones and cost benefits. Algorithm for mapping RTL structures onto FPGAs is given in section 4 and decomposition technique used is briefly described in section 5. An example illustrating the technique used is presented in section 6. In section 7, results of technology mapping on XILINX devices alongwith conclusions for some high level synthesis benchmarks are presented.

2 RTL COMPONENTS AND CELLS

The approach proposed in this paper is especially suited for implementing the data path of the design because most of the data path modules are generic modules with variable widths which cannot be mapped directly onto a single CLB. An important property of the RTL modules is that they are iterative structures of basic cells, where a *cell* is an indivisible part of a module that is iterated to form a module. Because of this, an RTL module can be partitioned into an array of single-cell or multi-cell slices, where a *slice* is an array of contiguous cells of a module. For example, a 16-bit adder can be expanded into an array of 16 one-bit cell slices or 8 two-bit cell slices or 6 three-bit cell slices and so on. Cells of different RTL components can be classified into two categories depending upon the nature of their inputs and outputs.

i) **Fixed Cell :** A fixed cell has fixed number of inputs and outputs. An important characteristic of these cells is that though the component to which it belongs may be generic, but the number of inputs and outputs of the cell always remains the same in all instances of the component. For example, we can expand an 8-bit or a 16-bit or a 64-bit adder into an array of single-bit cells but in all these cases, the basic 1- bit cell has fixed number of inputs and outputs i. e. , 3-inputs (including carry in) and 2-outputs(sum and carry out).

ii) **Generic Cell** : As the name implies a generic cell has variable number of inputs and outputs. For example, a basic 1-bit cell of an 8-bit 10-input mux and an 8-bit 16-input mux differ in the number of inputs.

Examples of RTL components with fixed cells are - adder, subtractor, alu, comparator, register, counter etc. Examples of RTL components with generic cells are - MUX, AND, OR, XOR, NAND, NOR, Decoder, Encoder etc.

An RTL component is composed of basic cells, arranged in the form of an array, a tree or a combination of these. For example -

- a 16-bit ALU is an array of 16 ALU cells,
- an 8-input AND is a tree of smaller (generic) AND cells,
- a 10-input 8-bit wide MUX is an array of trees of smaller (generic) 1-bit wide multiplexers,
- a 4 to 16 decoder is an array of 4 input partial decoders. Larger decoders may be considered as arrays of trees of smaller partial decoders.

Figure 1 illustrates some of these examples. It is clear that components like Mux or decoder can be composed in more than one way because the basic cells are generic. These choices are explored by our algorithm.

3   CONES AND COST BENEFITS

The input network to FAST is an RTL structure (Data path) obtained from High Level synthesizer and output is an optimized network of CLBs. The input network is represented as a directed graph G(V,E) where each node represents a register or a functional module (ALU, Mux, etc.,) and directed edges represent connections between the modules. In the FPGA architecture we have considered, flipflops appear at the output ends of CLBs. To facilitate mapping of register slices to these flipflops, each register node in G is split into two nodes - a register input (RI) node and a register output (RO) node, not connected with each other. After this splitting RO nodes become source nodes in addition to Primary Input (PI) nodes and RI nodes become sink nodes in addition to Primary Output (PO) nodes.

We use the term *width* to refer to the number of cells in a node or a slice. Let Width(n) and Width(s) represent width of a node n and slice s respectively. *Soft-slicing* is a process that dynamically determines slice width.

Let n_k represent slice of node n with width k, then

$$\texttt{max\_slice\_width(n)} \ = \ \texttt{max k} \ | \ n_k \ \texttt{is realizable in a CLB} \qquad (1)$$

For a slice s, node(s) denotes the corresponding node. Minimum number of instances of slice s required to cover node(s) is given by

$$\texttt{inst\_cnt(s)} = \left\lceil \frac{\texttt{width(node(s))}}{\texttt{width(s)}} \right\rceil \qquad (2)$$

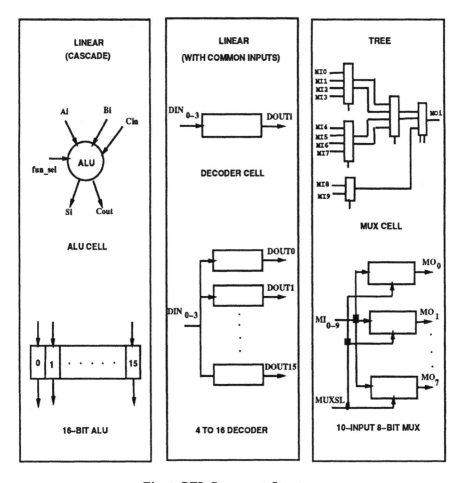

Fig. 1. RTL Component Structures

Minimum number of CLBs required to realize a node n is given by

$$\text{min\_CLB\_count}(n) = \left\lceil \frac{\text{width}(n)}{max\_slice\_width(n)} \right\rceil \qquad (3)$$
$$= \text{inst\_cnt}(n_k)$$

where k is maximum slice width

As our intention is to minimize the number of CLBs required to realize the graph, we start with an upper bound on CLBs required. This can be easily found by adding the minimum number of CLBs required for realizing each of the nodes.

$$\text{CLB\_upper\_bound}^1 = \sum_{n \in V} \text{min\_CLB\_count(n)} \qquad (4)$$

Our algorithm is based on packing slices from multiple nodes into a single CLB. This is achieved by identifying cones. A *Cone* is a set of slices of nodes which lie on paths converging on a particular node called apex of the cone. A *realizable cone* is a cone that fits in a CLB. Note that all slices are trivial form of cones and all slices of node n upto max_slice_width(n) are realizable cones. Naturally, in this context we consider only realizable cones. Further we consider only those cones which are *beneficial* i. e. , those which reduce the number of CLBs required.

As a cone could consist of slices of different widths, the number of instances of a cone required would be decided by the minimum inst_cnt of its slices.

$$\text{min\_inst\_cnt(c)} = \min_{s \in c} \text{inst\_cnt(s)} \qquad (5)$$

Let CA(c) denote the cost of realizing the nodes of cone c individually, i. e. , without forming the cone. It can be computed by simply summing the min_CLB_counts of the individual nodes that make up the cone c.

$$\text{CA(c)} = \sum_{n \in \text{node\_set(c)}} \text{min\_CLB\_count(n)} \qquad (6)$$

Let CB(c) denote the cost of realizing the nodes that comprise the cone c, with cone c formed. As each cone is realized by a CLB, min_inst_cnt(c) gives the number of CLBs realizing cone of type c. Due to differences in bit width of nodes as well as slice width of slices in c, the nodes may not be covered completely by cones. The remaining part of nodes are covered by the max_slice_width slices. We have observed that this assumption is mostly not restrictive. Therefore, CB(c) is given by the following formula -

$$\text{CB(c)} = \text{min\_inst\_cnt(c)} + \sum_{s \in c} \frac{\text{width(node(s))} - min\_inst\_cnt(c).\text{width}(s)}{max\_slice\_width(\text{node}(s))} \qquad (7)$$

Now we can quantify the benefit due to cone c as the difference between these two costs.

$$\text{Benefit}(c) = CA(c) - CB(c) \qquad (8)$$

We define a set of cones C as complete if it covers all the nodes in the graph. As per the above formulation, we consider only those cone sets in which non-trivial cones do not overlap. A non-trivial cone is one which contains slices from atleast two nodes. Cost of a cone set C denoted by CC is

$$\text{CC(C)} = \sum_{c \in C} \text{CB(c)} \qquad (9)$$

---

[1] In this paper we do not address the constraints imposed by limited number of interconnection resources available on the device.

4 FAST MAPPING ALGORITHM

The algorithm described in figure 2 shows the major steps involved in mapping RTL structure to FPGAs. Step 1 computes the CLB upper bound and step2 traverses the network and generates cones. We traverse the network backwards starting from register inputs/primary outputs and generate 'realizable' cones with non-negative 'cost-benefit' by considering various soft slicing options and merging them till no more merger is feasible or register outputs/primary inputs are reached. The feasibility of these cones are checked as they are generated and only 'realizable' ones are retained. Step 3 finds a cover which minimizes the CLB count following a greedy approach.

```
Algorithm FAST_MAP
1. Computation of CLB upper bound
      1.1  for each node n ∈ node set V of graph G
                1.1.1  compute  (i)  max_slice_width(n)
                        and  (ii)  min_CLB_count(n)
      1.2  compute CLB_upper_bound
2. Realizable Cone generation
      2.1  for all 'feasible slices' s of nodes in V do
                2.1.1  generate all 'realizable' cones
                       using FAST decomposition with
                       cost benefit ≥ 0 with s as apex
3. Minimal cone cover
      3.1  Generate complete cone sets with minimum
           cost in a 'greedy' manner
```

Fig. 2. Algorithm for Mapping RTL Structures onto FPGA's

The 'Realizable Cone generation' algorithm is shown in figure 3. The Cone-generation algorithm considers each node from the nodes of G and generates variable-width slices of width varying from 1 to max_slice_width and checks whether each slice of that node can be merged with a slice of the node at its fanin to form a cone. If the resulting cone is realizable and beneficial, then starting with this newly generated cone, it further checks whether it can be merged with slices of the node at its fanin. This process is repeated until no more slices can be packed into the cone. For each cone generated it computes the reduction in CLB count and rejects those for which no benefit occurs.

Minimal cone cover procedure given in figure 4 follows a greedy approach. It begins by sorting the list of conesets in the decreasing order of the benefit. Initially a coverset containing the first coneset of the cone_list is formed and the CLB_upper_bound is taken as the minimal cost for covering the entire net-

```
procedure Realizable_cone_generation()
  {
      Coneset =  φ
      for all  n ∈ V do
          for i = 1 to max_slice_width(n) do
              {
                  C = { nᵢ }
                  Coneset = Coneset + C
                  grow_cone(C)
              }
  }
procedure grow_cone(C)
  {
      for all u ∈ fanin_set(C) do
          if u is not a register output or a primary input
          then
              for j = 1 to max_slice_width(u) do
                  {
                      C¹ = merge(C, uⱼ)
                      if C¹ is realizable and beneficial
                        then
                          {
                              Coneset = Coneset + C¹
                              grow_cone(C¹)
                          }
                  }
  }
```

Fig. 3. Realizable cone generation

work. Next each coneset other than the first coneset is taken from the cone_list
and checked to see whether it overlaps with the coversets already generated. If
the coneset overlaps with all the coversets already generated, it creates a new
coverset with this coneset. Otherwise, it is added to all the non-overlapping coversets and minimal cost of realizing the network is made equal to the minimum
of CLB_upper_bound and cost of newly formed complete coversets. All coversets
exceeding this minimal cost are rejected.

5   DECOMPOSITION

During cone generation an important check to be performed is whether a cone is
realizable or not. A CLB is characterized by a fixed number of inputs, outputs
and flipflops. Every 'realizable' cone should have number of inputs, outputs and
flipflops less than or equal to those present in a CLB. But for realizability this

```
Procedure Minimal_cone_cover(coneset)
   {
        Cone_list = Sort_cones(coneset)
        Create a coverset = first(cone_list)
        UB = CLB_upper_bound
        for all C ∈ (cone_list except the first element) do
           {
                if(C overlaps with all coversets) then
                   Create a new coverset with C
                else
                   {
                        add C to the non_overlapping coversets
                        UB = min(UB,cost of newly formed
                                        complete coversets)
                   }
                Reject all coversets exceeding UB cost
           }
   }
```

Fig. 4. Minimal cone cover

check is not sufficient because all FPGA structures are characterized by one or more function generators which cannot realize any arbitrary function of the inputs. Typically the set of inputs have to be decomposed into two or three parts to be mappable onto a CLB.

Among the decomposition techniques employed by FPGA mapping systems, Roth-Karp[11] is the most versatile but suffers from high computation complexity. The complexity arises due to the fact that all possible combinations of variables have to be exhaustively checked for decomposition. Heuristics for fast decomposition have been developed and reported in [14]. The technique is based on checking some simple necessary conditions which candidate partitions have to satisfy for 'feasible' decomposition. Thus during the decomposition process a large number of candidate partitions are quickly rejected to achieve a speedup. An average speedup of 51.64% over Roth-Karp method has been achieved for decomposing MCNC logic synthesis benchmarks.

6   ILLUSTRATIVE EXAMPLE

The technique presented in this paper is unique in terms of its ability to map RTL structures onto FPGAs. We illustrate this technique using a GCD RTL structure obtained from IDEAS Data Part Synthesizer[13] which takes behavioral description of GCD(Greatest Common Divisor) High Level Synthesis Benchmark as input. Figure 5 a) shows the GCD RTL structure and figure 5 b) gives the CLB map of this structure for XILINX XC3000. In figure 5 b) the dotted rectangles enclosing the node(s) indicate that they can be realized using single CLBs and the number in the small square box associated with each node indicates the number of slices of that node packed to a CLB.

We traverse the GCD network starting from a register node *rega* and generate realizable cones by merging *rega* with slices of nodes at its fanin, i. e. , *muxa*. The table 1 shows the realizable cones rooted at *rega*, slices of nodes associated with these cones and CLB count. It is evident from the table that cone C_{22}

Cone	contents	CLB Count
C_{11}	1-bit slice of rega + 1-bit slice of muxa	16
C_{21}	2-bit slice of rega + 1-bit slice of muxa	16
C_{22}	2-bit slice of rega + 2-bit slice of muxa	8

Table 1. Partial Cone list

is beneficial as it requires only 8 CLBs whereas C_{11} and C_{22} both consume 16 CLBs and hence are rejected. Starting with this newly generated cone C_{22}, we further check whether it can be merged with the slices of the nodes at its fanin. Since no further merger is possible, this cone is added to the cone list. As it can be seen from the figure, that the 1-bit slice of the *alu* node has 4 inputs and 2 outputs and it cannot be merged with any other node and hence it forms a separate cone. Similarly, traversing the network from *regb* towards the primary inputs generates the next beneficial cone containing 2-bit slices of *regb* and *muxb*. Next the traversal is continued from primary outputs towards register inputs. The comparator node *cmp* has 5 inputs and 3 outputs and cannot be realized by a CLB, and hence it is decomposed into 3 sub-nodes, 1-bit slices of first two nodes occupy a single CLB whereas 2-bit slices of the third node get mapped onto one CLB. The register node *regc* is at the primary output and is realized using IOBs.

7   RESULTS AND CONCLUSION

The package FAST has been implemented on a workstation based on Motorola 68030 running at 25 MHz. We have synthesized five structures corresponding to High level synthesis benchmarks : GCD, Diff_eqn, AR_filter, Elliptic filter and Tseng structure. The mapping onto XC2000, XC3000 and XC4000 device CLBs has been performed and is reported in table 2. For benchmarks containing multipliers, we have assumed that multipliers are external to the design and are realized separately.

Table 2 lists the total number of CLBs required by FAST, XACT[2] and

---

[2] XACT is a proprietary product of XILINX and interfaces with a Schematic Capture tool for mapping onto XILINX devices.

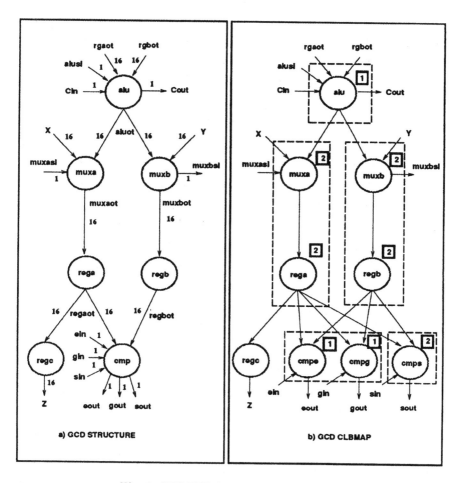

Fig. 5. GCD RTL Structure and CLB Map

| FPGA device class | Mapper | Total # of CLBs for HLS Benchmarks | | | | |
		GCD	Diif_eqn	AR_filter	Elliptic filter	Tseng
XC2000	FAST	96	320	240	656	630
	XACT	104	338	264	688	662
XC3000	FAST	56	160	120	328	317
	XACT	62	168	126	344	328
XC4000	FAST	40	120	104	256	294
	XACT	48	142	113	278	316
	XBLOX	40	136	104	272	304

Table 2. Total CLB count for HLS Benchmarks

XBLOX[3] for realizing the network after the mapping process. The synthesis technique reported in this paper results in a CLB count reduction of upto 16% over XACT and upto 12% over XBLOX. In addition to this, FAST has a much lower execution time. The details of the structure used are summarized in table 3.

RTL	HLS Benchmarks examples									
components	GCD		Diif_eqn		AR_filter		Elliptic		Tseng	
	width	no	width	no	width	no	width	no	width	no
ALU	16-bit	1	16-bit	1	16-bit	2	16-bit	1	16-bit	2
COMP	16-bit	1	-	-	-	-	-	-	16-bit	1
									1-bit	1
REG	16-bit	3	16-bit	6	16-bit	5	16-bit	12	16-bit	17
AND (16-bit)	-	-	-	-	-	-	-	-	2-inp	1
OR (16-bit)	-	-	-	-	-	-	-	-	2-inp	1
MUX (16-bit)	2-inp	2	5-inp	2	4-inp	3	16-inp	2	8-inp	1
			4-inp	2	2-inp	5	8-inp	1	4-inp	14
			2-inp	4			4-inp	4	2-inp	3
							2-inp	4		

Table 3. Component Details of HLS Benchmark Structures

To conclude, we have presented an approach for mapping RTL structures onto FPGAs. The technique is primarily meant for datapart of the design and effectively utilizes iterative structure of the data part components. The slices of well connected components are generated and are called cones. These functions form the inputs to the decomposition process. The approach is flexible and can handle a variety of look-up table based FPGAs and utilizes the architectural features of commercial devices(like intermediate outputs). The synthesized CLB boundaries correspond to RTL component boundaries which would imply ease in testability and simulation.

ACKNOWLEDGEMENTS

This work is partially supported by Department Of Electronics(DOE), Govt. of India under IDEAS project and Ministry of Human Resource Development (MHRD), Govt. of India under QIP programme.

[3] XBLOX is a recent product from XILINX and supports MSI level module library

References

1. R. Murgai, et al., "Logic Synthesis for Programmable Gate Arrays", Proc. 27th Design Automation Conf., June 1990, pp. 620-625.
2. Xilinx Programmable Gate Array Users' Guide, 1988 Xilinx, Inc.
3. R. J. Francis, J. Rose, K. Chung, "Chortle: A Technology Mapping Program for Lookup Table based Field Programmable Gate Arrays", Proc. 27th Design Automation Conf., June 1990, pp. 613-619.
4. R. J. Francis, J. Rose, Z. Vranesic, "Chortle-crf: Fast Technology Mapping for Look-up Table-Based FPGAs", Proc. 28th Design Automation Conf. 1991, pp.227-233.
5. D. Filo, J. C. Yang, F. Malihot, G.D. Micheli, "Technology Mapping for a Two-output RAM-based Field Programmable Gate Arrays", European Design Automation Conf., February 1991, pp. 534-538.
6. K. Karplus, "Xmap: A Technology Mapper for Table-Lookup Field Programmable Gate Arrays", Proc. 28th Design Automation Conf., June 1991, pp. 240-243.
7. Nam-Sung Woo, "A heuristic Method for FPGA Technology Mapping Based on Edge Visibility", Proc. 28th Design Automation Conf.,June 1991, pp. 248-251.
8. R. J. Francis, J. Rose, Z. Vranesic, "Technology Mapping of Look- up Table- Based FPGAs for performance", Proc. Int. Conf. on CAD, 1991, pp.568-571.
9. Kuang-Chien Chen et al., "DAG-Map: Graph-Based FPGA Technology Mapping for Delay Optimization", IEEE Design & Test , September 1992, pp. 7-20.
10. R. Murgai et al., "Performance-Directed Synthesis for Table Look-up Programmable Gate Arrays", Proc. Int. Conf. on CAD, 1991, pp. 572-575.
11. P. J. Roth and R. M. Karp, "Minimization over Boolean graphs", IBM Journal of Research and Development vol. 6/No.2/April 1962 pp. 227-238.
12. Anshul Kumar et al. "IDEAS : A Tool for VLSI CAD ", IEEE Design and Test, 1989, pp.50-57.
13. M. V. Rao, M. Balakrishnan and Anshul Kumar, "DESSERT : Design Space Exploration of RT Level Components", Proc. IEEE/ACM 6th Int. Conf. on VLSI Design'93, January 1993, pp. 299-303.
14. A. R. Naseer, M. Balakrishnan and Anshul Kumar , "FAST : FPGA targeted RTL structure Synthesis Technique", Proc. IEEE/ACM 7th Int. Conf. on VLSI Design'94 January 1994, pp. 21-24.
15. A. R. Naseer, M. Balakrishnan and Anshul Kumar , "A technique for synthesizing Data Part using FPGAs", Proc. IEEE/ACM 2nd Int. Workshop on Field Programmable Gate Arrays, Berkeley, February 1994.

A Testbench Design Method Suitable for FPGA-based Prototyping of Reactive Systems

Volker Hamann

Institut für Technische Informatik
Vienna University of Technology
Treitlstrasse 3 - 182/2
A-1040 Wien, AUSTRIA

email: *volker@vlsivie.tuwien.ac.at*

Abstract. As reactive systems are growing more and more complex, costs stemming from a misconception in early design phases like requirements analysis and system specification show a tendency to explode. This makes rapid prototyping inevitable for extensive simulation in early design phases. In this article, we investigate into the usability of field programmable logic for designing a "real-world" testbench suitable for simulating the environment of a prototype. We present a top-down design method which significantly reduces effort, especially when FPGAs are also used as implementation technology for the system prototype.

1 Introduction

Each environment of a system can be seen as a system of its own. When verifying a system under development (SUD), it suffices to show that the interaction between the system itself and the system called "environment" at no time leaves the specification as long as certain constraints are fulfilled. Most of the time, this interaction is performed according to a state-oriented protocol.

Statecharts [Har87] have proven to be a good capture and simulation basis for designing state-based reactive systems. Statecharts are not only suitable for designing such systems themselves but can also be used for specifying their working environment and thus may serve as a basis for testbench development. Instead of being bothered with the development of case studies and/or test vectors, the designer creates a model of the environment and steps down the design process in parallel with the SUD itself.

Since the introduction of the FPGA technology, their power and consequently their field of usage have increased to a high extent. Several applications of medium to high complexity have been realized and lots of tools have been created supporting top-down synthesis from higher design levels like hardware description languages (e.g. VHDL [IEEE88]) or logic level encoding based on schematics [View91]. However, verification techniques have not been able to keep up with this development. Most simulation tools accept test vectors as input and produce either a kind of waveform or a "go-nogo" statement when comparing simulation results with expected results.

Section 2 will describe a possible design flow supporting codesign of a system and its testbench. In Section 3 this method is being discussed and some results are presented.

2 Design Flow

The design flow is shown in Fig. 1. The testbench as well as the SUD are modelled in quite the same way, possibly even in parallel. As a front end, a Statecharts tool like SPeeDCHART [Spd93] can be used. This tool offers a VHDL generation facility. Thus, the SUD can be simulated directly in Statecharts or in an VHDL environment, e.g. Synopsys [Syn92]. After refining the VHDL code with hand-written entities, which has proven to be useful when designing modules of higher complexity, it can be synthesized to logic level using e.g. the Xilinx [Xil93] backend available for Synopsys and downloaded into FPGAs. Here it is possible to make the connections between SUD and testbench by the internal routing channels of an FPGA, thus using the same chip for testbench and SUD. If, however, the design is too large, the connections are realized by external wiring between the various FPGA chips.

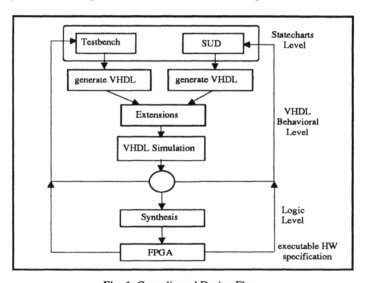

Fig. 1: Co-ordinated Design Flow.

3 Discussion and Results

The proposed method has several advantages:

1.) The designer is obliged to spend more thoughts on the environment of the system, almost as much as in a design method based on formal semantics. As he does so in an early design phase, costs for searching for the optimum solution can be kept small.

2.) High level design tools like Statemate or SPeeDCHART encourage the usage of statechart models not only for the design itself but also for simulation testbenches, so why throw them away when stepping down to the next level?

3.) When reaching the FPGA prototype level, it is much cheaper to test the system prototype by connecting it to another FPGA prototype of the system environment. Especially in an industrial environment where expensive machines are controlled by electronic devices, erroneous behavior in the integration test stage may have disastrous consequences.

4.) When a system consists of several submodules, a single submodule may see the rest of them as its testbench. By having a thoroughly designed testbench, at least parts of it can be reused for modelling other components of the system under development. This is especially true for interacting replications of equally structured entities.

Of course, this method - if used stand-alone - will not be feasible for optimizations necessary for ASIC development based on processes different from FPL technologies. This is due to the fact that it views system and environment (or subsystem and other subsystems) very much as separate entities, neglecting behavior that they might have in common. Anyway, this is not really a need as long as a system is only prototyped. Also, the method will not be helpful when statemachine based design is out of question unless different front-end tools are empoyed.

The method is currently being tested for feasibility in the design of a fuzzy controller. As far as we can see now, there is a significant improvement in design time and fault coverage compared to standard testing methods. Especially when designing the rulebase, the interactive testing and observation of the external behavior is eased in a considerable way [SH94]. The rulebase is first modelled with Statecharts and tested without fuzzifier and defuzzifier. Appropriate fuzzified input values are presented by the testbench, and output values are evaluated for soundness giving an idea how stable the system is and how sensible it reacts to changes of input stimuli.

Afterwards, fuzzifier and defuzzifier models are created in VHDL and tested together with the already designed rulebase. Logic synthesis and FPGA implementation finish the prototyping cycle. Along with the model, the testbench is refined and made more and more similar to the real operating environment and is finally also brought to FPGA level, allowing for "real-time" observations.

References

[IEEE88] IEEE, *IEEE Standard VHDL Language Reference Manual*, IEEE, 1988.

[Har87]   David Harel: *"Statecharts: A Visual Formalism for Computer Systems"*, in: Science of Computer Programming, Vol. 8, North Holland, 1987.

[SH94]   Valentina Salapura, Volker Hamann: *"Using Statecharts and Embedded VHDL for Fuzzy Controller Design"*, Proc. of VHDL-Forum for CAD in Europe, 1994.

[Spd93]   Speed: *"SPeeDCHART Reference Manual"*, Speed SA Neuchatel (Switzerland), 1993.

[Syn92]   Synopsys, *"Design Compiler Reference Manual"*, Version 3.0., Synopsys Inc., 1992.

[View91] VIEWlogic: *"Workview Reference Manual"*, Version 4.1, VIEWlogic Systems Inc., 1991.

[Xil93]   Xilinx: *"The Programmable Logic Data Book"*, Xilinx Inc., 1993.

Using Consensusless Covers for Fast Operating on Boolean Functions

Eugene Goldberg   Ludmila Krasilnikova

Institute of Engineering Cybernetics, the Academy of
Sciences of Belarus, Surganov str.6, Minsk 220012, Republic of
Belarus, fax: ++7 0172 31 84 03, phone: ++ 7 0172 39 51 71,
e-mail: katkov@adonis.iasnet.com

Abstract. The paper presents a method for fast operating on covers of Boolean functions. The method develops the one based on the unate paradigm (UP)[1]. The proposed method differs from the UP one in two aspects (1) the initial cover is decomposed into a set of prime rather than unate subcovers, (2) prime covers are obtained by applying to each reched not prime subcover either branching by the Shannon expansion or a procedure of making the subcover consensusless in a variable by the consensus operation. Experiments on MCNC-91 two-level logic benchmarks and random functions show that operations based on the proposed method are less laborious than their UP based counterparts.

1 Basic Definitions and Propositions

To present the proposed method we need to introduce some definitions and propositions.

A subset $C = S_1 \times .. \times S_n$ of the Boolean n-space $\{0,1\}^n$ is said to be a *cube* . A subset S_i we shall call the i-th component of C. A cube C is called an *implicant* of a completely specified single-output Boolean function f if $C \subseteq ON\_SET(f)$ where $ON\_SET(f)$ is the vertices of the n-space in which f evaluates to 1. An implicant C is said to be a *prime* if any cube strictly containing C is not an implicant. A set of implicants of f which contain any vertex of $ON\_SET(f)$ is called a cover of f. We shall call a cover *prime* if all primes of the Boolean function specified by the cover are in the cover.

Cubes C' , C'' are said to be *orthogonal* in the k-th variable if $S'_k \cap S''_k = \emptyset$. Let cubes C' and C'' be orthogonal only in the j-th variable. Then a cube $S'_1 \cap S''_1 \times .. \times S'_j \cup S''_j \times S'_n \cap S''_n$ is said to be produced by the *consensus* operation [2]. We shall call a pair of cubes which are orthogonal only in one variable a *consensus pair* in the variable. We shall call a cover *consensusless in a variable* if any cube produced by the consensus operation from a pair of cubes from F which are consensus pair in the variable is contained in some cube from F.

Denote by F_{x_j} and $F_{\overline{x}_j}$ subcovers (sometimes called cofactors with respect to x_j and $\overline{x}_j$ [2]) of subfunctions $f(x_1, .., 1, ..x_n)$ and $f(x_1, .., 0, ..x_n)$ of function $f(x_1, .., x_j, .., x_n)$ formed from a cover F of f.

Proposition 1 *If a cover F is consensusless in all variables, F is prime.*

Proposition 2 *Let a cover F be consensusless in the j-th variable. Then the covers F_{x_k} and $F_{\overline{x}_k}$, $k \neq j$ obtained by Shannon expansion or any cover F'*

obtained by adding to F a set of cubes produced by the consensus operation are consensusless in the j-th variable too.

Proposition 3 *Let F be a cover and F'_j be a set of all cubes produced from consensus pairs of cubes from F orthogonal in the j-th variable. Then the equivalent to F cover $F^*_j = F \cup F'_j$ is consensusless in the j-th variable .*

The proofs of the propositions are omitted for short.

2 Formulation of the Method

The UP method is to apply the Shannon expansion to the initial cover to decompose the cover into a set of unate subcovers. This allows one to substitute operating on the cover for doing on the unate subcovers. The main property of unate covers that makes performing many operations trivial is that a unate cover is prime [2]. The key point of the method presented in the paper is to decompose the operated cover into prime subcovers which, generally speaking, may not be unate.

The method consists in recursive performing the following algorithm. (1) If there is a variable in which a subcover F is not unate and an ancestor of F was not made consensusless in the variable then step 2 is performed. Otherwise F is prime.(2) The subcover F is either decomposed by branching in a "not processed" variable or made consensusless in one of such variables.

By making the subcover F consensusless in variable x_j is meant the described in proposition 3 procedure of substituting F for cover F^*_j. To choose between the two alternatives in step 2, the values $a = min(2|F| - (n^j_1 + n^j_0))$ and $b = min(|F| + n^j_1 * n^j_0)$ are calculated where n^j_1 and n^j_0 is the number of cubes from F the j-th component of which is equal to $\{0\}$ and $\{1\}$ respectively. The value of $2|F| - (n^j_1 + n^j_0)$ is equal to $|F_{x_j}| + |F_{\overline{x}_j}|$ and so a describes the most effective way of branching. The value of $|F| + n^j_1 * n^j_0$ is the upper bound of $|F^*_j|$ and so b describes the most effective way of making the cover F consensusless in a variable.If $a \leq b$ the branching in a variable minimizing $2|F| - (n^j_1 + n^j_0)$ is chosen. Otherwise from F cover F^*_j is obtained where j is the index of variable for which $|F| + n^j_1 * n^j_0$ is minimum.

Justification of the method is based on propositions 1-3.

3 Experimental Results

To evaluate the efficiency of using the proposed method (further referred to as the EPC- (expansion plus consensus) method) programs EPC-Reduce and EPC-Decomposition have been written. EPC-Reduce implements Reduce operation used in the two-level logic minimizer Espresso [2]. The program was applied to a number of MCNC-91 two-level examples (table 1). When implementing the operation the extension of the EPC-method to the case of multi-output Boolean functions was made. The program EPC-Decomposition is intended just to decompose the cover into a set of prime covers. The program was applied to single-output covers obtained by the pseudorandom number generator (table 2).

Both programs were compared with their UP based counterparts written in accordance with [2]. To compare the performance of operations based on the UP- and EPC- methods it is reasonable to use the total number of the variables processed when reaching prime subcovers. For the UP-based operations the number is equal to the number of branchings and for the EPC-method based ones the number is equal to the sum of the number of branchings and the number of variables in which subcovers were made consensusless. Programs EPC-Reduce and UP-Reduce were applied to covers obtained after performing Irredundant_Cover procedure during the first iteration of the minimization loop of Espresso [2]. When constructing a random cover the number of components of a cube different from {0,1} was uniformly distributed in the range $[r_1, r_2]$ shown in table 2.

Table 1

Examp.	n_c	n_i	n_o	n_{exp}	n_{con}	n_{epc}	n_{up}
apex2	459	39	3	6749	2219	8968	13561
apex5	308	117	88	1210	339	1549	1566
alu4	360	14	8	4595	2624	6219	6310
cordic	1082	23	2	2824	2694	5518	51474
cps	106	24	109	216	57	273	268
ex4	141	128	28	584	344	928	1033
seq	262	41	35	821	209	1030	1056
misex3	70	14	14	1807	384	2191	2792

Table 2

Examp.	n_c	n_i	r_1	r_2	n_{exp}	n_{con}	n_{epc}	n_{up}
1	100	10	3	4	310	311	621	851
2	100	10	6	7	222	41	263	268
3	50	20	4	4	403	249	652	5692
4	50	20	3	5	446	305	751	7584
5	60	20	3	5	1493	846	329	15241
6	60	20	4	4	1173	576	749	14253
7	100	30	8	18	2289	1556	845	3209
8	100	30	6	7	15154	11027	2181	95789

n_c, n_i, n_o - the number of cubes, inputs and outputs respectively, n_{exp} , n_{con} - the number of variables processed in EPC-based programs by expansion and consensus operation, n_{epc}, n_{up}- the total number of variables processed by EPC- and UP-based programs.

References

[1] Brayton R.K., e.a. Fast recursive Boolean function manipulation , *Proc. Int. Symp. Circuits and Syst. Rome, Italy, May 1982*,pp.58-62.

[2] Brayton R.K., e.a. *Logic minimization algorithms for VLSI synthesis.* Norwell, MA: Kluwer Academic Publishers, 1984.

Formal Verification of Timing Rules in Design Specifications

Tibor Bartos, Norbert Fristacky

Abstract:

An algorithm for formal verification of the set of timing rules that express timing discipline in digital systems is described. It is based on a digital system specification model and notation transferrable to VHDL and concerns formal consistency verification at the design level of system specification development procedure.

1. Introduction

In the process of top-down synthesis, developed system specifications have to be verified even during the specification refinement process, for it must be assured that the derived specification is consistent and correct, and that towards it any synthesised implementation can be formally verified.

This paper concentrates on the formal verification of specified timing discipline that should hold in the system and its environment (i.e. it ignores the functional verification). The specifications are expressed in frame of a higher-level specification model we employ.

2. Specification model

We will represent time as a finite real interval TI and variables as functions of type $X: TI \rightarrow DX \cup \{u\}$ where DX is a finite domain and u denotes an unspecified value. A finite number of changes in TI is supposed. Variables are described as sequences of *events* (changes of values). Two types of events exist: *up(X,v,i)* (X changes from any value to v; i is an index used to distinguish between different changes to the same value) and *down(X,v,i)* (X changes from v to any other value). For every up(X,v,i), down(X,v,i) exists. Symbol *tm(e)* denotes the time when event e occurs.

A *digital system* S is described by input, output and state variables. Let V, H, Q be sets of vectors of input, output and state variable values, respectively. The vectors are called *input vectors, output vectors* and *states* of S. Let v(t), h(t) and q(t) denote their values indicated in time point t. A *timed input/output word* (shortly i/o word) of system S is a finite sequence $w = (v_1, e_1, h_1) (v_2, e_2, h_2) \ldots (v_N, e_N, h_N)$. Every (v_i, e_i, h_i) is a *timed input/output vector* (shortly i/o vector), where e_i is an event, $v_i = v(tm(e_i))$ and $h_i = h(tm(e_i))$, $tm(e_1) < tm(e_2) < \ldots < tm(e_N)$. Events e_i are called *timing events*. The last timing event e_N is called *final timing event* ef.

The model [1] is based on entities named *agents* that specify a partial behaviour of the system S over finite time interval. An agent specifies a particular set of i/o communications (given by i/o words and initial states) taking place in a time interval and the final state at the end of the interval. All agents reflect deterministic finite state machine behaviours. The timing discipline of an agent is specified by a set of predicates called *timing rules* [1], [2], [3]. To achieve the behaviour specified by an agent, all timing rules have to be fulfilled. The following types of rules exist: *Delays* specify that a difference between two event occurence times is equal to a given timing parameter, *constraints* specify that a difference between two event occurence times is not less than a given timing parameter. *Stability rules* specify that the value of a given variable is stable in an interval containing a timing event. They can be converted to constraints. *Or-rules* are disjunctions of previous types of rules.

3. Verification algorithm

As the timing of events is "closed" within the agents, we can verify the timing of every agent separately. The timing of an agent is correct if the timing rules are true for every i/o word w specified by the agent. Change of a variable in w generates an event. Let $E(w)$ be the set of all these events and $TR(w)$ the set of timing rules that deal only with events from $E(w)$. The timing of w is consistent if for every event $e \in E(w)$ exists an occurence time $tm(e)$ such that all timing rules from $TR(w)$ are true and event e_1 occurs in w before event e_2 if and only if $tm(e_1) < tm(e_2)$.

An approach to verification of timing rules over a graph representing *timing diagrams* was developed in [5], [3], [2]. A timing diagram is semantically equal to a sequence of events together with the set of timing rules. We have concentrated on the question how the set of i/o words and timing rules can be transformed to a graph, in order to use the known approach. As a result, timing rules in an agent are verified if:

1. All possible i/o words are generated from a regular expression that describes the set of i/o communications.

2. For every i/o word, all events resulting from changes of input and output variables are added to the sequence of timing events at appropriate positions, creating so-called *precedence graph*. The precedence graph is a graph G where nodes represent events and oriented weighted edges represent the precedence (or relative time order and "distance") of events. The mapping from nodes to events is a one-to-one function, so we will use the names of events as names of nodes. The edges have two weights. There is an edge from e_1 to e_2 with weights p and s in G if either $s = 0$ and $tm(e_2) \geq tm(e_1) + p$ or $s = 1$ and $tm(e_2) > tm(e_1) + p$.

Whenever an "up" event is added to the graph, also the corresponding "down" is added and they are connected by an edge with weights $p = 0$ and $s = 1$. If an "up" event describing a change of variable X is added to the graph, an edge with weights $p = 0$ and $s = 0$ from last "down" event for X to this event is added. If an edge with weights p_0 and s_0 is being added, but another edge with weights p_n and s_n connecting given nodes already exists, only the weights of the existing edge are updated. The resulting weights will be p_n, s_n if $p_n > p_0$; p_0, s_0 if $p_n < p_0$ or ($p_n = p_0$ and $s_n = 0$); p_0, 1 if $p_n = p_0$ and $s_n = 1$. The graph G is initially empty.

The precedence graph is created as follows: Let $w = (v_1, e_1, h_1) \dots (v_N, e_N, h_N)$ be the input i/o word, where $e_1, \dots, e_N$ are timing events, $v_1, \dots, v_N$ and $h_1, \dots, h_N$ are vectors of values of input and output variables. For every $i = 1, \dots, N$:

a) the node representing the timing event e_i is added along with an edge with weights $p = 0$ and $s = 1$ from previously added timing event to this event.

b) Events generated by changes of values of every variable X $(X(tm(e_i)) \neq u)$ in vectors v_i and h_i are added. If X does not change its value, only an edge with weights $p = 0$ and $s = 1$ from e_i to the last "down" event for X is added. If the value of X changes, corresponding "up" and "down" events are added together with two edges. The first one connects the "up" event with e_i and has weights $p = 0$ and $s = 0$ and the second one connects e_i with the "down" event and has weights $p = 0$ and $s = 1$.

3. Edges representing the set of timing rules are added to the graph. Edges are added only if there exist both nodes they connect (so that only rules from TR(w) are added for i/o word w). For constraints, one edge with weight p equal to the timing parameter is added. For delays, two edges with weight $s = 0$ between given events e_1, e_2 are added: an edge from e_1 to e_2 with weight $p = x$ and an edge from e_2 to e_1 with weight $p = -x$, where x is the given timing parameter.

It is possible to prove [3] that the set of timing rules is inconsistent iff the precedence graph contains a positive cycle (a cycle in which either the sum of p-weights of edges is positive, or the sum of p-weights is zero and there is at least one edge with s-weight equal to 1 in the cycle).

Satisfying one of the subrules contained in an or-rule suffices for the entire or-rule to be satisfied. If one of the subrules causes existence of a positive cycle in the graph, adding another one cannot cancel it. As a result, the consistence of an or-rule can be verified by searching all graphs where exactly one of subrules contained in the or-rule is added. Therefore, after all "simple" rules have been added to the graph, one of subrules from every or-rule is added, and the graph is searched for positive cycles. The subrules are then removed from the graph and the process is repeated until all combinations of subrules are checked. The whole set of rules is consistent, if at least one consistent combination is found.

4. Conclusion

This algorithm is intended to be used in the process of specification design to ensure that the developed specification is correct. It was implemented as a program in C. A simple language was defined as input notation for that program. Definition of the input notation and implementation details can be found in [4].

References:

[1] Fristacky, N., Cingel, V.: A functional and timing specification model for digital systems. Proc. of the 7th Symp. on Microcomp. and Microproc. App., Budapest, 1992, pp. 185-190.
[2] Cingel, V.: A graph based method for timing diagrams representation and verification. In Correct Hardware Design Methodol. CHARME 93, Arles France, Springer Verlag, 1993.
[3] Cingel, V.: Specification and Verification of Timing in Digital Systems. Ph.D. Thesis, Dept. of Comp. Science and Eng., Slovak Techn. Univ., Bratislava, 1991 (in Slovak).
[4] Bartos, T.: Program for Verification of Timing Rules in Digital System Specifications. Diploma Thesis, Faculty of El. Eng., Slovak Techn. Univ., Bratislava, 1993 (in Slovak).
[5] Jahanian, F., Mok, A. K.: A Graph-theoretic Approach for Timing Analysis and its Implementation. IEEE Tr. on Computer, Vol. 8, 1987, pp. 961-975.

Optimized Synthesis of Self-Testable Finite State Machines (FSM) Using BIST-PST Structures in Altera Structures

Andrzej Hlawiczka-Senior Member, IEEE[1], Jacek Binda[1]

Technical University of Gliwice, Poland email:<hlawiczoss.iele.gliwice.edu.pl>

Abstract. The testing of PCBs containig ASICs, e.g., Altera FPGA is an important problem which needs consideration. One of the ideas of solving this problem is using BIST architecture for each ASIC. With the use of built-in testers, the additional cost, in the form of overhead of macrocells is added. A certain idea of built-in tester structures is BIST-PST [1]. The disadventage of this idea is, that the FSM memory block in form of MISR with a given characteristic polynomial may be realized only in form of: IE-MISR and EE-MISR. In our paper, the new kind of MISR registers consisting of D and T flip-flops has been used in BIST-PST. They make it possible for a given characteristic polynomial to achieve a wide range of possible realizations of MISR type memory block, ranging from tens to thousands. In effect, it is possible to choose the minimal excitation function saving a considerable number of Altera FPGA macrocells.

1 Optimized Synthesis

The idea of BIST-PST structures given by Wunderlich in [1], and presented in Fig. 1, didn't include the solution of problems concerning synthesis of self-testable FSMs using minimum cost, e.g., standard cells in Altera 7000 family. The main drawback of the theory proposed in [1] was used, for a given primitive polynomial, only the Internal EXOR gates based Multi Input Signature Registers (IE-MISR) and External EXOR gates based MISR (EE-MISR). On the basis of the Wunderlich theory, the primitive polynomial, e.g., $p(x) = 1 + x + x^3$ has only two following realizations: D⊕DD and DD⊕D (see Fig. 2). In effect the number of

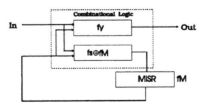

Fig. 1. The BIST-PST architecture

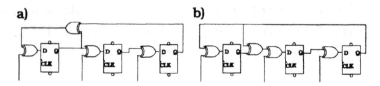

Fig. 2. Realizations of MISR: a – D⊕DD, b – DD$^{\oplus}$D

different excitation functions f_M of above mentioned MISRs is very small. So the chance of choosing the optimized new excitation function $f_M \oplus f_S$ (f_S – excitation function of sequential circuits) was too poor. This chance rapidly grows using new class of MISRs presented in [2]. The paper [2] deals with a uniform algebraic description of operations of MISRs consisting of D and T flip-flops (DT) and their combinations with XOR gate based linear feedback path. Structures based on the internal (external) linear feedback path with new kind of D or T flip-flops have a huge number of possible realizations of MISR registers named IEDT-MISR and EEDT-MISR. The main point of the new theory is discovery the possibility of designing DT type linear registers using XOR gates contained in the T flip-flops (JK f/fs) instead of XOR gates placed in the feedback path. The example of $p(x)$ mentioned above obtain now new five realizations of IEDT-MISRs and EEDT-MISRs: DTT; TDT; TTD; T$^{\oplus}$TT; TT$\oplus$T. The DT type registers operate faster than any of their equivalents with XOR gates in the feedback path and what's more are less complicated and use less number of cells in some of PLDs, FPGAs etc. The more is the number of MISR's stages the more rapidly increases the number of IEDT-MISR and EEDT-MISR realizations for given polynomial. For example some of ten degrees polynomials have hundreds and some of seventeen degrees polynomials have thousands or more realizations of IEDT-MISRs and EEDT-MISRs. Discovering the rules of finding out IEDT-MISR (EEDT-MISR) with primitive polynomial quaranteing the minimal or close to (quasi) minimal cost of realization of the function $(f_S \oplus f_M) \oplus f_M$, was the main point of this new theory. According to the rules of creating the excitation functions of i-th stage of register of the BIST-PST memory block structure presented in [2] ($f'_{Si} = f_{Si} \oplus f_{Mi}$), the new excitation function for i-th stages of the sequential circuit has been created. Owing to its compensation by the f_{Mi} in the MISR type memory block, the truth table of the sequential circuit is unchanged. The realization of such a function is $f_{Si} = (f_{Si} \oplus f_{Mi}) \oplus f_{Mi} = f'_{Si} \oplus f_{Mi}$. The main task of the optimization relied on finding out the type of IEDT-MISR register, having primitive polynomial and which f_{Mi} formula for each (i-th) stage warranty the minimal realization for every $f_{Si} \oplus f_{Mi}$ functions. Depending on values of p_i and k_i factors (Fig.6 in [2]), the function $f'_{Si} = f_{Si} \oplus f_{Mi}$ may be for n-bit register stated in four formulas (1)

$$f'_{Si} = f_{Si} \oplus q_{i-1}$$
$$f'_{Si} = f_{Si} \oplus q_{i-1} \oplus q_n \qquad (1)$$
$$f'_{Si} = f_{Si} \oplus q_{i-1} \oplus q_i$$

$$f'_{Si} = f_{Si} \oplus q_{i-1} \oplus q_i \oplus q_n$$

Every of (1) formulas carry in the different cost factors of the i-th stage of IEDT-MISR register's realistation. To achieve the minimal f'_{Si} function the minimization process based on Karnaugh maps must be done. Their number is stated by the expression: $N_n = 4(n-1)$ where n denotes the length of the register. As a result is the complexity (the number of terms) of the i-th stage of IEDT-MISRs. Having the minimal complexity of the f'_{Si} functions of the IEDT-MISRs i-th stages, the choice of the minimal cost IEDT-MISR register may be realized. This set of minimal f'_{Si} excitation functions of i-th stages, makes possible to determine the characteristic polynomial of the pointed IEDT-MISR register. If achieved in this way polynomial is the primitive one, the designed register based on it is regarded as proper to realize the minimal BIST-PST structures. In other way, the next from the set of the f'_{Si} minimal realizations of the i-th MISRs stage is chosen. This process is repeated until the primitive polynomial is reached. Comparing the way of the IEDT-MISR realization to the IE-MISR ones, the following conclusions have been created:

- there are only 2^{n-2} registers consisting of the D flip-flops,
- there are $2^{2(n-1)} = 4^{(n-1)} = 2^{(n-2)}2n$ possible to achieve IEDT-MISR registers based on the D and T flip-flops,
- the number of realizations of the IEDT-MISRs is 2^n times more than the number of realizations of the IE-MISRs.

So the chance of finding out the possible minimal realizations of IEDT-MISRs in comparison to IE-MISRs grows rapidly, together with the length of the register (2^n). Based on the concept of the new class of IEDT-MISR registers, a new family of FPGA circuits - Altera 7000 was taken to realize this theory. Thanks to existence of the XOR gates connected to programmable flip/flop and used to build internal (external) feedback paths, the idea of constructing the new kind of MISR registers based on the given primitive polynomial was realized. Owing to this new theory, one can reduce not only the number of macrocells used to realize project, but also to change the proportions between the number of macrocells used in realization of FSM and the number of macrocells used in realization of the BIST.

References

1. B. Eschermann, H.J. Wunderlich: Optimized Synthesis of Self Testable Finite State Machines. Proc. 20th Int. Symp.Fault-Tolerant Computing, pp.390-397,1990
2. A. Hławiczka: D or T flip-flop based linear registers. Archives of Control Sciences Volume 1(XXXVII) No.3-4, pp.249-268, 1992

A High-Speed Rotation Processor

Jan Lichtermann, Günter Neustädter

Department of Computer Science, University of Kaiserslautern
D-67653 Kaiserslautern, PO-Box 3049, Germany

Abstract. We present a high-speed rotation processor for rotating digital images based on the backrotation algorithm. The design is part of our research project on real time volume visualization architectures. The processor is implemented on an Actel FPGA and the solution is compared to a Xilinx implementation. Performance measurements show a throughput of more than 360 images per second with 256^2 pixels per image. Each pixel is represented with 12 bit data.

1 Introduction

Modern imaging techniques in medicine like computer tomographie or magnetic resonance imaging produce digital images, called slices, of the human anatomy. Combining adjacent slices of one scene results in a volumetric data set. Volume visualization methods, eg. raytracing directly through the sampled data points, make it possible to compute images that show a perspective (3D) view of these volumetric data sets. To achieve real time computation speed (20 pictures per second) we separate the entire visualization process into data volume rotation and raytracing with fixed observer position [2]. Therefore we can use a pipeline architecture for these computations. In this paper we present a high-speed rotation processor for the first pipeline step [3]. It is capable of rotating 320 slices with 256 x 256 data points with 12 bit intensity resolution in one second, equivalent to 16 slices at the required rotation speed.

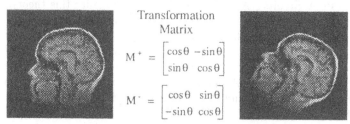

$$M^+ = \begin{bmatrix} \cos\theta & -\sin\theta \\ \sin\theta & \cos\theta \end{bmatrix}$$

$$M^- = \begin{bmatrix} \cos\theta & \sin\theta \\ -\sin\theta & \cos\theta \end{bmatrix}$$

Fig. 1. MR picture of a head before and after 35^0 rotation

2 Image Rotation

The sample points in the slices are on a rectangular, evenly spaced grid. They can be described by integer coordinates in a source coordinate system. 2D-Rotation by angle θ can be described by matrix M^+ for counterclockwise and M^- for clockwise rotation. In the rotated slice, the new sampling points are described in a destination coordinate system.

The *backrotation algorithm* enumerates all points in the destination slice and transforms their coordinates back into the source coordinate system by applying the inverse matrix of the rotation matrix. Enumerating these points in a regular fashion leads to simplifications in the algorithm because line drawing DDA algorithms can be used to easily compute the points in the source coordinate system [1]. The following algorithm (figure 2) enumerates the points in a slice column by column.

/* transform the coordinates of the first point of first column p'(x',y') from the
   destination coordinate system into source coordinates p(x,y): */
$x_{bak} = x = x' \cdot \cos\theta + y' \cdot \sin\theta, \ y_{bak} = y = y' \cdot \cos\theta - x' \cdot \sin\theta$
for all columns of the destination slice {
   for all points p'(x',y') in the column {
      compute the intensity at point p'(x',y') from the intensities of the neighborhood
      of point p(x,y) in the source slice (resampling)
      $x = x + \sin\theta, \ y = y + \cos\theta$ }
   $x = x_{bak} = x_{bak} + \cos\theta, \ \ y = y_{bak} = y_{bak} - \sin\theta$ }

Fig. 2. Backrotation algorithm using line drawing

The new sampling points p(x,y) generally do not exactly meet the integer coordinates in the source coordinate system. In a *resampling* step we use bilinear interpolation to compute the intensity i(p) at a point p(x,y) between grid points from the intensities of the four surrounding grid points.

3 Rotation Processor

The architecture of our rotation processor is shown in figure 3. The source memory consists of four memory banks and is able to store 16 slices. Partitioning the source memory into four banks allows parallel access to all four values necessary for bilinear interpolation (memory interleaving). Data read from source memory is routed through a crossbar switch to the inputs of the bilinear interpolator. The result of the interpolation is stored in destination memory which is of the same size as the whole source memory. Crossbar switch and bilinear interpolator are realized by a standard cell design (die size 40 mm^2) with ES2 1.5μ cell library.

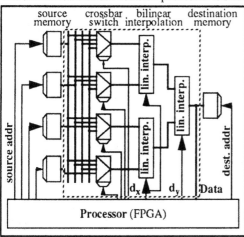

Fig. 3. Architecture of backrotation processor

The backrotation algorithm (figure 2) has been implemented in an Actel FPGA A-1280-1. The address generator traverses a slice column by column and computes the source coordinates of the four points necessary for bilinear interpolation. The source coordinates are split up into integer parts which give the addresses to access the intensity values at the grid points and fractional parts which are the weights (dx, dy) for bilinear interpolation. Computation of the first point in the first slice takes seven clock cycles. After that, we are able to rotate one pixel within each 42 ns clock cycle.

4 FPGA-Design Experiences

The Actel A-1280-1 contains 1232 logic modules (8000 gate array equivalent gates). In our design we have a module utilization of about 80%. Therefore it was difficult to achieve our timing requirements. Our limit was a clock period of at most 47ns. Actel design tools predicted a probability of 94% for complete wiring. Only about half of the 50 placement and routing attempts succeeded. Initial critical path analysis showed that the design would not reach the goal (30% too slow). Avoiding large fanouts by duplicating parts of the logic in critical paths and using additional buffers and gates with few pins improved the critical path by about 5ns. Another 5-10ns could be saved by marking nets more or less critical. We observed that the delay of one net differs between 8ns and 15ns after distinct place and route attempts. Nets not marked critical driven by modules with a fanout of 8 or more can easily reach delays up to 50ns.

Finally, after all improvements, critical path analysis tools reported a delay of 55ns for worst case process (chip fabrication) and operating conditions 75^0C ambient temperature at 4.75V supply voltage. For typical process conditions and 25^0C ambient temperature at 5.0V supply voltage the result was 46ns.

The design was retargeted to the Xilinx XC4000 family. This was done by giving a VHDL description of the processor to the Mentor AutoLogic synthesis tool. Since the synthesized adders turned out to be much too slow (120 ns) we replaced them by hand designed adders. Xilinx place and route tools suggested an XC4008 with 324 configurable logic blocks (approx. 8000 gates). After placement and routing the tools reported a CLB utilization of 82% and a critical path length of 52.9ns for an XC4008 with speed grade -5 (5ns CLB delay). The critical path analysis reflects worst case values over the recommended operating conditions. The current version of AutoLogic is not able to optimize the critical path length during technology mapping for Xilinx designs. Several place and route attempts led to comparable results. Even the attempt to use an XC4010 with more CLBs and routing resources did not improve the result.

Thus we can conclude that both FPGA families, Actel and Xilinx , are suitable for our application and result in designs (Actel 55ns worst case versus Xilinx 52.9ns worst case) with similar speed.

References

1.   Joel H. Dedrick: Transforming Digital Images in Real Time. The Electronic Design Magazine, 27-30, August 1987

2.   Jan Lichtermann, Gangolf Mittelhäußer: Eine Hardwarearchitektur zur Echtzeitvisualisierung von Volumendaten durch 'Direct Volume Rendering', GI Workshop Visualisierungstechniken, Stuttgart, June 1991 (in German)

3.   Jan Lichtermann, Günter Neustädter: A High-Speed Rotation Processor for the PIV^2-Architecture, SFB 124 Report, University of Kaiserslautern, Germany, 1994

The MD5 Message-Digest Algorithm in the XILINX FPGA

P. Gramata[1] P. Trebatický and E. Gramatová[2]

[1] INFOTRANS,
Dúbravská cesta 9, 842 20 Bratislava, Slovak Republic
[2] Institute of Computer Systems of Slovak Academy of Sciences,
Dúbravská cesta 9, 842 37 Bratislava, Slovak Republic

Security algorithms are very often implemented by the FPGA technology because only a limited amount of circuits with these algorithms are necessary for production.

This paper describes some possibilities for implementation of the MD5 Message-Digest Algorithm [1] by means of XILINX FPGA's. A message digest algorithm (one-way hash function generator) takes as input a message of arbitrary length and produces as output say 128-bit "fingerprint" or "message digest" of the inputs. Such algorithms are often usefull for digital signature applications.

Several algorithms of this kind are known. SNEFRU, FFT-Hash II, MD4, MD5 algorithms, or the I.S.O. proposal are among them [1], [2], [3].

We chose the MD5 algorithm for implementation for several reasons:

- it was good accepted by experts,
- it needs minimum ROM memory, so it can be fully implemented in one circuit (higher security),
- it offers good trade-offs for implementation of hash functions with respect of development time and price,
- only one small change (new initial constants) allows to distinguish different users.

The algorithm begins with the initial constant MD0 and processes n blocks of words by means of four round functions - FF, GG, HH and II as follows:

$$MD_i = MD_{i-1} + II(M_i, HH(M_i, GG(M_i, FF(M_i, MD_{i-1})))). \qquad (1)$$

Each round consists of 16 steps, where in each step must be values of four registers A,B,C,D calculated according to formula:

$$A = B + ((A + f(B,C,D) + x[s] + t[i]) << k). \qquad (2)$$

The XILINX XC4000 FPGA family was chosen for implementation of this algorithm [5]. Several experiments have been executed to find optimal architecture. Results of them are described in this paper. The main goal was to achieve a circuit, which can be "fed" by words of a message practically with the speed of the PC bus. Then, we can use the circuit as a peripheral device of a PC computer.

XC4000 FPGA's were chosen because of their possibility to implement very effectively ROM and RAM memories. ROM memory was used as a table instead of computing of function t [i] = 4294967296 * abs (sin (i)). Dimension of the table is 64x32 bits. Another ROM tables contain values for shift- ing (32x5 bits) and initial constants (4x32 bits). As the architecture of such circuit is "heavy multiplexed", we have preferred XC4000 family to XC3000 one.

RAM memory contains 16 32-bit words. Some experiments have shown, that this part of the circuit and substantial part of control block could be saved, when sequence of steps in the MD5 algorithm is modified by sequence of input words. But such modification of the algorithm gives lower security.

RAM's and ROM's dimensions don't allow to implement them with a small amount of packages on board. Therefore for the full implementation of the algorithm, they can be put into one XC4004 package with 82 active i/o pins. This circuit is interfaced to another one, where all computational steps of the MD5 algorithm are done. Content of the 16x32 data RAM, from which the fingerprint is being received, is defined by the controlling PC computer, i.e. final padding of the RAM is its task. Structure of the circuit is on Fig.1.

The most critical part of the algorithm from the point of view of its performance was the summing of two 32 bit words. Implementation of dedicated carry logic in CLB's of the XC4000 FPGA's was really substantial advantage. All other sub-functions were implemented as synchronous with a clock signal.

Trade-offs between price and performance of the implemented algorithm forced us to propose "modified MD5 algorithm". The main cause of the modification is based on the fact, that the original algorithm has 32-bit structure, which gives 128-bit fingerprint but it needs 32-bit buses which are limiting factor for routing. We have experimented with 8-bit and 16-bit structures, which offer 32-bit nad 64-bit fingerprints. Such fingerprints can be used in specific applications, where CFB mode of operation for ciphering is used.

The main part of the control logic consists of 13-bit state shift register, which generates impulses for the basic function, described by equation (2). Control signals for 64 steps in 4 rounds, described by (1), are generated by another state machine.

Table 1 shows possible implementation of these structures in XC4000 FPGA's without RAM and ROM memories. In these implementations was the limiting factor not the number of gates, but number of interconnections among blocks. As a final result, we received a class of circuits, containing MD5 algorithm or its modification with respect of their complexity and price. The simplest version is implemented into XC 4004A FPGA.

Some experiments are running now to receive a circuit for message digest with fingerprint larger than 128 bits, based on a modification of the MD5 algorithm. In such circuit can be also other arithmetical functions used, to achieve better confusion for cryptoanalysts. Such circuit will be useful e.g. in an archiving system with high performance, where security and throughput are determined also by chosen hash function.

References

1. R. Rivest.: "The MD4 Message Digest Algorithm", *RFC 1320, MIT and Data Security, Inc.*, April 1992.
2. R. Rivest.: "The MD5 Message Digest Algorithm", *RFC 1321, MIT and Data Security, Inc.*, April 1992.

3. "ISO/IEC Draft 10118 Information Technology - Security Techniques - Hash functions," ISO 1992
4. C.P. Schorr, "FFT-Hash II, Efficient Cryptographic Hashing," *Proc. of EURO-CRYPT*, Budapest 1992, pp. 44-47.
5. "The Programmable Logic Data Book," *XILINX, 2100 logic Drive, San Jose*, CA 95124, 1993.

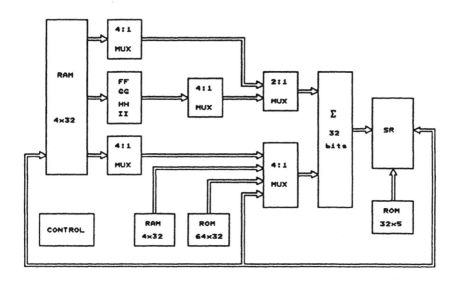

Figure 1.
Block Structure of circuit

Structure	Number of gates	Implementation
8 bits	2 550	XC4004A
16 bits	4 600	XC4005A
32 bits	8 600	XC4010

Table 1.
Implementation of different structures
of the MD5 algorithm

A Reprogrammable Processor
for Fractal Image Compression

Barry Fagin

Department of Computer Science
United States Air Force Academy
2354 Fairchild Drive Suite 6K41
USAF Academy, Colorado 80840-6234 USA

Pichet Chintrakulchai

Thayer School of Engineering
Dartmouth College
Hanover, NH 03755-8000 USA

Abstract. Fractal image compression appears to be a good candidate for implementation with a reprogrammable processor. It is computationally intensive, slow on existing technology, and employs a few basic, well-defined functions that are clear candidates for hardware implementation. We discuss our implementation of a reconfigurable processor for fractal image compression, used to evaluate the utility of different compression methods faster than a software-only approach.

1 Introduction

Fractal image compression has recently been proposed as an alternative to JPEG and other compression techniques [1,2,3]. By identifying an appropriate set of affine transformations, images can be stored as mathematical functions which can then be iterated on an arbitrary initial data set to approximate the original image. The challenge is to construct the transformation set. Because the search space is large, fractal image compression is computationally intensive. Fractal image compression of a 128 x 128 color image can take over 4 hours on an RS/6000.

The long turnaround time for a software-only implementation enhances the difficulty of experimentation with different search and comparison functions. We require both 1) improved performance of frequently executed functions, and 2) the ability to evaluate different functions as candidates for use in fractal image compression. This suggests the use of a reconfigurable coprocessor.

2 Architecture

Our coprocessor system is designed to work with the Apple Power Macintosh computer using the NuBus interface [4], as shown in Figure 1:

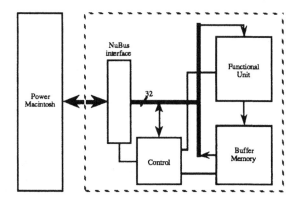

Fig. 1. Coprocessor Architecture

The coprocessor functions in either in compute mode or program mode. In program mode, the user feeds configuration data for a particular candidate function through the NuBus into the FPGAs. In compute mode, the host computer pumps input data through the NuBus which is then fed through the programmable functional unit. Output results are stored in the buffer memory. The functional unit is user-programmable and set up according to the directed acyclic graph (DAG) of the candidate operation. This DAG is the data flow pipeline of the candidate operation where many basic units can operate in parallel. Data dependency is the only limit for the degree of parallelism in each stage; results dependent upon the outputs of the current stage become the next stage of the pipeline. This way, we are able to exploit the parallelism both within a single stage of the pipe and across the entire pipeline where results are pumped out of the pipe every cycle after the pipe is filled.

3 Candidate Operation

Profile results gathered from running a fractal compression program on sample images on IBM RS6000/340 workstation show that the program spends more than 80% of total run time computing the absolute error between image blocks. This suggests that this function is a good candidate for hardware implementation.

Figure 2 shows the DAG for absolute error computation. Our calculations indicate implementation requirements of 123 CLB's and 64 I/O pins, easily within the limits of a Xilinx 4000-series FPGA [5]. Our results indicate that this routine can be sped up by approximately a factor of 50. Applying Amdahl's Law reduces the speedup to an approximate factor of 4, still a significant improvement. We are looking to incorporate other candidate functions into hardware, including genetic algorithms for searching the problem space and user-defined functions for determining similarity between portions of images.

This work was supported by the National Science Foundation under award # MIP - 9222643.

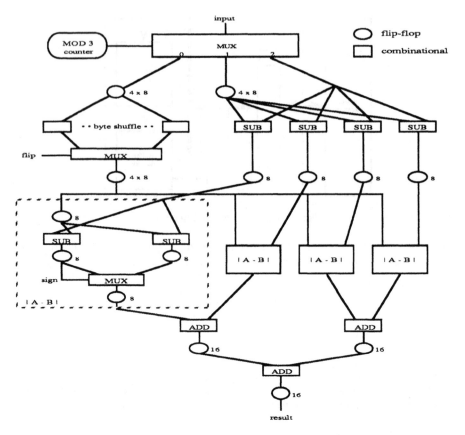

Fig. 2. DAG for 2x2 Pixel Error Computation

References

1. Barnsley, M.F. and Hurd, L.P., *Fractal Image Compression*, A.K. Peters, 1993.

2. Fisher, Y., Jacobs, E.W. and Boss, R.D., *Fractal Image Compression using Iterated Transforms*, Image and Text Compression, J. A. Storer Editor, Kluwer Academic Publishers, 1992.

3. Jacquin, A.E., *Image Coding Based on a Fractal Theory of Iterated Contractive Image Transformations*, IEEE Trans. on Image Processing, Vol. 1 (1), 1992, p. 18-30.

4. Apple Computer, Inc., *Designing Cards and Drivers for the Macintosh Family (3rd Edition)*, Addison-Wesley, 1992

5. Xilinx Inc., *The Programmable Logic Data Book*, San Jose, CA, 1993.

Implementing GCD Systolic Arrays on FPGA

Tudor Jebelean

RISC-Linz, A-4040 Linz, Austria
jebelean@risc.uni-linz.ac.at

Abstract. We implement on Field Programmable Gate Arrays from Amtel (old CLi) three systolic algorithms for the computation of greatest common divisor of integers. The experiments show that elimination of global broadcasting significantly reduces both area and time consumption. We eliminate broadcasting by using a novel technique which is more suitable to arithmetic algorithms than Leiserson conversion lemma.

This report is part of an on-going research aimed at speeding-up exact-arithmetic systems by adding a systolic dedicated coprocessor. Computation of the greatest common divisor (GCD) of long integers is an important and expensive subalgorithm in exact arithmetic and in some cryptography schemes. Practically, the systolic approach has been so far the most successful in parallelizing this algorithm - see [BrKu85], [YuZh86], [Guyo91].

We compare three GCD systolic implementations using field programmable logic from Amtel (old CLi): a **semisystolic** parallelization of a variant of Brent-Kung plus-minus algorithm, using signed arithmetic and global broadcasting; a purely **systolic** algorithm previously described in [Jebe93], which is obtained from the first by eliminating global broadcasting; an **improved** version of the purely systolic algorithm by halving the number of cells.

We start from a simplified version of the plus-minus algorithm introduced by [BrKu85]. In the description in figure 1, A, B are the input integers and a_0, a_1, b_0, b_1 denote their least-significant bits. A termination and correctness proof of this algorithm are presented in [Jebe94].

1 Systolic Algorithms

In order to parallelize the plus-minus algorithm systolically, the operands are represented using signed-digits, for avoiding carry propagation. Each cell contains 9 registers (variables): the 3-bit signed-digit representation of each operand; one-bit "tags" ta, tb which show the sign-bits of each operand; sa: the sign of A (finally of the result). Cell 0 (the rightmost) determines the "instruction" to be executed, which is than broadcasted to all the other cells (there are 5 "instructions"). The tags ta, tb are essential for termination detection in the presence of variable-length operands. The tags are set to 1 in those processors containing

---

* Supported by Austrian FWF, project P10002-PHY.

while $B \neq 0$ do
    if $a_0 = 0$ and $b_0 = 0$ then [shift both]
        $(A, B) \leftarrow (A/2, B/2); (A_s, B_s) \leftarrow (A, B)$
    if $a_0 = 0$ and $b_0 = 1$ then [interchange and shift]
        $(A, B) \leftarrow (B, A/2)$
    if $a_0 = 1$ and $b_0 = 0$ then [shift B]
        $(A, B) \leftarrow (A, B/2)$
    if $a_0 = 1$ and $b_0 = 1$ then [plus-minus and shift]
        if $a_1 = b_1$ then $(A, B) \leftarrow (B, (A - B)/2)$
              else $(A, B) \leftarrow (B, (A + B)/2)$
return A [the pseudo-GCD], A_s, B_s

Fig. 1. The plus-minus algorithm.

non-significant bits, and shifted rightward when possible. When tb reaches cell 0, than we know B is null. Also, the value of the lowest tagged bit of A is shifted rightward using sa. At the end this will indicate the sign of the result.

For the elimination of global broadcasting we do not use Leisersons systolic conversion scheme [Leis82]. This method would triplicate the number of registers and the new circuit will inherit the complications of signed-redundant arithmetic. Rather, we us a novel technique, which is more suitable to least-significant bits first arithmetic algorithms: the carries/borrows are rippled along together with the "instruction" signal (which has now 8 values), and the operands are represented in the classical fashion. This reduces the number of registers needed for operands by a factor of 3, compensating the registers added for buffering the instruction signal: only 8 registers are used now instead of 9. The details of this systolic algorithm are presented in [Jebe93].

Similarly to what happens in Leiserson systolic conversion method, our broadcasting-elimination scheme also introduces a slow-down of 2: the array is used with only 50% efficiency. In order to improve this situation, we "pack" two cells in one: a new "double" cell will contain the registers of two old neighboring cells, but only one implementation of the transition function. The function is multiplexed alternatively between the two sets of registers. This brings the theoretical efficiency to 100%, however from the practical point of view only a small improvement in area consumption can be noted.

Nevertheless our systolic algorithms represent a significant improvement over Brent-Kung scheme [BrKu85]. While the old device needed 4n cells, for n-bit operands, each with 24 registers, our devices need n cells with 8 registers (respectively $n/2$ cells with 16 registers) - hence a reduction of 12 times. The running time of the old algorithm is $4n$ for any inputs, the running time of our algorithm is $4.34n$ in the average.

2   Practical Experiments and Conclusions

The three designs were implemented for 8-bits operands using Amtel FPGA development system (successor of Concurrent Logic), which includes CAD from Viewlogic. The layout was done automatically on a 6005 CLi chip (has $56 * 56 = 3,136$ primitive cells), and was successful only for the purely systolic designs. For the first (semi-systolic) design, 97 nets out of 694 could not be routed. Figure 2 presents the resulting data.

	semisystolic	systolic	improved
macros	710	476	426
registers	98	89	92
gates	607	377	313
equivalent gates	2,928	2,167	2,210
cells used			
before layout	704	466	405
after layout	2,649	1,889	1,576
time/clock (ns)			
before layout	60.8	56.4	64.0
after layout	650	450	440
increase (times)	11	8	7

Fig. 2. Comparison of the three designs.

The most important conclusion of the experiments is that elimination of global broadcasting using our novel technique is benefic from all points of view: the purely systolic algorithm was **simple** enough for the layout software to be able to process it successfully; the **area consumption** of the semi-systolic device is bigger by 32% (equivalent gates), 73% (logic cells), 68% (layout cells); the **timing** of the semi-systolic device is bigger by 8% (logic), 47% (layout).

References

BrKu85 R. P. Brent, H. T. Kung, *A systolic algorithm for integer GCD computation*, 7th IEEE Symp. on Computer Arithmetic.

Guyo91 A. Guyot, *OCAPI: Architecture of a VLSI coprocessor for the GCD and extended GCD of large numbers*, 10th IEEE Symposium on Computer Arithmetic.

Jebe93 T. Jebelean, *Systolic normalization of rational numbers*, ASAP'93.

Jebe94 T. Jebelean, *Systolic algorithms for long integer GCD computation*, Conpar94.

Leis82 C. E. Leiserson, *Area-efficient VLSI computation*, PhD Thesis, Carnegie Mellon University, MIT Press, 1982.

YuZh86 D. Y. Y. Yun, C. N. Zhang, *A fast carry-free algorithm and hardware design for extended integer GCD computation*, ACM SYMSAC'86.

Formal CAD Techniques for Safety-Critical FPGA Design and Deployment in Embedded Subsystems

R.B. Hughes[1], G. Musgrave[2]

[1] Abstract Hardware Limited, No. 1 – Brunel Science Park,
Kingston Lane, Uxbridge, Middlesex,
UB8 3PQ, U.K.
[2] Department of Electrical Engineering and Electronics,
Brunel University, Uxbridge, Middlesex,
UB8 3PH, U.K.

Abstract. In this short paper we describe the formal specification of interface chips which are used in embedded subsystems. The typical applications come from the areas of mission critical systems which are most commonly found in the avionics and space industries. Our application, by which we illustrate our formal techniques for the design of an embedded FPGA controller is that of an ABS (anti-lock braking system) as used by the automotive industry. We describe our innovative technological approach for ASIC design and show that it may equally well be applied to the area of FPGA design which are more cost-effective for small production runs or where the system specification may need to be changed at short notice.

1 Introduction

This paper describes part of our on-going work[1, 2, 3] to formally specify interface chips which are used in embedded subsystems. The typical applications with which we illustrate our technique are the use of FPGAs in safety-critical applications in avionics and space, which require extremely high levels of mission reliability, extended maintenance-free operation, or both. The automotive industry is also increasing its use of control and interface chips in engine management, cruise control and ABS subsystems. The need for design assurance, increasingly of a contractual nature, has led to the increasing use of formal methods in this area[4]. Our use of formal methods is industrial, our industry-provided example being developed on a commercially available formal CAD toolset (LAMBDA), and not purely academic. We show how the LAMBDA (Logic and Mathematics Behind Design Automation) system, which very successfully exploits many years of academic research and whose logical core is based on HOL[5, 6], can be used to address these problems.

2 Overview of Technological Approach

At the core of our approach is a theorem-proving tool in which a specification[7] can be transformed through a series of rule transformations into a design which is correct by construction[8]. The design decisions are made, interactively, by the engineer and the system automatically introduces constraints (e.g. on timing, connections of inputs and outputs and wiring) as a result of the partitioning decisions made by the engineer.

The current design state is represented as a rule which must keep track of formal relationship between four things, viz. the original specification, the implementation developed so far, the work which remains to be done and environmental constraints introduced by the system. Initially the rule is a tautology, which is valid, and this is transformed by the theorem prover as the designer makes implementation decisions.

IF	current_design + further_work	ACHIEVES	task_n
AND	...		
AND	current_design + further_work	ACHIEVES	task_1
THEN	current_design + further_work + constraints	ACHIEVES	specification

We do not go into detail about the rule transformations, as these are not what the engineer needs, or wants, to work with. Such mathematical details need to be, and are, hidden from the engineer; all that the engineer sees is a graphical view of the current state of the implementation and a view of what design work remains to be done to provide an implementation which satisfies the specification. If details of the logical transformation process are of interest then see[1, 3].

The engineer makes a series of partitioning decisions(c.f. [9]), some of which are aided by the system, and transcends a design hierarchy. We advocate that this approach, which is currently being commercially used for ASICs, can be applied to the design FPL. The primitive elements, i.e. the leaves of the design tree, are the cells of the FPGA, the connections between them having been generated automatically in a mathematically rigorous formal manner This highly novel approach is illustrated by example for the design of an FPGA for deployment in an anti-lock braking system. An FPGA is worthwhile when the ideal specification of the FPGA system is not known. By this, we are referring to the possibility of a braking system in which the brakes may be independent or one in which front and rear are grouped together. These systems have different mechanical properties relating to yaw movement of the car. In some systems, including the phase II version of our development, the wheel angle is taken into account in the specification so that increased braking force may be applied to the wheels on the inside of the steering curve, thus increasing the car's ability to corner while braking. To change the specification and reproduce a formal interconnect to the FPGA cells is much easier, and hence less costly, than redesigning and refabricating a dedicated ASIC.

3 Conclusions

In general, the theorem-proving research emphasis towards microprocessors has left the formal development of FPGAs relatively unexplored. This is extremely unfortunate since, as mentioned in the introduction, these are now being designed into safety-critical systems; it can also be argued that, because the user base for an FPGA is so much smaller than for a mature commercial microprocessor, a design design flaw in an FPGA is more likely to find its way into a deployed critical system. Clearly, this situation needs to be addressed. What we have shown is a thoroughly practical approach for the formal specification and formal design of FPGAs. The automatic generation of firmware for embedded microcontrollers has also been demonstrated, and provides a small step in the direction of hardware/software codesign yet addresses a very large sector (60–70%) of the software requirements of the codesign market. Further work in this area is certainly required. Our approach is also suitable for highly-distributed systems but more "real-world" problems need to be tackled by industrial designers adopting our methods in-house; it is only from feedback gained by such experiences that further refinement and customisation of our technique for the particular problems of various niche areas can be achieved.

References

1. R. B. Hughes, M. D. Francis, S. P. Finn, and G. Musgrave. Formal tools for tri-state design in busses. In L.J.M. Claesen and M.J.C. Gordon, editors, *IFIP Transactions: Higher Order Logic Theorem Proving and Its Applications (A-20)*, pages 459–474, Amsterdam, The Netherlands, 1993. Elsevier Science Publishers B.V. (North-Holland). ISSN 0926-5473.
2. G. Musgrave, S. Finn, M. Francis, R. Harris, and R. Hughes. Formal Methods in the Electronic Design Environment. In *Proceedings of the NORCHIP Conference*, Finland, October 1992.
3. R.B. Hughes and G. Musgrave. Design-Flow Graph Partitioning for Formal Hardware/Software Codesign. In J.W. Rozenblit and K. Buchenrieder, editors, *Codesign: Computer-Aided Software/Hardware Engineering*, chapter 10. (to be published by IEEE Computer Society Press), September 1994.
4. Fura, Windley, and Cohen. Towards the formal specification of the requirement and design of a processor interface unit. NASA Contractor 4521, Boeing Space and Defense Systems, 1993.
5. M. Gordon. Why Higher-Order Logic is a good conclusion for specifying and verifying hardware. In G. Milne and P.A. Subrahmanyam, editors, *Formal Aspects of VLSI Design*. North-Holland, 1986.
6. Gordon and Melham. *Introduction to HOL: A Theorem Proving Environment for Higher Order Logic*. Cambridge University Press, 1993.
7. K.D. Müller-Glaser and J. Bortolazzi. An approach to computer aided specification. *JSSC*, 25(2):45–47, April 1990.
8. G. Musgrave, S. Finn, M. Francis, R. Harris, and R.B. Hughes. Formal Methods and Their Future. In F. Pichler and R. Moreno Díaz, editors, *Computer Aided Systems Theory – EUROCAST'93*, pages 180–189. Springer-Verlag, Heidelberg, January 1994.
9. E. D. Lagnese and D. E. Thomas. Architectural partitioning for system level synthesis of integrated circuits. *Transactions on Computer-Aided Design*, 10(7):847–860, July 1991.

Direct Sequence Spread Spectrum Digital Radio DSP Prototyping Using Xilinx FPGAs

T.Saluvere, D.Kerek, H.Tenhunen

Royal Institute of Technology
KTH-Electrum, ESD-lab, Electrum 229, 164 40 Kista, Sweden

Abstract. Spread spectrum digital radio receivers and transmitters are very difficult to simulate for overall system performance evaluation. Reliable estimates
for Bit Error Rates and effect of indoor and outdoor fading radio channels can be
best studied via practical hardware measurements. In this work we propose a
flexible CDMA spread spectrum radio architecture structure well suited for
FPGA prototyping. FPGA based rapid system prototyping techniques provide
complementary information than simulations and also facilities earlier system
integration activities across different project groups.

1 Introduction

The design of future mobile communication systems requires thorough performance analysis before the hardware can be built. Many of the design options can only be
evaluated and characterised in real working environment making the analytical or
completely simulation based approaches infeasible. Due to real radio environment
with multipath fading, interference and interactions with natural noise sources, rapid
system prototyping techniques need to be adopted as an integral part of the design
cycle. Prototyping will not replace simulations, but will complement and identify specific problem areas which need to be characterised in more detail with modelling and
simulations.

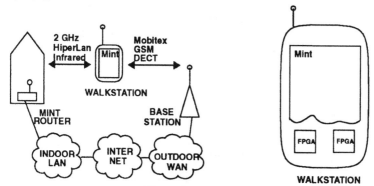

Fig. 1. System overview

Our work relates to testing system solutions in mobile high data rate communication network and building the first testbenches in order to demonstrate the feasibility and application integration[1].

Mobile users expect global network connectivity, mobility transparent applications, and quality of services comparable to that of fixed networks. This will require the integration of fixed, and mobile indoor local area and outdoor wide-area networks in order to provide global connectivity. In practice this will require from portable radio subsystems flexibility to handle multiple radio air interfaces based on availability and quality of services and air interfaces. Appropriate radio interfaces will be controlled and selected by a system management function (implemented as a software) to meet criteria such as achievable throughput and delay, real time requirements, usage cost, and impact of selected communication link battery lifetime and currently available radio transmission power.

2 System overview

2.1 Xilinx FPGA based digital receiver

In order to have a configurable and flexible radio interface to the host-MINT [1] computer the main signal processing tasks for encoding transmitted and extracting received data are performed digitally using Xilinx FPGAs. For communication with analog world, 8-bit AD and DA video speed converters are used and, whole DSP itself is implemented on 4 XC4000 series PLCC84 chips to achieve also flexibility of available logic on the board (see fig.2).

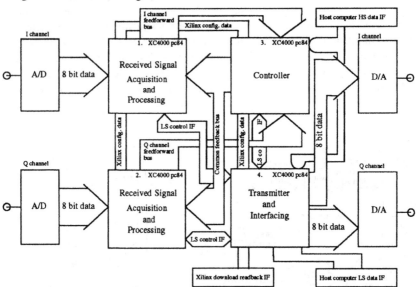

Fig. 2. Xilinx XC4000 based flexible DSP for DS CDMA digital radio

In general, there is 2 AD interfaced inputs to the board for I and Q channel and also

2 DA interfaced outputs. For communication with host computer DSP board is equipped with Xilinx download-readback and also with low - and high speed control interfaces. System partitioning is done in a way that the most extensive incoming signal processing will normally be done in first chips directly connected to ADs thereafter extracted control information is passed to third chip which acts as a controller and is passing needed correction information back to the first two chips. The fourth chip is dedicated for transmitting and interfacing functions. Such an architecture is turned out to be universal enough to test out several different design solutions without need to alter PCB. Different modulation and access schemes, data transmission speeds and coding algorithms can be utilized just by downloading different chip configuration. Xilinx configuration data is possible to download in a daisychain way or every chip individually.

The homodyne direct conversion receiver is used in the radio frontend. After mixing down and amplifying the received signal, the DSP is fed with[1]:

$$I(t)=data(t)*pn(t)*cos(\omega t) \text{ and } Q(t)=data(t)*pn(t)*sin(\omega t)$$

The actual data extraction and synchronization and received power estimation are done digitally. One of the non-trivial problems one must face with a DS CDMA system is the synchronization of the PN sequence in the receiver. Different solutions have been proposed from which Noncoherent Tracking Loop is used due to inherent simplicity for VLSI (Xilinx FPGA) implementation.

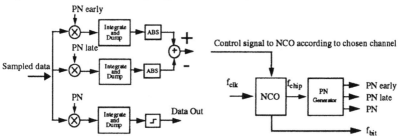

Fig. 3. Functional diagram of the digital receiver

The received signal is multiplied with three shifted replicas of the PN sequence. Each of the PN sequence is shifted by half a chip in time from each other. Correlated input signals are integrated over the one bit arrival time T_b and thereafter dumped.

As shown in fig. 3, the data is extracted from the in time integrator, and by taking the difference of the late and early we get an error signal which is feed back to a digital variable oscillator. It is implemented with the help of an NCO and a PN generator.

3 References

1. D.Kerek, H.Tenhunen, G.Q.Maguire, F.Reichert, "Direct Sequence CDMA Technology and its Application to Future Portable Multimedia Communication Systems," IEEE 3. International Symposium on Spread Spectrum Techniques & Applications, (Oulu, Finland), July 1994.

FPGA Based Reconfigurable Architecture for a Compact Vision System

R.Nguyen[1], P.Nguyen[2]

Lab. Système de Perception, ETCA,
16bis, Av. Prieur de la Côte d'Or, 94114 Arcueil Cedex, FRANCE
E-mail address : [1]bob@etca.fr, [2]pn@etca.fr   Fax Number : (33) (1) 42319964

Abstract. The EMFRI board is a set of hardware resources (FPGA, memories, communication links) for the rapid prototyping of control units dedicated to our home-made programmable artificial retinas. Its full on-line reconfigurability enables to load different architectures and softwares. We now project to use it as an "active" architecture.

1 Introduction

The original goal was to validate and compare different ideas of control unit architectures for the "NCP retina", a SIMD matrix of 5000 boolean processors (65x76) on a single chip, each of them including a photosensitive device, an A/B converter and 3 bits of memory. The NCP retina was extensively described in [1]. Previous researches [2] on the top level architecture of a vision system

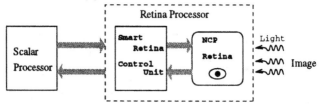

Fig. 1. Bi-Processor Vision System based on NCP retina

based on the NCP retina led us to associate an undedicated scalar unit (80186 or Sparc) to the retina chip (Fig 1). In this framework, rather than building a new electronic board for each control unit we had in mind, we decided to design an unique FPGA-based re-programmable board. This platform proved itself so flexible that it led us to use it for other purposes, in particular as a co-processor for compilation of communication inside the P.E. network. Eventually, the concept of "active architecture" can be supported, that is, the down-loaded architecture is changed according to the context, by the vision process itself.

2 Global Structure

A set of fully re-programmable resources: The particular architecture of the NCP retina and its operating mode lead us to take into account hardware or software parameters: *Bandwidth between control unit and smart retina data path (about 20 Mo/s), Latency delay between both processor request and execution, "Real-time" control (due to the image flow), Control of the instruction flow, Execution model (CISC, RISC, independent thread), Diversity and specificity of operators.* They might be strongly algorithm dependent, so we privileged flexibility and re-programmability features by interconnected resources: *Memories,*

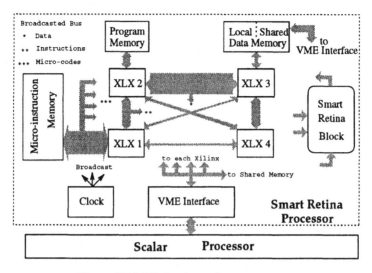

Fig. 2. E.M.F.R.I. schematic structure

N.B.: The width of grey arrows are proportional to the number of interconnections

Programmable Logic, NCP Retina, Fixed Communication links (wire). The most complex control unit requires 3 distinct memory spaces (ie, the C.I.S.C. HAR-VARD architecture). Besides, specific requirements for each memory space imply different bus widths and sizes. The FPGA chips are specialized and have privileged links according to their allocation of a memory space (see Fig 2). Three buses are also broadcasted to all chips. Cohabitation of operative and routing part in each FPGA avoids re-wiring. Unlike e.g. the B.O.R.G. system [3], and for speed reasons we have not used a specific FPGA for routing.

All these fixed specifications relative to FPGA,memory, fixed wiring, access protocols are derived into *a platform skeleton, that defines the minimal description of any new control unit.* The platform can be swapped between two operation modes: configuration and user mode.

Inter-processor communications: Processors communicate by sending messages through two resources : the shared memory with a dual port (extending the local data memory, cf top-left of Fig 2) and the FPGA internal registers (providing an easy hardware solution for access conflict on shared data).

3   Creation and debugging of a vision system

The methodology: The hardware and software co-description of the processor (ie: program, instruction set and material architecture for a C.I.S.C. based system) is compiled by an extended chain (50 tools) of processes. The software description is based on our retina specific extended C language (RC). The hardware functionalities are described by logical schematics or by a behavioral language, even if it has proved possible to automatically deduce a data path directly from an algorithm (e.g. P.R.I.S.M. [4]). Our type of description enables us full control of control structures and signal paths. A specific loader (VxLoad) down-loads each resource with all hardware and software result objects. Fig 3 summarizes the whole creation flow chart.

Debugging facilities: The debugging of an algorithm of vision and of its dedicated processor use both software and hardware facilities: *debug procedures, specific debug instructions, added data paths.* Like the "make" UNIX facility, only upgraded description needs to be compiled again.

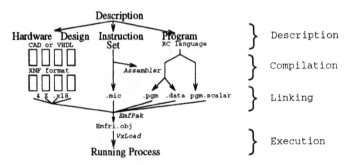

Fig. 3. Design Flow for an EMFRI's vision system

4 Developed Applications

The following application examples emphasize on the diversity of implementable architectures on the EMFRI platform. Even if some operators or data paths are added to the design to improve program efficiency, the basements of an architecture won't be altered. So we have identified and studied three types of architecture which seem particularly interesting:

- a C.I.S.C. like Control Unit
- an Independent Thread Control Unit
- an architecture for Compilation of R.I.S.C. smart retina code

Further research should address the fact that some vision application involve distinct modes: e.g. target detection and target tracking need very different functionalities. The swapping between mode is performed by on-line reprogramming of our control platform. This gives birth to the concept of "active" architecture as a reminder of "active" vision for perception system.

5 Conclusion

The main feature of the "EMFRI" platform is its great flexibility. It is aimed at providing a material support for a quick implementation of a control unit and additional data path for the NCP retina. It could be considered as a step for abolishing the dependences to hardware (see a nice example in [5]). Hardware design and test become fully resolved by software. Retargetting facilities from Xilinx to silicon designs is an other advantage of the platform. In this way, "EMFRI" provides a reconfigurable workspace to conceive and debug future control unit ASICs for upcoming versions of programmable artificial retinas.

The authors are indebted to T.Bernard, F.Devos and B.Zavidovique for their support and fruitful discussions.

References

1. T. Bernard, B. Zavidovique, F. Devos, *A programmable artificial retina*, IEEE J. Solid-State Circuits, Jul 93, vol. 28 pp 789-798.
2. Ph. Nguyen, T. Bernard, F. Devos, B. Zavidovique, *A Vision Peripheral Unit Based On A 65x76 Smart Retina*. SICICA'92, Malàga Spain, May 20-22 1992, pp 643-648.
3. Pak K. Chan, Martine D.F. Schlag, and Marcelo Martin *B.O.R.G. : A Reconfigurable Prototyping Board using FPGA*. Tech. report UCS-CRL-91-45, Computer Engineering, University of California, Santa Cruz, California 95064
4. Peter M. Athanas, Harvey F. Silverman, *Processor Reconfiguration Through Instruction-Set Metamorphosis (P.R.I.S.M.)* Computer, March 1993, pp 11-18.
5. Dr. D. E. Van den Bout, O. Kahn, D. Thomae, *The 1993 AnyBoard Rapid-Prototyping Environment*. RSP'93, North Carolina, June 28-30 1993, pp 31-39.

A New FPGA Architecture for Word-Oriented Datapaths

Reiner W. Hartenstein, Rainer Kress, Helmut Reinig

University of Kaiserslautern
Erwin-Schrödinger-Straße, D-67663 Kaiserslautern, Germany
Fax: ++49 631 205 2640, email: abakus@informatik.uni-kl.de

Abstract. A new FPGA architecture (reconfigurable datapath architecture, rDPA) for word-oriented datapaths is presented, which has been developed to support a variety of Xputer architectures. In contrast to von Neumann machines an Xputer architecture strongly supports the concept of the "soft ALU" (reconfigurable ALU). Fine grained parallelism is achieved by using simple reconfigurable processing elements which are called datapath units (DPUs). The word-oriented datapath simplifies the mapping of applications onto the architecture. Pipelining is supported by the architecture. It is extendable to almost arbitrarily large arrays and is in-system dynamically reconfigurable. The programming environment allows automatic mapping of the operators from high level descriptions. The corresponding scheduling techniques for I/O operations are explained. The rDPA can be used as a reconfigurable ALU for bus-oriented host based systems as well as for rapid prototyping of high speed datapaths.

1 Introduction

Word-oriented datapaths are convenient for numerical computations with FPGAs. A recent trend in FPGA technology moves toward the support of efficient implementation of datapath circuits. The Xilinx XC4000 series [9] provides fast 2-bit addition at each logic cell by a special carry circuit. AT&T's ORCA [4] supports even 4-bit arithmetic operations. A 16 bit adder requires only four function blocks for example. Word-oriented datapaths are not directly supported by FPGAs currently available since these circuits are designed for both random logic control and datapath applications. Word-oriented datapaths in reconfigurable circuits have the additional advantage of operators being mapped more efficiently.

The reconfigurable datapath architecture (rDPA) provides these word-oriented datapaths. It is suitable for evaluation of any arithmetic and logic expression. Statement blocks in inner loops of high performance applications can be evaluated in parallel. The rDPA array is in-system dynamically reconfigurable, which implies also partial reconfigurability at runtime. It is extendable to almost arbitrarily large arrays. Although the rDPA has been developed to support Xputer architectures it is useful for a wide variety of other applications for implementation of numerics by field-programmable media.

First, this paper gives an overview on the rDPA. Section 3 explains a support chip which allows the efficient use of the rDPA in bus-oriented systems. Section 4 presents the programming environment for the automatic mapping of operands and conditions to the rDPA. The scheduling algorithm is described. Section 5 shows the utilisation of the rDPA within the Xputer hardware environment. Finally some benchmark results are shown and the paper is concluded.

2 Reconfigurable Datapath Architecture

The reconfigurable datapath architecture (rDPA) has been designed for evaluation of any arithmetic and logic expression from a high level description. It consists of a regular array of identical processing elements called datapath units (DPUs). Each DPU has two input and two output registers. The dataflow direction is only from west and/or north to east and/or south. The operation of the DPUs is data-driven. This means that the operation will be evaluated when the required operands are available. The communication between the neighbouring DPUs is synchronized by a handshake. This avoids the problems of clock skew and each DPU can have a different computation time for its operator. A problem occurs with the integration of multiple DPUs into an integrated circuit because of the high I/O requirements of the processing elements. To reduce the number of input and output pins, a serial link is used for data transfer between neighbouring DPUs on different chips as shown in figure 1. The DPUs belonging to the converters are able to perform their operations independent of the conversion. Using a serial link reduces the speed of the communication, but simulation results showed that by using pipelining, the latency is increased whereas the throughput of the pipeline is decreased only slightly. Internally the full datapath width is used. For the user this serial link is completely transparent.

A global I/O bus has been integrated into the rDPA, permitting the DPUs to write from the output registers directly outside the array and to read directly from outside. This means, that input data to expressions mapped into the rDPA do not need to be routed through the DPUs. The communication between an external controller, or host, and the DPUs is synchronized by a handshake like the internal communications.

An extensible set of operators for each DPU is provided by a library. The set includes the operators of the programming language C. Other operators such as the parallel pre-

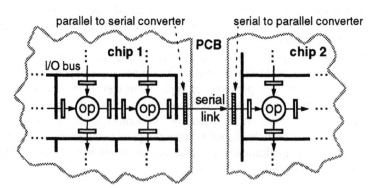

Fig. 1. The extendable rDPA architecture between chip boundaries

fix operator are provided [3]. For example a queue of 'scan-max' operators can be used for easy implementation of a hardware bubble sort [7]. The 'scan-max' computes the maximum from the input variable and the internal feedback variable and gives the maximum as result and stores the other value internally. In addition to expressions, the rDPA can also evaluate conditions. Each communication channel has an additional condition bit. If this bit is true, the operation will be computed, otherwise not. In each case the condition bit is routed with the data using the same handshake. The 'false' path is evaluated very quick, because the condition bit has to be routed only. With this technique also nested if_then_else statements can be evaluated (see also figure 4). The then and the else path can be merged at the end with a merge operator (m). This operator routes the value with the valid condition bit to its output.

The operators of the DPUs are configurable. A DPU is implemented using a fixed ALU and a microprogrammed control, as shown in figure 2. This means, that operators such as addition, subtraction, or logical operators can be evaluated directly, whereas multiplication or division are implemented sequentially. New operators can be added by the use of a microassembler.

As mentioned before the array is extendable by using several chips of the same type. The DPUs have no address before configuration since all rDPA chips are identical. A DPU is addressed by its x- and y-location, like an element in a matrix. The x- and y-location are called addresses later for convenience. A configuration word consists of a configuration bit which distinguishes the configuration data from computational data. Furthermore it consists of the x- and the y-address, the address of the DPU's configuration memory, and the data for this memory.

Each time a configuration word is transferred to a DPU, the DPU checks the x- and the y-address. Four possible cases can occur:
- the y-address is larger than zero and the x-address is larger than zero
- the y-address is larger than zero and the x-address is zero
- the y-address is zero and the x-address is larger than zero
- both, the y-address and the x-address are zero

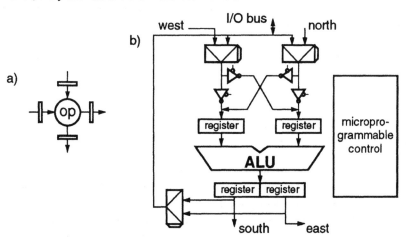

Fig. 2. A datapath unit (a) and its implementation (b)

In the first case the DPU checks if the neighbouring DPUs are busy. If the neighbouring DPU in y-direction is not busy, the y-address will be decreased by one and the resulting configuration word will be transferred to this DPU. If the DPU in y-direction is busy and the DPU in x-direction is not busy the x-address will be decreased by one and the resulting configuration word will be transferred to this DPU. If both neighbouring DPUs are busy, the DPU waits until one finishes. With this strategy an automatic load distribution for the configuration is implemented. Internally the configuration words are distributed over the whole array and several serial links are used to configure the rest of the chips. An optimal sequence of the configuration words can be determined since these can be interchanged arbitrarily.

In the second case, the y-address will be decreased by one and the configuration word will be transferred to the next DPU in y-direction. In the third case when the y-address is zero and the x-address is larger than zero, the x-address will be decreased by one and the configuration word will be transferred in x-direction. In the last case when both addresses are zero, the target DPU is reached, and the address of the DPU's configuration memory shows the place where the data will be written.

Because of the load distribution in the rDPA array, one serial link at the array boundary is sufficient to configure the complete array. The physical chip boundaries are completely transparent to the user. The communication structure allows dynamic in-system reconfiguration of the rDPA array. This implies partial reconfigurability during runtime [6]. Partial reconfigurability is provided since all DPU can be accessed individually. The configurability during runtime is supported because each DPU forwards a configuration word with higher priority than starting with the next operation. The load distribution takes care of that most of the configuration words avoid the part of the rDPA array which is in normal operation. Further the configuration technique allows to migrate designs from a smaller array to a larger array without modification. Even newer generation rDPA chips with more DPUs integrated do not need a recompilation of the configuration data. The configuration is data-driven, and therefore special timing does not have to be considered.

With the proposed model for the DPA, the array can be expanded also across printed circuit board boundaries, e. g. with connectors and flexible cable. Therefore it is possible to connect the outputs of the east (south) array boundary with the west (north) one, to build a torus.

3 Support Chip for Bus-Oriented Systems

With the rDPA, a programmable support chip for bus-oriented systems is provided. Together they form a data-driven reconfigurable ALU (rALU). The support chip consists of a control unit, a register file, and an address generation unit for addressing the DPUs (figure 3).

The register file is useful for optimizing memory cycles, e. g. when one data word of a statement will be used later on in another statement. Then the data word does not have to be read again over the external bus. In addition, the register file makes it possible to use each DPU in the rDPA for operations by using the internal bus for routing. If different expressions have a common subexpression, this subexpression has to be computed only once. If the rDPA does not provide the routing capacity for this reduction,

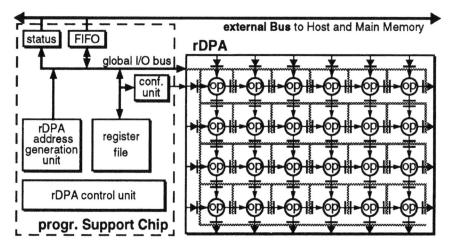

Fig. 3. The reconfigurable datapath architecture (rDPA) with the programmable support chip

e. g. if three or more subexpressions are in common, the interim result can be routed through the register file.

The address generation unit delivers the address for the DPU registers before each data is written into the rDPA over the bus. Usually the address is increased by one but it can also be loaded directly from the rDPA control unit.

The rDPA control unit holds a program to control the different parts of the data-driven rALU. The instruction set consists of instructions for loading data into the rDPA array to a special DPU from the external units, for receiving data from a specific DPU, or branches on a special control signal from the host. The rDPA control unit supports context switches between three control programs which allows the use of three independent virtual rALU subnets. The control program is loaded during configuration time. The reconfigurable data-driven ALU allows also pipelined operations.

A status can be reported to the host to inform about overflows, or to force the host to deliver data dependent addresses. The input FIFO is currently only one word deep for each direction. The datapath architecture is designed for an asynchronous bus protocol, but it can also be used on a synchronous bus with minor modifications of the external circuitry.

4 Programming Environment

Statements which can be mapped to the rDPA array are arithmetic and logic expressions, and conditions. The input language for programming the rALU including the rDPA array is the rALU programming language, called ALE-X (arithmetic & logic expressions for Xputers). The syntax of the statements follows the C programming language syntax. A part of an ALE-X example is shown in figure 4.

A data dependency analysis is performed to recognize possible parallelization and to find dependencies between the statements. The statements are combined to larger expressions and a data structure which is a kind of an abstract program tree is built (figure 5). Then the data structure is mapped onto the rDPA array structure. The map-

```
a = b + c * d;  (1)
if (e < 10)     (2)
   f = g + h;   (3)
else            (4)
   f = g - h;   (5)
i = c + f;      (6)
```

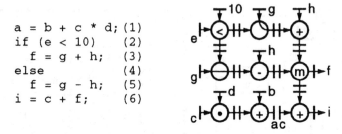

Fig. 4. Part of an ALE-X program example mapped onto the rDPA array

ping algorithm starts at the leaf cell nodes of the data structure for each expression. These nodes are assigned to DPUs in a first line of the rDPA array. A second line starts if there is a node of higher degree in the data structure. The degree of a node increases if both sons of the node are of the same degree. After that the mapped structure is shrunk by removing the nodes which are used only for routing. There are several possibilities for the mapping of each expression. Finally the mapped expression with the smallest size is chosen. Figure 5 shows an example of the mapping. Now the mapped expressions are allocated in the rDPA array, starting with the largest expression. If the expressions do not fit onto the array, they are split up using the global I/O bus for routing. If the number of required DPUs is larger than the number of DPUs provided by the array, the array has to be reconfigured during operation. Although this allocation approach gives good results, future work will be done in the optimization of this algorithm to incorporate the scheduling process for advance timing forecast.

Due to the global I/O bus of the rDPA array, the loading of the data and the storing are restricted to one operation per time. An optimal sequence of these I/O operations has to be determined. For the example in figure 4, starting with loading the variables c and d is better than starting with h. The operators do not have to be scheduled, since they are available all the time. The operands have to be scheduled with the additional restriction that operands used at multiple locations have to be loaded several times at succeeding time steps. For example, when the variable c is scheduled, the c of the multiplication and the c of the last addition have to be loaded in direct sequence. To find the time critical operations, first an 'as soon as possible' schedule (ASAP) and an 'as late as possible' schedule (ALAP) are performed. No other resource constraints

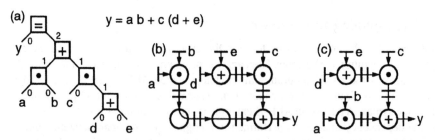

Fig. 5. Example of the mapping process: a) data structure with the degree of the node, b) mapped structure, c) shrunk mapped structure

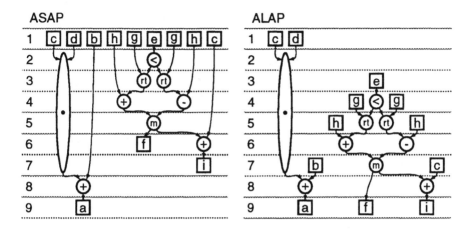

Fig. 6. ASAP and ALAP schedules for the program example

(like only a single I/O operation at a time) are considered at this moment. For simplicity in our example, all operations, including the route operations (rt) are assumed to need a single time step for finishing in the worst case. The multiplication is assumed to need six time steps. The rALU compiler considers the real time delays in the worst case. Due to the self timing of the data-driven computation, no fixed time step intervals are necessary. Figure 6 shows the ASAP and ALAP schedules for the program example.

Comparing the ASAP with the ALAP schedule, the time critical path is found. It is the multiplication of c and d with the succeeding addition of b. The range of this operation is zero. A priority function is developed from these schedules which gives the range of the I/O operations of the operands. This is the same as in a list based scheduling [2]. The highest priority i. e. the lowest range have the variables c and d. Since c has to be loaded twice, d is loaded first. The complete priority function is listed in figure 7b.

When the variable c is scheduled twice in direct sequence the ASAP and the ALAP schedule may change because of the early scheduling of c in the addition operation. Then the schedule of d, c, and c is kept fixed and a new priority function on the remaining variables is computed to find the next time critical operation. For simplicity this is not done in the illustration. Figure 7a shows the final schedule of the program example.

In time step 10 no I/O operation is performed. If the statement block of the example is evaluated several times, the global I/O bus can be fully used by pipelining the statement block. The pipeline is loaded up to step 9. Then the variable d from the next block is loaded before the output variables a, i and f are written back. The statement block is computed several times (step 10 to 21, figure 7c) until the host signals the rALU control to end the pipeline. Step 22 to the end is performed, and the next operators can be configured onto the rDPA array.

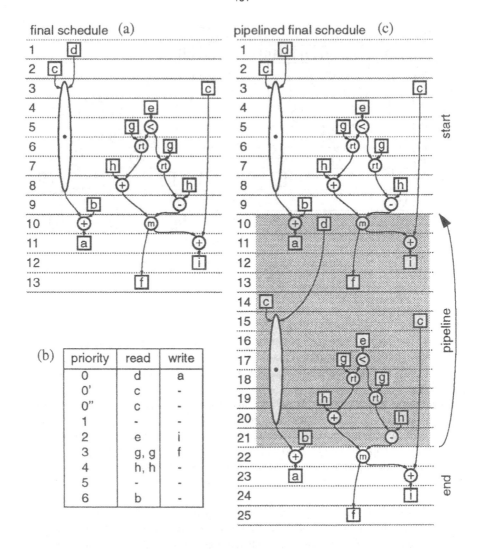

Fig. 7. The final schedule (a), the priority function (b) and the pipelined final schedule (c) for the program example

The rDPA configuration file is computed from the mapping information of the processing elements and a library with the microprogram code of the operators. The configuration file for the rALU control unit is extracted from the final schedule of the I/O operators.

5 Utilisation with the Xputer Hardware Environment

Although the proposed rALU can be used for any bus-oriented host based system, it is originally build for the Xputer prototype Map-oriented Machine 3 (MoM-3). The Xputer provides a hardware and a software environment for a rALU. The rALU has to compute a user defined compound operator only. A compiler for the Xputer supports the high level language C as input [8]. The rALU programming environment has to compile arithmetic and logic expressions as well as conditions onto the rALU.

Many applications require the same data manipulations to be performed on a large amount of data, e. g. statement blocks in nested loops. Xputers are especially designed to reduce the von-Neumann bottleneck of repetitive decoding and interpreting address and data computations. In contrast to von Neumann machines an Xputer architecture strongly supports the concept of the "soft ALU" (rALU). The rALU allows for each application a quick problem-oriented reconfiguration. High performance improvements have been achieved for the class of regular, scientific computations [5], [1].

An Xputer consists of three major parts: the data sequencer, the data memory and the rALU including multiple scan windows and operator subnets. Scan windows are a kind of window to the data memory. They contain all the data words, which are accessed or modified within the body of a loop. The data manipulations are done by the rALU subnets, which provide parallel access to the scan windows. The scan windows are updated by generic address generators, which are the most essential part of the data sequencer. Each generic address generator can produce address sequences which correspond to nested loops under hardware control. The term data sequencing derives from the fact that the sequence of data triggers the operations in the rALU, instead of a von-Neumann instruction sequence. Generally, for each nesting level of nested loops a separate rALU subnet is required to perform the computations associ-

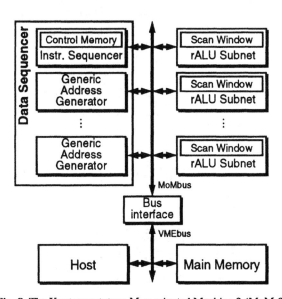

Fig. 8. The Xputer prototype Map-oriented Machine 3 (MoM-3)

ated with that nesting level. The rALU subnets perform all computations on the data in the scan windows by applying a user-configured complex operator to that data. Pipelining across loop boundaries is supported. The subnets need not to be of the same type. Subnets can be configured for arithmetic or bit level operations.

The Xputer prototype MoM-3 has direct access to the host's main memory. The rALU subnets receive their data directly from a local memory or via the MoMbus from the main memory. The MoMbus has an asynchronous bus protocol. The datapath architecture is designed for the asynchronous bus protocol of the MoMbus, but it can also be used by a synchronous bus with minor modifications. Figure 8 shows our prototype MoM-3.

A complete rALU programming environment is developed for the rALU when using it with the Xputer prototype. The input language for programming the rALU is the ALE-X programming language. The syntax of the statements follows the C programming language syntax (see also figure 4). In addition, the language provides the size of the scan windows used and the next handle position which is the lower left corner of the boundary of the scan window. Providing the handle position gives the necessary information for pipelining the complete statement block in the rALU.

The ALE-X programming language file is parsed and a data structure like an abstract program tree is computed. Common subexpressions are taken into consideration. The

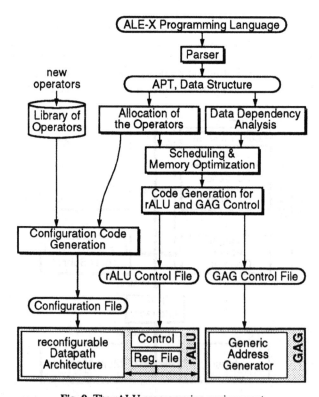

Fig. 9. The rALU programming environment

operators of each statement are associated to a DPU in the rALU array as described in section 4. Memory cycles can be optimized using the register file when the scan pattern of the GAGs works with overlapping scan windows.

The rDPA configuration file is computed from the mapping information of the DPUs and a library containing the code of the operators. The configuration file for the rALU control unit and the configuration file for the GAGs is extracted from the final schedule of the I/O operators. The programming environment of the rALU is shown in figure 9.

6 Results

The prototype implementation of the rDPA array works with 32 bit fixed-point and integer input words. Currently the host computer's memory is very slow. The clock frequency of the system is 25 MHz. In a single chip of the rDPA array fits at least a 3×3 matrix of DPUs. In many applications the coefficients in e. g. filter implementations are set up in such a way that shift operations are sufficient and multiplications are not necessary. If high throughput is needed the DPU processing elements can be linked together to implement a pipelined multiplier for example. Benchmark results are given in table 1. The performance figures are a worst case estimation of our prototype. They give the duration of the operation time per data word. The speed of the examples 2 to 5 does not depend on the order of the filter as long as the necessary hardware (number of DPUs) is provided. The same applies for example 6.

#	Algorithms	Opera-tions	number of active DPUs	number of necessary chips	Time Steps per Operation	Perfor-mance
1	1024 Fast Fourier Transform	*, +, -	10	2	$16 \cdot 10240$	20 ms
2	FIR filter, n^{th} order	*, +	2(n+1)	$\left\lceil \dfrac{n+1}{3} \right\rceil$ [a]	15	1800 ns / data word
3	FIR filter, n^{th} order	shift, +	2(n+1)	$\left\lceil \dfrac{n+1}{3} \right\rceil$	4	500 ns / data word
4	n × m two dim. FIR filter	*, +	2(n+1)(m+2)-1	$\left\lceil \dfrac{n+1}{3}(m+2) \right\rceil$	15	1800 ns / data word
5	n × m two dim. FIR filter	shift, +	2(n+1)(m+2)-1	$\left\lceil \dfrac{n+1}{3}(m+2) \right\rceil$	4	500 ns / data word
6	Bubblesort, length n	scan-max	n-1	$\left\lceil \dfrac{n-1}{9} \right\rceil$	2	240 ns / data word

Table 1. Benchmark results

a. $\lceil x \rceil$ = the smallest integer, which is greater or equal to x

7 Conclusions

An FPGA architecture (reconfigurable datapath architecture, rDPA) for word-oriented datapaths has been presented. Pipelining is supported by the architecture. The word-orientation of the datapath and the increase of the fine granularity of the basic operations extremely simplifies the automatic mapping onto the architecture. The extendable rDPA provides parallel and pipelined evaluation of the compound operators. The rDPA architecture can be used as reconfigurable ALU for bus-oriented host based systems as well as for rapid prototyping of high speed datapaths. It suits very well for the Xputer prototype MoM-3. The architecture is in-system dynamically reconfigurable, which implies also partial reconfigurability at runtime.

A prototype chip with standard cells has been completely specified with the hardware description language Verilog and will be submitted for fabrication soon. It has 32 bit datapaths and provides arithmetic resources for integer and fixed-point numbers. The programming environment is specified and is currently being implemented on Sun SPARCstations.

References

1. A. Ast, R. W. Hartenstein, H. Reinig, K. Schmidt, M. Weber: A General purpose Xputer Architecture derived from DSP and Image Processing; in M. A. Bayoumi (Ed.): VLSI Design Methodologies for Digital Signal Processing Architectures; Kluwer Academic Publishers, Boston, London, Dordrecht, pp. 365-394, 1994

2. D. D. Gajski, N. D. Dutt, A. C.-H. Wu, S. Y.-L. Lin: High-Level Synthesis, Introduction to Chip and System Design; Kluwer Academic Publishers, Boston, Dordrecht, London, 1992

3. S. A. Guccione, M. J. Gonzalez: A Data-Parallel Programming Model for Reconfigurable Architectures; IEEE Workshop on FPGAs for Custom Computing Machines, FCCM'93, IEEE Computer Society Press, Napa, CA, pp. 79-87, April 1993

4. D. Hill, B. Britton, B. Oswald, N.-S. Woo, S. Singh, C.-T. Chen, B. Krambeck: ORCA: A New Architecture for High-Performance FPGAs; in H. Grünbacher, R. W. Hartenstein (Eds.): Field-Programmable Gate Arrays, Lecture Notes in Computer Science, Springer-Verlag, Berlin, 1993

5. R. W. Hartenstein, A. G. Hirschbiel, M. Riedmüller, K. Schmidt, M. Weber: A Novel ASIC Design Approach Based on a New Machine Paradigm; IEEE Journal of Solid-State Circuits, Vol. 26, No. 7, July 1991

6. P. Lysaght, J. Dunlop: Dynamic Reconfiguration of Fieldprogrammable Gate Arrays; Proceedings of the 3rd International Workshop on Field Programmable Logic and Applications, Oxford, Sept. 1993

7. N. Petkov: Systolische Algorithmen und Arrays; Akademie-Verlag, Berlin 1989

8. K. Schmidt: A Program Partitioning, Restructuring, and Mapping Method for Xputers; Ph. D. Thesis, University of Kaiserslautern, 1994

9. N. N.: The XC4000 Data Book; Xilinx, Inc., 1992

Image Processing on a Custom Computing Platform

Peter M. Athanas and A. Lynn Abbott

The Bradley Department of Electrical Engineering
Virginia Polytechnic Institute and State University
Blacksburg, Virginia 24061-0111

Abstract. Custom computing platforms are emerging as a class of computing engine that not only can provide near application-specific computational performance, but also can be configured to accommodate a wide variety of tasks. Due to vast computational needs, image processing computing platforms are traditionally constructed either by using costly application-specific hardware to support real-time image processing, or by sacrificing real-time performance and using a general-purpose engine. The SPLASH-2 custom computing platform is a general-purpose platform not designed specifically for image processing, yet it can cost-effectively deliver real-time performance on a wide variety of image applications. This paper describes an image processing system based on the SPLASH-2 custom computing engine, along with performance results from a variety of image processing tasks extracted from a working laboratory system. The application design process used for these image processing tasks is also examined.

1. Introduction

Many of the tasks associated with image processing can be characterized as being computationally intensive. One reason this is true is because of the vast amount of data that requires processing -- several million pixels need to be processed per second for images with respectable resolution. Another reason is that for many tasks, several operations need to be performed on each picture element within the image, and a typical image may be composed of more than a quarter of a million picture elements. To keep up with these capacious data rates and demanding computations in real-time, the processing engine must provide specialized data paths, usually application-specific operators, creative data management, and careful sequencing and pipelining.

A typical design process necessitates extensive behavioral testing of a new concept before proceeding with a hardware implementation. For any task of reasonable complexity, simulation of a VHDL model with a representative data set on a respectable workstation is prohibited due the enormous simulation processing time. Days, or even weeks, of processing time are sometimes needed to simulate the processing of a single image. In many instances several seconds, or even minutes of image data, which may consist of hundreds or thousands of images are needed to make a fair subjective analysis, or to exercise the design sufficiently. Because of this, the designer is often forced to into a trade-off on how much testing can be afforded verses an acceptable risk of allowing an iteration in silicon.

An alternative approach discussed in this paper is the automated transformation of the structural representation (or the transformation of a behavioral model) into a real-time implementation. Using this approach, the prototype image processing platform would not only serve as a means to evaluate the performance of an experimental algorithm/architecture, but also may serve as a working component in the development and testing of a much larger system. The platform used to provide this capability is an experimental general-purpose custom computing platform called SPLASH−2[1]. SPLASH is a reconfigurable attached processor featuring programmable processing elements and programmable communication paths as the mechanism for performing computations. The SPLASH−2 system utilizes arrays of RAM-based FPGAs, crossbar networks, and distributed memory as a means of accomplishing the above goals. Even though SPLASH was not designed specifically for image processing, this platform possesses architectural properties that make it well suited for the computation and data transfer rates that are characteristic of this class of problems. Furthermore, the price/performance of this system makes it a highly competitive alternative to conventional real-time image processing systems.

There are several aspects of image processing which distinguish it as being computationally challenging; these are identified in Section 2. Sections 3 and 4 provide a synopsis of the pertinent architectural features of the SPLASH processor, along with a description of the laboratory image processing system. Section 5 provides a narration of the application design process. Descriptions of some of the applications implemented on the laboratory system can be found in Section 6. Performance results are given in Section 7.

2. Architectural Aspects of Image Processing

Conventional general-purpose machines fail to manage the distinctive I/O requirements of most image processing tasks, nor are they equipped to take advantage of the opportunities for parallel computation that are present in many vision-related tasks. Parallel processing systems such as mesh computers or pipelined processors have been successfully applied to some image processing applications. Mesh architectures often provide very large speedup after an image is loaded, but overall performance often suffers severely from I/O limitations. Pipelined machines can accept image data in real time from a camera or other source, but historically they have been difficult to reconfigure for different processing tasks.

Image processing is characterized by being computationally intensive, and often by the repeated application of a single operator. A typical edge detector, for example, may be implemented as a 3x3 (or 5x5, etc.) operator which is applied at every picture element (*pixel*) in an image, producing a new image as the result. Other examples of such neighborhood operations are template matching and morphological processing.

Other forms of image processing do not produce new images, but instead compute statistical information from the image data. Examples of this include histogram generation and the computation of moments. Furthermore, many applications require that *sequences* of such images be processed quickly. Image compression and motion compensation applications typically operate on sequences of two or more images.

Image data are typically available in *raster order* -- pixels are presented bit-, byte-, or word-serially, left-to-right for each image row, beginning with the top row. If a typical image frame is 512 rows by 512 columns of 8-bit pixels, then the total data in a single frame is 256 kilo-pixels, or 2 megabits of data. The discussions in this paper will assume that data represent monochrome light intensity values.

3. The SPLASH-2 Custom Computing Machine

SPLASH-2 is a second generation custom computing attached processor designed by the Supercomputing Research Center in Bowie, Maryland. SPLASH is intended to accelerate applications by reconfiguring the hardware functionality and processing element interconnections to suit the specific needs of individual applications.

SPLASH is classified as an *attached processor* since it is intended to append a host machine through an expansion bus. It differs from a *coprocessor* in that it does not reside directly on the host processor bus. SPLASH-2 has been designed with an SBus interface, and currently serves a Sun SPARC-2 host. A SPLASH attached processor is comprised of an interface board (for formatting and buffering input and output data), and from one to fifteen processor boards. Each processor board contains 17 *processing elements* and a crossbar network. A Xilinx XC4010 and a fast 256K×16 static RAM together make one processing element. The crossbar network contains sixteen 36-bit bidirectional ports for augmenting communications between processor elements. The crossbar switches present on SPLASH can be dynamically adjusted to support complex interconnection topologies.

The SPLASH-2 system offers an attractive alternative to traditional architectures. With this computing platform, not only can the specific operations be custom designed (for function and size), but the data paths can also be customized for individual applications. Furthermore, these platforms can be completely reconfigured in just a few seconds. The reconfigurable nature of SPLASH provides the performance of *application-specific* hardware, while preserving the *general-purpose* nature of being able to accommodate a wide variety of tasks. A more complete description of SPLASH hardware and software development environment can be found in [1,2].

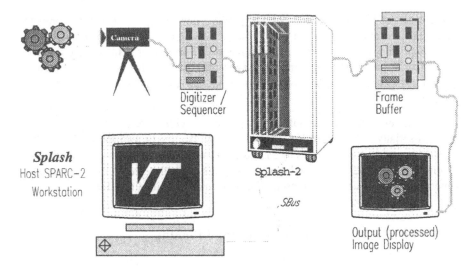

Fig. 1: Components in the VTPLASH *laboratory system.*

4. A Real-Time Image Processing Platform

Performing the computations of a vision related task within the time permitted of a live video data stream is a challenging task, mainly because of:

- the quantity of data involved -- 256 Kbytes for a single 512×512 image frame,
- the input / output data requirements (30 frames per second for RS-170), and
- the high computational requirements (per pixel) for many image applications.

The adaptive nature of the SPLASH architecture makes it well suited for the computational demands of image processing tasks, even though it was not specifically designed for such tasks. Furthermore, SPLASH features a flexible interface design which facilitates customized I/O for situations which cannot be accommodated by the host SPARC processor. There is sufficient memory distributed on each processor board to buffer several images for the processing of two or more frames simultaniously (if needed). A real-time image processing custom computing system has been constructed at Virginia Tech based on SPLASH-2. The VTSPLASH laboratory system is depicted in Figure 1.

A monochrome video camera or a VCR is used to create an RS-170 image stream. The signal produced from the camera is digitized with a custom built frame grabber card. This card was designed not only to capture images, but also to perform any needed sequencing or simple pixel operations before the data are presented to SPLASH. The frame grabber card was built with a parallel interface which can be connected directly to the input data stream of the SPLASH processor. The SPLASH system used in this work consists of a modified interface board and two processor array boards.

The output of SPLASH, which may be a real-time video data stream, overlay information, or some other form of information, is first presented to another custom board for converting the data (if necessary) to an appropriate format. Once formatted, the data are then presented to a second frame grabber card (a commercial card: Data Translation DT2867LC). A new RS-170 signal is formed and presented to a video monitor. A Sun SPARC 2 serves as the SPLASH host, and is responsible for configuring the SPLASH arrays and interjecting run-time commands within the video stream.

5.   Application Development on SPLASH-2

While the programming environment for SPLASH-2 is one of the most advanced and automated for its class, there are a number of difficulties that exist that must be addressed before this type of machine can become accepted into mainstream computing. In this section, a brief summary of the application development process is given.

Figure 2 illustrates the basic design flow for the development of a typical SPLASH application. This figure is somewhat simplified, and may not depict all of the

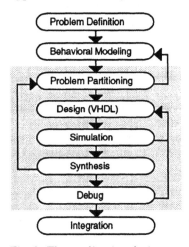

Fig. 2: The application design process.

possible iteration paths in the design process. The first step in the process is the definition of the problem. As in all hardware and software system design, a sound *problem definition* will facilitate the design process. Step two is the *behavioral modeling* of the problem. Typically, a VTSPLASH programmer models the problem using C or with a behavioral VHDL model. Not only is the model verified to comply with the problem definition, but sample images are run through the model (when possible) for comparison with the results of the synthesized implementation.

The next step in the process, and often a difficult step, is *partitioning* the model into a form that is suitable for the final implementation on SPLASH. The model is first mapped onto processor boards, and then partitioned more finely into individual processing elements. The three main factors that drive a partition are *time, area* and *communication complexity*. The time and area factors are familiar problems that are discussed in the high-level synthesis and silicon compiler literature[3]. *Time* relates to how much computation is desired per clock cycle. *Area* relates to how much of the reconfigurable resources should be allocated to a given computation, and to the total available reconfigurable resources within each processor board and within each of the 17 processing elements on each board. Even though SPLASH contains ample hardware support to aid signal propagation between processing elements, not all communications are equal in cost and in bandwidth (*communication complexity*). There are finite limitations on the available communication resources that the designer must comply with in SPLASH-2. Some of these are:[1]

> • a maximum of 36 bits of input data and 36 bits of output data per processing element (72 bits if the crossbar network is used to augment these paths).
> • a 16 bit data path between a processing element and its 2 megabyte RAM
> • several single-bit signals for global communications and broadcasts.
> • a 36 bit data path between processing boards, along with several 1-bit global signals.

While these numbers are quite generous, they represent realistic design trade-offs versus cost, and any upper bounds on these data paths will eventually disgruntle some designers. Not all applications easily map to these communications limitations, and tough design trade-offs must be considered. As it stands now, there are few quantitative up-front measures available to gauge partitioned alternatives. A designer must often wait until after the synthesis step before it is known whether a given problem partition is feasible.

After the design is partitioned, a detailed structural design is produced and verified. Many different alternatives are available to the designer for "programming" an application, including FPGA design tools like XBLOX[4]; however, the best supported design environment for SPLASH is with the Synopsys VHDL simulation and synthesis tools. There is only so much one can simulate within a reasonable amount of time. The simulations for many of the image processing tasks discussed in this paper consumed several days of CPU time per run on a SPARC-10 -- in many cases, for just a small fraction of an image. Therefore, the stimulation input for a simulation run must be considered judiciously.

The application simulations are based on VHDL models developed prior to placement and routing; hence, are barren of signal propagation annotation. Actual propagation delays in the Xilinx FPGAs are highly sensitive to the outcome of the placement and routing process, and can have a disturbing effect on the application behavior. To counter these problems, and to help cope with the limited functional coverage that can be achieved by the simulation tools, a powerful debugging environment has been built for SPLASH-2. The *T2* interactive debugger [5] provides the power of conventional high-level language debuggers by allowing such features as monitoring internal state variables and tracing. Debugging a hardware/software design adds new conceptual difficulties not found in traditional debugging environments. Once the image operations are performing satisfactorily, they must be

---

[1]These numbers are rather simplistic. A better understanding of the SPLASH architecture is required to get a full appreciation of the available communication resources.

integrated within the body of application. A rich **C** library has been developed by SRC[5] which facilitates communication between a host program and the attached processor.

As stated before, SPLASH is representative of the state-of-the-art in custom computing processors -- both in hardware capabilities and software support -- yet it requires a substantial time investment to develop an application. The authors observed that graduate students well versed in VHDL and hardware design required from one to four months to develop their first SPLASH application. This time decreased by half for the development of the second. To make this class of machinery more widely accepted and cost-effective, methods must be developed to reduce application development time. There are many promising endeavors that focus on this issue [6,7,8]; the main emphasis of these are to automate to some degree the portions of the shaded region of Figure 2.

6. Image Processing Tasks

Common image processing tasks can be classified into the following categories: neighborhood operations (both linear and nonlinear), statistical computations, and transformations. All of these types of operations have been implemented (with varying degrees of difficulty) on the VTSPLASH laboratory system, and are briefly discussed in this section to illustrate the types of computations each require.

6.1. Linear and Nonlinear Neighborhood Operations

Two-dimensional filtering techniques are very common in image processing. The most common methods process small neighborhoods in an input image to generate a new output image. The resulting image is often a smoothed or enhanced version of the original, or may comprise a 2D array of *features* that have been detected. Neighborhood-based filtering is characterized by the repeated application of identical operations, and often serves as a preprocessing step that is followed by higher-level image analysis.

Neighborhood operations typically use a 2D template, usually rectangular, which is applied at every pixel in the input image. (The template is often called a *mask* operator, or *filter*.) In the linear case, applying a template means centering the template at a given pixel of the input image, multiplying each template pixel by the associated underlying image pixel, and summing the resulting products. The sum is used as the pixel value (for this template position) in the output image.

In addition to the linear filtering described above, template operations can be nonlinear. For example, a median filter can be implemented by using a template. For every position of the template, the median value is chosen from the image pixels covered by the template, and is used as the new pixel value for the output image. In this case, the template simply serves as a window, and has no cell values.

Another form of nonlinear image processing is based on *mathematical morphology*[9]. This is an algebra which uses multiplication, addition (subtraction), and maximum (minimum) operations to produce output pixels. The filtering operations are known as *erosion* and *dilation*, and can be used to perform such tasks as low-pass or high-pass filtering, feature detection, etc. One advantage of this nonlinear approach is reduced blurring, as compared with linear filtering.

6.2. Statistical Computations

Unlike the previous types of processing, statistical analysis typically does not result in a new output image. Instead, the goal is to extract descriptive statistics of the input image. For example, the mean and standard deviation of pixel values in the image are often of interest. These and similar statistics can be computed using simple multiply-accumulate processing, where one such operation is required for each input pixel.

Real-time histogram generation is another useful operation which is often used as an initial step for other applications, such as region detection and region labeling. In generating a histogram, the processor must maintain and update a one-dimensional array which records the number of occurrences of particular pixel values. In addition, histograms are often analyzed further and used to adjust parameters for image enhancement.

6.3. 1-D and 2-D Transformations

The 2-D discrete Fourier transform (DFT) is an extremely useful operation which is often avoided because of its large computational requirements. Although it is a linear operation, it differs from the neighborhood operations described above since every transformed output pixel depends on every pixel of the input image. The problem can be simplified somewhat, since the 2-D Fourier transform can be decomposed into multiple 1-D Fast Fourier transforms. For example, a 512×512 DFT can be implemented as 512 one-dimensional FFT computations (one for each row) followed by 512 additional one-dimensional FFTs (one per column). This application was implemented using floating point arithmetic.

The Hough transform, another 2-D transformation, can be appended after an edge detection task for the purpose of determining if a set of points lie on a curve of specified shape, namely a straight line. The coordinates of high-intensity points in the transform domain correspond to the position and orientation of best-fit lines in the original image. A more complete discussion on the Hough transform can be found in [12].

6.4. Other Image Transformations

After an image has been appropriately low-pass filtered, the image can be subsampled without fear of violating the Nyquist criterion. If an image is recursively filtered and subsampled, the resulting set of images can be considered a single unit and is called a *pyramid*. This data structure shows promise in applications which require complex image analysis, because analysis which begins at the lower-resolution portion of the pyramid can be used to guide processing at higher-resolution levels. For some tasks (such as surveillance and road following) this approach can greatly reduce the overall amount of processing required.

In addition to low-pass pyramids, it is possible to generate band-pass pyramids, in which each level of the pyramid contains information from a single frequency band. A popular technique for generating these pyramids (known as Gaussian and Laplacian pyramids) is described in [10]. Reconfigurable data paths through the crossbar networks are used as the mechanism for dynamically restructuring and reconnecting the pyramid processing elements.

7. Performance

A diversity of image processing tasks have been completed on the VTSPLASH laboratory system. This section provides a quantitative summary of a representative number of these. Qualitative evaluation of the real-time visual results are absent; the interested reader can refer to contemporary texts on the subject[11][12]. Furthermore, example (stationary) pictures of the processed results are not included since they do not contribute to the major theme of this paper.

Application	Description	Class
Fourier Transform	2-D transformation to / from spatial domain. Implemented using floating point arithmetic.	Repeated 1-D transformation.
Convolution	8×8 window operation for linear filtering.	Neighborhood operator.
Pyramid Transform	Repeated application of a Gaussian filter, Laplacian filter, and decimation.	2-D transform, filtering, decimation, and reconstruction.
Morphological Operators	Non-linear 3×3 window operator.	Neighborhood operator.
Median Filter	Non-linear 3×3 window operator.	Neighborhood operator.
Hough Transform	Transformation of x-y coordinates into angle and displacement. Useful for line finding.	2-D statistical operation and transformation.
Region Detection and Label	Uses point statistics to determine regions, and then assigns a unique number to the region.	Mixture of window operations, point statistics, and pixel manipulation.
Histogram	Determines image intensity distribution.	Statistical point operation.

Table 1: A representative list of image processing tasks.

Table 1 summarizes a number of tasks that were discussed in the previous section, which have been implemented in the laboratory. Table 2 provides an estimate of the computational performance of each of these tasks. In Table 2, the application name is listed in the first column. The second column provides a rough estimate of the number of general-purpose operations (operations that are likely to be found in the repertory of most common RISC processors) performed on average each pixel clock cycle. In the third column, an estimate of the number of (equivalent) storage references are given. The purpose of these two columns is to provide a basis for quantifying the computational load of each of the tasks. These numbers are used to produce a very rough estimate of the "MIPS" rating given in the fourth column of Table 2.

Application	Arithmetic/logical operations per second	Memory operations per second	Effective number of operations per second
Median Filter	3.9×10^8	2×10^7	4.1×10^8
Hough Transform	2.6×10^8	8×10^7	3.4×10^8
Region Detection and Labeling	1.8×10^8	4×10^7	2.2×10^8
Fast Fourier Transform (forward & reverse)	2.2×10^8 (floating point) 2.0×10^8 (fixed point)	2.0×10^8	6.6×10^8
Pyramid Generation	3.8×10^8	6×10^7	4.4×10^8
Histogram	1.2×10^8	2×10^7	1.4×10^8
Morphological Operators	4.8×10^8	2×10^7	5.0×10^8
8×8 Linear Convolution	6.4×10^8	1×10^7	6.5×10^8

Table 2: Estimated performance of image processing tasks.

In many of the applications developed, a pipeline architecture was used. The pipeline accepts digitized image data in raster order, often directly from a camera, and produces output data at the same rate, possibly with some latency.

8. Summary

With the addition of input/output hardware, the SPLASH platform has proven to be well suited for many meaningful image processing tasks. Reconfigurable computing platforms, such as SPLASH, can readily adapt to meet the communication and computational requirements of a variety of applications. Real-time processing of image data is an effective approach for demonstrating the potential processing power of adaptive computing platforms.

Acknowledgments

The authors are indebted to the graduate students who have contributed to the VTSPLASH program which include the VTSPLASH application developers Luna Chen, Robert Elliott, James Peterson, Ramana Rachakonda, Nabeel Shirazi, Adit Tarmaster, and Al Walters, along with the VTSPLASH hardware support duo of Brad Fross and Jeff Nevits. In addition, the authors appreciate the support and guidance from Jeffrey Arnold and Duncan Buell from the Supercomputing Research Center.

References

[1] J. M. Arnold, D. A. Buell, E. G. Davis, "Splash 2," in *Fourth Annual ACM Symposium on Parallel Algorithms and Architectures,* San Diego, CA, pp. 316-322, 1992.

[2] J. M. Arnold, "The Splash 2 software environment," in *IEEE Workshop on FPGAs for Custom Computing,* Napa, CA, pp. 88-93, Apr 1993.

[3] D. Gajski, *Silicon Compilation,* Addison-Wesley, Reading, Massachusetts, 1988.

[4] *The Programmable Gate Array Data Book,* Xilinx Inc. San Jose, California.,1994.

[5] J. M. Arnold, M. A. McGarry, "Splash 2 Programmer's Manual," Supercomputing Research Center,Tech. Rep. SRC-TR-93-107, Bowie, Maryland, 1993.

[6] M. Gokhale, R. Minnich, "FPGA Computing in a Data Parallel C," in *IEEE Workshop on FPGAs for Custom Computing,* Napa, CA, pp. 94-101, Apr 1993.

[7] L. Abbott, P. Athanas, R. Elloitt, B. Fross, L. Chen, "Finding Lines and Building Pyramids with Splash-2" in *IEEE Workshop on FPGAs for Custom Computing,* Napa, CA, pp. 155-163, Apr 1994.

[8] P. Athanas, H. Silverman, "Processor Reconfiguration through Instruction-Set Metamorphosis: Architecture and Compiler," *IEEE Computer,* vol. 26, no. 3, pp. 11-18, Mar 1993.

[9] A. L. Abbott, R. M. Haralick, X. Zhuang, "Pipeline Architectures for Morphologic Image Analysis," *Machine Vision and Applications,* vol. 1, no. 1, pp. 23-40, 1988.

[10] P. J. Burt, E. H. Adelson, "The Laplacian Pyramid as a Compact Image Code," *IEEE Transactions on Communications,* vol. COM-31, no. 4, pp. 532-540, April 1983.

[11] B. Jahne, *Digital Image Processing,* Springer-Verlag, New York, 1991.

[12] A. Rosenfeld, A. Kak, *Digital Picture Processing, 2nd Edition,* Academic, New York, 1982.

A Superscalar and Reconfigurable Processor

Christian Iseli and Eduardo Sanchez

Laboratoire de Systèmes Logiques, École Polytechnique Fédérale, CH–1015
Lausanne, Switzerland *

Abstract. Spyder is a processor architecture with three concurrent, reconfigurable execution units implemented by FPGAs. This paper presents the hardware evolution of the Spyder processor and its evolving
software development environment.

1 Introduction

The performance (P) of a processor is usually measured as a function of the
time (T) necessary for the execution of a given benchmark. This execution time
is itself a function of three parameters: the number of instructions executed (N_i),
the average number of cycles per instruction (C_i) and the clock frequency (F) [1]:
Performance $= f\left(T^{-1}\right)$ where $T = N_i \cdot C_i \cdot F^{-1}$.

Assuming that the clock period is mostly a technological parameter (even
if it also depends on the processor organization), the designer is left with two
parameters to optimize, so as to realize the highest-performance processor in the
world. Unfortunately, however, these two parameters are in direct conflict: the
optimization of one implies the deterioration of the other, and vice-versa. Processor designers are thus divided into two opposing schools: the CISC (Complex
Instruction Set Computer) school emphasizes the optimization of the number of
instructions, while the RISC (Reduced Instruction Set Computer) school, which
at the moment is dominant, emphasizes the optimization of the number of cycles
per instruction [1].

In a conventional scalar processor, at least three cycles (fetch, decode and
execute) are necessary to execute an instruction: special techniques are thus
needed to improve this value. At the moment, the two most common techniques
are pipelining (decomposition of the execution in independent phases, so as to be
able to execute more than one instruction at the same time, each in a different
phase) and superscalar design (entirely parallel execution of multiple instructions, thanks to multiple processing units).

However, some drawbacks are associated with these two techniques:

- Pipelining introduces hazards of many types, which are difficult to handle,
  with repercussions notably on the compiler and on the handling of excep-
  tions. Moreover, if the number of stalls in the pipeline is very high, the
  performance gain is minor, given the added complexity of hardware and
  software.

---

* FAX: +41 21 693 3705, E-mail: Christian.Iseli@di.epfl.ch

- The multiple execution units of a superscalar processor, which are sure to cause an increase of the surface of the silicon and a deterioration of the clock frequency, are rarely used in full: it is very difficult for the fetch unit to find enough independent instructions capable of being executed in parallel. Moreover, a large bandwidth is necessary to deliver the data to all the units which request them. And, because of the fixed size of the data in a conventional processor, a sizable portion of this bandwidth can remain unused. For these reasons, superscalar architectures very often imply an appreciable waste of silicon.

Spyder (Reconfigurable Processor DEvelopment SYstem) is an alternative to these two techniques, proposing an improvement of superscalar design, called SURE (SUperscalar and REconfigurable):

- the instruction is very large (128 bits), allowing, as with microprogramming, the direct control of the different execution units;
- only three execution units are available in parallel, but, thanks to the use of FPGA circuits in their implementation, they are entirely configurable by the user, according to the application. The waste in the number of units and in the bandwidth (the configuration of the units includes not only the functionality but also the size of the handled data) is thus avoided;
- to increase the bandwidth, the large register bank shared by the three execution units is multiple-access (four accesses: the three units plus the data memory).

Right now, the configuration of the execution units is done "by hand" by the user, but the final goal is to endow Spyder with an "intelligent" compiler, capable of producing, in addition to the executable code, the configurations best suited for a given problem.

The architecture of Spyder, together with a first implementation, has been described in [2]: here we plan to show mostly the evolution of its development software, as well as a second, extended, implementation.

2 Processor Architecture

The overall architecture of Spyder, shown in figure 1, derives mainly from the architecture of VLIW processors [3] with some features of RISC processors. It is register-based and uses only load and store operations to communicate with the data memory. The data and program memory are separate (Harvard-type). The data memory is dual-port, 16-bit wide and the program memory is 128-bit wide. The program consists of horizontal microcode that drives all the components of the processor in parallel. There are three execution units that can work in parallel and two register banks. Each execution unit has one separate bidirectional data bus connected to each register bank and can perform one read and one write operation during each clock cycle. Both register banks are connected to the data memory and one data transfer from/to each register bank can occur during each clock cycle.

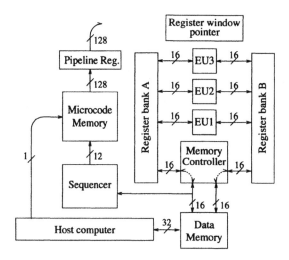

Fig. 1. Overall architecture

A detailed description of the structure and execution of an instruction on Spyder can be found in [2]. The architecture of Spyder has undergone some modifications, described below, since it was presented in [2].

The structure of the sequencer is shown in figure 2. It can perform the usual jump, call and return operations, both conditionally and unconditionally. It now also includes a stack of counters. We soon discovered, by experimenting with the first version of Spyder, that an execution unit often had to be used to implement a loop counter. The counter also had to be stored in some data register. We felt it was better to have the counting done where it was actually needed: in the sequencer. There is a stack of 32 16-bit counters to handle nested loops.

The registers are accessed using a windowing system similar to the SPARC architecture [4]: four windows of 4 to 16 registers are accessible at any given time: the global window which is always accessible (at the highest index), the current window, the previous window, and the next window. The size of the register banks and the number of windows has been increased, compared to the previous version of Spyder. Each bank consists of 2048 16-bit registers and there can be 128, 256 or 512 windows of, respectively, 16, 8 or 4 registers. The current window index is incremented by the subroutine call instruction and decremented by the return from subroutine instruction. It now can also be incremented, decremented and reset at will. Indeed, the windowing mechanism is very valuable to temporarily store a few lines of pixels from an image or the state of the cells of a cellular automata. For example, if we configure Spyder to have 512 windows, we can store at most 8 lines (there are 2 register banks) of a 8192-pixel-wide black-and-white image in the registers.

The operation of the sequencer and the increment, decrement and reset of the window index are controlled by the same 4 bits of the microcode, as would be expected since the window index is modified by call and return instructions.

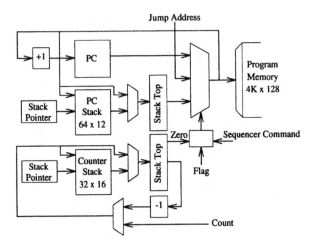

Fig. 2. Sequencer

It is interesting to note that there is no explicit increment operation for the sequencer: it is replaced by an explicit jump to the next instruction. No problem is caused by this approach since we use horizontal microcode where the jump address field is always available, and since it lowers the number of bits needed to control the sequencer.

The execution units are now also connected in a ring by two 16-bit wide data buses. This allows the use of the execution units in a pipelined mode and increases the data transfer bandwidth available to the execution units.

3 Implementation

The sequencer and the register windowing controller are each implemented in a Xilinx 4005 chip [5]. Each execution unit is implemented in a Xilinx 4008 chip; a Xilinx 4010 chip could also be used, should the need arise, without having to change the board. The whole Spyder processor is being realized on a double-Europe VME board.

4 Software

In classical superscalar processors, it is very hard to keep the n execution units available busy at all time. To keep them all busy, the dispatch unit would have to be able to read a large number of instructions in parallel and find among them n independent instructions of different types. For example, the PowerPC 601 can execute up to 3 operations in parallel (an integer operation, a floating-point operation and a branch), chosen among 8 instructions stored in the execution queue, and its dispatch unit is very complex [6].

```
void
Add(short ai, short bi,
    Phase ph,
    short &ao, short &bo)
{
  static short temp;

  switch (ph) {
  case PH_1:
    temp = ai + bi;
    break;
  case PH_2:
    ao = bo = temp + ai + bi;
    break;
  }
}
```

```
operator Add {
  input i1, i2, i3, i4;
  output sum;
  phases {
    1: (i1,i2) -> ( , );
    1: (i2,i1) -> ( , );
    2: (i3,i4) -> (sum,sum);
    2: (i4,i3) -> (sum,sum);
  }
}
memory {
  short d[4];
  short result;
}
main()
{
  Add.i1 = d[0]; Add.i2 = d[1];
  Add.i3 = d[2]; Add.i4 = d[3];
  result = Add.sum;
}
```

Fig. 3. Example of operator   **Fig. 4.** Example of microcode

The Spyder architecture tries to provide two solutions to this waste of resources:

1. The execution units are configured according to the application. They implement the operations actually used by the application in an optimal way.
2. There is no dispatch unit. Each instruction commands the execution units and all the other components of the processor in parallel.

To summarize, the Spyder architecture shifts the complexity from the hardware to the compiler. Ideally, the compiler should be able to produce the object code, the instruction set and the hardware to implement this instruction set.

In other words, the functions of the ideal Spyder compiler are:

— to analyze the source code for a given application, written in a standard high-level language like C++, and extract the hardware operators necessary to the application;
— to classify these operators in three independent groups to be implemented each in one of the three execution unit;
— to generate the configuration of the FPGA circuits implementing the execution units;
— to generate the object code, maximizing the parallel use of the three execution units.

```
class Spyder {
public:
  unsigned long *ramPtr;
  unsigned char *ctrlPtr;

  virtual int WriteData(u_int start, u_int length, void *data) = 0;
  virtual int ReadData(u_int start, u_int length, void *data) = 0;
  virtual void Start(void) = 0;
  virtual void Stop(void) = 0;
  virtual void WaitAndStop(void) = 0;
  virtual int RunningP(void) = 0;
};
```

Fig. 5. The Spyder interface class

The design of such a compiler, if at all possible, is unfortunately beyond our reach. The added complexity of parallelizing a sequential algorithm and generating the configuration of FPGA circuits from a behavioral description of their functionality, all described in a single high-level language, seems overwhelming.

Moreover, considering the fact that Spyder is a coprocessor, and thus is unable to handle input/output, another program must run on the host computer to complete the development system. This program, which can be called a monitor, feeds data to Spyder, reads back the results, displays them, etc.

So, for the time being, a Spyder application is decomposed in several parts where more intelligence is provided by the programmer than by the tools. In a first phase, the programmer must decide which operators to implement in the execution units and describe them using a subset of C++. Figure 3 shows the description of an example operator implementing the addition of 4 numbers in 2 phases. This description contains the operator interface and the operations to be performed in each phase. From this description, a compiler generates a netlist which is then fed to the Xilinx placement and routing tools which in turn produce the configuration for the execution units.

In a second phase, the programmer uses the operators defined in the first phase to write the program which solves a given problem. Again a subset of C++, with a few extensions, is used to describe the algorithm. As shown in figure 4, first the available operators are described, followed by the data memory organization and the program itself. The compiler handles the register allocation and schedules the operations to try to maximize the use of all the execution units in parallel. The compiler can also translate the source program in standard C++, in order to be able to simulate the program on the host computer, and thus facilitate the debugging.

Eventually, the programmer has to write the monitor program running on the host computer which handles the communication between Spyder and the host. A C++ interface class, shown in figure 5, is provided to ease this task. All the details of the hardware implementation of Spyder are thus hidden. Accesses to the data memory of Spyder are performed through the `ramPtr` pointer or the `ReadData` and `WriteData` methods. This class also allows to hide from the monitor whether the Spyder application is being simulated or run on the actual hardware.

5 Conclusion

A first attempt at implementing the Abingdon Cross Benchmark (an image processing benchmark) [7] shows that the performance of Spyder for this benchmark is roughly an order of magnitude worse than the performance of the connection machine CM-1. But it seems our algorithm could be improved. It also turns out that the performance of Spyder could very easily be improved by using wider data words (32 or 64-bit wide data word). It is interesting to note that the ability of Spyder to handle more that one data element per data word (i.e., 16 black-and-white pixels per 16-bit data word) in a meaningful manner allows an easy scaling of the computing power with the width of the data words. This is usually not the case with regular (non-reconfigurable) processors. Who needs 64-bit integers?

References

1. David A. Patterson and John L. Hennessy, *Computer Organization & Design, The Hardware/Software Interface*, Morgan Kaufmann Publishers, San Mateo, 1994.
2. Christian Iseli and Eduardo Sanchez, "Spyder: A reconfigurable VLIW processor using FPGAs", in *IEEE Workshop on FPGAs for Custom Computing Machines*, Napa, April 1993.
3. B. Ramakrishna Rau and Joseph A. Fisher, "Instruction-level parallelism: History, overview, and perspectives", in *Instruction-Level Parallelism*, B. Ramakrishna Rau and Joseph A. Fisher, Eds., pp. 9–50. Kluwer Academic Publishers, Boston, 1993.
4. Sun Microsystems, *The SPARC Architecture Manual*, Sun Microsystems, Inc, Mountain View, 1987.
5. Xilinx, *The Programmable Logic Data Book*, Xilinx, San Jose, 1993.
6. Motorola, Phoenix, Arizona, *PowerPC 601 RISC Microprocessor User's Manual*, 1993.
7. Kendall Preston Jr., "The abingdon cross benchmark survey", *IEEE Computer*, vol. 22, no. 7, pp. 9–18, July 1989.

A Fast FPGA Implementation of a General Purpose Neuron

Valentina Salapura, Michael Gschwind, Oliver Maischberger
{ *vanja, mike, oliver* } *@vlsivie.tuwien.ac.at*

Institut für Technische Informatik
Technische Universität Wien
Treitlstraße 3-182-2
A-1040 Wien
AUSTRIA

Abstract. The implementation of larger digital neural networks has not been possible due to the real-estate requirements of single neurons. We present an expandable digital architecture which allows fast and space-efficient computation of the sum of weighted inputs, providing an efficient implementation base for large neural networks. The actual digital circuitry is simple and highly regular, thus allowing very efficient space usage of fine grained FPGAs. We take advantage of the re-programmability of the devices to automatically generate new custom hardware for each topology of the neural network.

1 Introduction

As conventional computer hardware is not optimized for simulating neural networks, several hardware implementations for neural networks have been suggested ([MS88], [MOPU93], [vDJST93]). One of the major constraints on hardware implementations of neural nets is the amount of circuitry required to perform the multiplication of each input by its corresponding weight and their subsequent addition:

$$n_i\left(x_1, \ldots, x_{m_i}\right) = a_i\left(\sum_{1 \leq j \leq m_i} w_{ji} * x_j\right),$$

where x_j are the input signals, w_{ji} the weights and a_i the activation function.

The space efficiency problem is especially acute in digital designs, where parallel multipliers and adders are extremely expensive in terms of circuitry [CB92]. An equivalent bit serial architecture reduces this complexity at the cost of net performance, but still tends to result in large and complex overall designs.

We decided to use field-programmable gate arrays (FPGAs) to develop a prototype of our net [Xil93]. FPGAs can be reprogrammed easily, thus allowing different design choices to be evaluated in a short time. This design methodology also enabled us to keep overall system cost at a minimum. Previous neural network designs using FPGAs have shown how space efficiency can be achieved [vDJST93], [GSM94], [Sal94]. The neuron design proposed in this paper makes a compromise between space efficiency and performance.

We have developed a set of tools to achieve complete automatization of the design flow, from the network architecture definition phase and training phase to the finished hardware implementation. The network architecture is user-definable, allowing the implementation of any network topology. The chosen network topology is described in an input file. Tools automatically translate the network's description into a corresponding net list which is downloaded into hardware. Network training is performed off-chip, reducing real estate consumption for two reasons:

- No hardware is necessary to conduct the training phase.
- Instead of general purpose operational units, specialized instances can be generated. These require less hardware, as they do not have to handle all cases. This applies especially to the multiplication unit, which is expensive in area consumption terms. Also, smaller ROMs can be used instead of RAMs for storing the weights.

As construction and training of the neural net occurs only once in an application's lifetime, namely at its beginning, this off-chip training scheme does not present a limitation to a net's functionality. Our choice of FPGAs as implementation technology proved beneficial in this respect, as for each application the best matching architecture can be chosen, trained on the workstation and then down-loaded to the FPGA for operational use.

2 Related Work

The digital hardware implementations presented in literature vary from bit-stream implementations, through bit-serial and mixed parallel-serial implementations to fast, fully parallel implementations.

The pulse-stream encoding scheme for representing values is used in an analog implementation by Murray and Smith [MS88]. They perform space-efficient multiplication of the input signal with the synoptic weight by intersecting it with a high-frequency chopping signal.

van Daalen et al. [vDJST93] present a bit-stream stochastic approach. They represent values v in the range $[-1, 1]$ by stochastic bit-streams in which the probability that a bit is set is $(v + 1)/2$. Their input representation and architecture restrict this approach to fully interconnected feed-forward nets. The non-linear behavior of this approach requires that new training methods be developed.

In [GSM94], we propose another bit-stream approach. Digital chopping and encoding values v from the range $[0, 1]$ by a bit stream where the probability that a bit is set is v are used. Using this encoding, an extremely space efficient implementation of the multiplication can be achieved. In this design, only 22 CLBs [Xil93] are required to implement a neuron. This method enables the construction of any network architecture, but constrains applications to those with binary threshold units.

The approach in [Sal94] is based on the idea to represent the inputs and synaptic weights of a neuron as delta encoded binary sequences. For hardware implementation delta arithmetic units are used which employ only one-bit full adders and D flip-flops. The performance of the design is improved and some real-estate savings are achieved. The design can be used for assembling of feed-forward and recursive nets.

GANGLION [CB92] is a fast implementation of a simple three layer feed forward net. The implementation is highly parallel achieving performance of 20 million decisions per second. This approach needs 640 to 784 CLBs per neuron, making this implementation extremely real estate intensive.

3 The Neuron

Each processing unit computes a weighted sum of its inputs plus a bias value assigned to that unit, applies an activation function, and takes the result as its current state. The unit performs the multiplication of 8 bit unsigned inputs by 8 bit signed integer weights forming a 16 bit signed product. The eight products and a 16 bit signed unit-specific bias are accumulated into a 20 bit result. The final result is computed by applying an arbitrary activation function. This process scales the 20 bit intermediate result stored in the accumulator to an 8 bit value (see figure 1).

We use the fact that multiplication is commutative, and instead of multiplying the input values with the weight, we multiply the (signed) weight with the (positive) input values. Thus, multiplication is reduced to multiplying a signed value by an unsigned value. This can be implemented using fewer logic gates.

Multiplication is performed by using the well-know shift and add algorithm. The first synapse weight is loaded into the 16 bit shift register from the weight ROM, and the synapse input in the 8 bit shift register. Then, the shift and add multiplication algorithm is performed, using a 20 bit accumulator.

After eight iterations, the first multiplication $w_{ji} * x_j$ has been processed. To process the next neuron input, the input and weight values for the next multiplication are loaded into their respective shift registers and the process starts over. At the same time, the accumulator is used for implementing the accumulation of the multiplication result and adding the results of all eight multiplications.

After the result $\sum_{1 \leq j \leq m_i} w_{ji} * x_j$ has been computed, the activation function is applied to this intermediate result. Depending on the complexity of the activation function, this can take 0 or more cycles. This activation function also scales the intermediate result to an unsigned 8 bit output value. This output value is either the final result or fed to a next layer neuron.

As the constructed unit can have at most eight inputs and as the multiplication of one input requires eight cycles, a new computation cycle is started every 64 cycles (plus the time used for computing the activation function). This

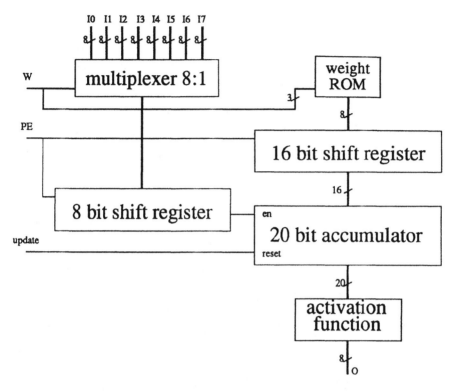

Fig. 1. Schematic diagram of a neuron.

condition is checked by a global counter, and distributed to all neurons. Upon receiving this signal, the neurons will latch their input state into an output register, load the bias into the accumulator and start a new computation.

51 CLBs are used for implementing the base neuron. Depending on the complexity of the activation function used, additional CLBs may be necessary to implement look-up tables or other logic. The **ppr** tool [Xil92] reports the following design data for a single neuron:

Packed CLBs	51
FG Function Generators	102
H Function Generators	16
Flip Flops	44
Equivalent "Gate Array" Gates	1458

4 The Overall Network Architecture

The design of the neurons is such that any neural architecture can be assembled from single neurons. Users can choose an optimal interconnection pattern for their specific application, as these interconnections are performed using FPGA routing. This neuron design can be used to implement a wide range of different models of neural networks whose units have binary or continuous input and unit state, and with various activation functions, from hard-limiter to sigmoid. The implementation of both feed-forward networks and recursive networks [Hop82], [Koh90] is possible.

Any network can be implemented using the proposed units. The design includes a global synchronization unit which generates control signals distributed to the whole network. Figure 2 shows a feed-forward network with four neurons in the input layer, four neurons in the hidden layer and two neurons in the output layer.

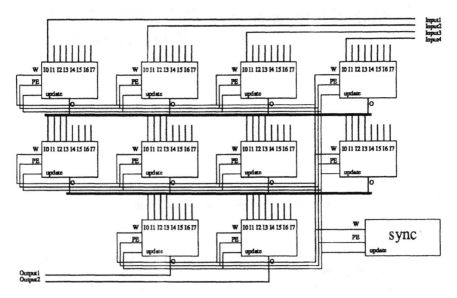

Fig. 2. Example architecture: a feed-forward network with ten neurons.

Several neurons can be placed on one FPGA. The exact number of neurons fitting on one FPGA depends on the exact FPGA type and the complexity of the activation function. By using multiple FPGAs, arbitrarily large, complex neural nets can be designed cheaply and efficiently. Having neurons as indivisible functional units allows absolute freedom in choosing any topology required.

5 Automation of the Design Process

To design a network for a new application, a new network topology is selected. On this network, the training process is performed, yielding a set of new weights and biases. These new connections, weights and biases have to be mapped to the

logic of the LCAs. Embedding these parameters into the LCAs alters the routing
within the LCAs. To customize the base LCA design for each new application, we
have developed tools that enable the fully automation of the designing process.
The arbitrarily network topology with trained weights is described in an input
file. Complete translation into LCAs and design optimization is then performed
automatically, entirely invisible to the user.

```
I    I1
I    I2
N    SON0        0
C    I1   GND  GND  GND  GND  GND  GND  GND
W    126  0  0   0   0   0   0   0
N    SON1        0
C    I2   GND  GND  GND  GND  GND  GND  GND
W    126  0  0   0   0   0   0   0
N    S1N0        192
C    SON0 SON1 GND  GND  GND  GND  GND  GND
W    63  63  0   0   0   0   0   0
N    S2N0        64
C    SON0 SON1 GND  GND  GND  GND  GND  GND
W    63  63  -126  0   0   0   0   0
O    S2N0
```

Fig. 3. Example input file: a feed-forward network with four neurons.

The input file contains all parameters needed. For illustration, a simple input
file is shown in figure 3. It describes a small network with two inputs, two neurons
in the first, one neuron in the second and third layers and one output. At the
beginning of the file inputs are specified (denoted with I), assigning a name to
every input. Then, the neurons are described. The order of neurons in the file
is irrelevant. Every neuron is defined with four parameters. Firstly, a name is
assigned to every unit. Then, the bias value assigned to the unit is given. After
that, the connections are specified: for each of the eight neuron inputs, the name
of the input to the network or the name of the unit with which to connect is
given. If an input of the unit is unused, it is connected to GND. Finally, the
corresponding weights (signed integers) are given. At the end of the file, the list
of the outputs is defined, containing the names of the units whose output should
be used as outputs of the network.

After the network has been defined and trained, our tool set generates a
configuration net list for the FPGA board. The configuration bit-stream is used
to initialize the Xilinx FPGAs. Figure 4 shows the phase model for the design
of a neural net from training to hardware operation.

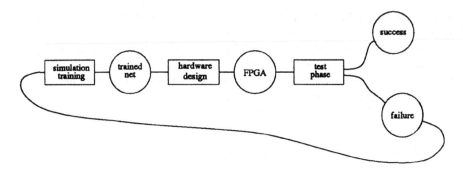

Fig. 4. Phase model of net development

6 Conclusion

We propose a space-efficient, fast neural network design which can support any network topology. Starting from an optimized, freely interconnectable neuron, various neural network models can be implemented.

The simplicity of the proposed neuron design allows for the massive replication of neurons to build complex neural nets. FPGAs are used as hardware platform, facilitating the implementation of arbitrary network architectures and the use of an off-chip training scheme.

Tools have been developed to completely automate the design flow from the network architecture definition phase and training to the final hardware implementation.

References

[CB92] Charles E. Cox and W. Ekkehard Blanz. GANGLION – a fast field-programmable gate array implementation of a connectionist classifier. *IEEE Journal of Solid-State Circuits*, 27(3):288–299, March 1992.

[GSM94] Michael Gschwind, Valentina Salapura, and Oliver Maischberger. Space efficient neural net implementation. In *Proc. of the Second International ACM/SIGDA Workshop on Field-Programmable Gate Arrays*, Berkeley, CA, February 1994. ACM.

[Hop82] John J. Hopfield. Neural networks and physical systems with emergent collective computational abilities. In *Proceedings of the Academy of Sciences USA*, volume 79, pages 2554–2558, April 1982.

[Koh90] Teuvo Kohonen. The self-organizing map. *Proceedings of the IEEE*, 78(9):1464–1480, September 1990.

[MOPU93] Michele Marchesi, Gianni Orlando, Francesco Piazza, and Aurelio Uncini. Fast neural networks without multipliers. *IEEE Transactions on Neural Networks*, 4(1):53–62, January 1993.

[MS88] Alan F. Murray and Anthony V. W. Smith. Asynchronous VLSI neural networks using pulse-stream arithmetic. *IEEE Journal of Solid-State Circuits*, 23(3):688–697, March 1988.

[Sal94]   Valentina Salapura. Neural networks using bit stream arithmetic: A space
             efficient implementation. In *Proceedings of the IEEE International Sympo-
             sium on Circuits and Systems*, London, UK, June 1994.

[vDJST93]   Max van Daalen, Peter Jeavons, and John Shawe-Taylor. A stochastic
             neural architecture that exploits dynamically reconfigurable FPGAs. In
             IEEE Workshop on FPGAs for Custom Computing Machines, Napa, CA,
             April 1993. IEEE CS Press.

[Xil92]   Xilinx. *XACT Reference Guide*. Xilinx, San Jose, CA, October 1992.

[Xil93]   Xilinx. *The Programmable Logic Data Book*. Xilinx, San Jose, CA, 1993.

Data-procedural Languages
for FPL-based Machines

A. Ast, J. Becker, R.W. Hartenstein, R. Kress, H. Reinig, K. Schmidt

Fachbereich Informatik, Universität Kaiserslautern
Postfach 3049, D-67653 Kaiserslautern, Germany
fax: (+49 631) 205-2640, e-mail: abakus@informatik.uni-kl.de

ABSTRACT. This paper introduces a new high level programming language for a novel class of computational devices namely data-procedural machines. These machines are by up to several orders of magnitude more efficient than the von Neumann paradigm of computers and are as flexible and as universal as computers. Their efficiency and flexibility is achieved by using field-programmable logic as the essential technology platform. The paper briefly summarizes and illustrates the essential new features of this language by means of two example programs.

1 Introduction

Usually procedural machines are based on the von Neumann machine paradigm. (Data flow machines are no procedural machines, since the execution order being determined by an arbiter is indeterministic.) Both, von Neumann machines, as well as von Neumann languages (Assembler, C, Pascal, etc.) are based on this paradigm. We call this a *control-procedural paradigm*, since execution order is control-driven. Because in a von Neumann machine the instruction sequencer and the ALU are tightly coupled, it is very difficult to implement a reconfigurable ALU supporting a substantial degree of parallelism.

By turning the von Neumann paradigm's causality chain upside down we obtain a data sequencer instead of an instruction sequencer. We obtain a new machine paradigm called a *data-procedural machine paradigm*. This new paradigm is the root of a new class of procedural languages which we call *data-procedural languages*, since the execution order is deterministically data-driven. This new data-procedural paradigm [1], [4], [5] strongly supports highly flexible FPL-based reconfigurable ALUs (rALUs) permitting very high degrees of intra-rALU parallelism. That's why this paradigm opens up new dimensions of machine architecture, reconfigurability, and hardware efficiency [4].

This paper introduces this new class of languages by using a data-procedural example language. The language MoPL-3 used here is a C extension. Such data-procedural languages support the derivation of FPL-based data path resource configurations and data sequencer code directly from data dependences. The usual detour from data dependences via control flow to data manipulation, as practiced by von Neumann program-

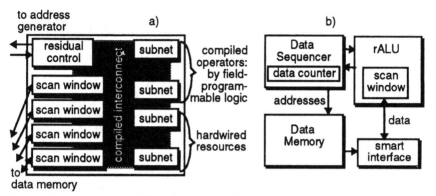

Fig. 1. Basic structures of Xputers and the MoM architecture: a) reconfigurable ALU (rALU) of the MoM, b) basic structure of Xputers

ming, is almost completely avoided. The paper illustrates data-procedural language usage and compilation techniques as well as their application to FPL-based hardware.

2 Summarizing the Xputer

For convenience of the reader this section summarizes the underlying machine paradigm having been published elsewhere [1], [4], [5], [6], [9]. Main stream high level control-procedural programming and compilation techniques are heavily influenced by the underlying von Neumann machine paradigm. Most programmers with more or less awareness need a von-Neumann-like abstract machine model as a guideline to derive executable notations from algorithms, and to understand compilation issues. Also programming and compilation techniques for Xputers need such an underlying model, which, however, is a data-procedural machine paradigm, which we also call *data sequencing paradigm*. This section summarizes and illustrates the basic machine principles of the Xputer paradigm [9]. Later on simple algorithm examples will illustrate MoPL-3, a data-procedural programming language.

2.1 Xputer Machine Principles

The main difference to von Neumann computers is, that Xputers have a *data counter* (as part of a data sequencer, see figure 1b) instead of a program counter (part of an instruction sequencer). Two more key differences are: a *reconfigurable ALU* called *rALU* (instead of a hardwired ALU), and *transport-triggered operator activation* ([11], instead of the usual control-flow-triggered activation). Operators are preselected by an *activate* command from a *residual control* unit. Operator activation is transport-triggered. Xputers are data-driven but unlike data flow machines, they operate deterministically by *data sequencing* (no arbitration).

Scan Window. Due to their higher flexibility (in contrast to computers) Xputers may have completely different processor-to-memory interfaces which efficiently support the exploitation of parallelism within the rALU. Throughout this paper, however, we use an Xputer architecture supported by smart register files, which provide a 2-dimensional scan windows (e.g. figure 2b shows one of size 2-by-2). A scan window gives

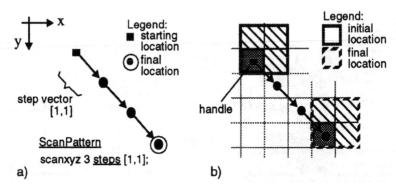

Fig. 2. Simple scan pattern example: a) source text and illustration, b) scan pattern moves a scan window by its handle

rALU access to a rectangular section of adjacent locations in data memory space. Its size is adjustable at run time.

Scan Pattern. A scan window is placed at a particular point in data memory space according to an address hold by a data counter (within a data sequencer). A data sequencer generates sequences of such addresses, so that the scan window controlled by it travels along a path which we call scan pattern. Figure 2a shows a scan pattern example with four addresses, where figure 2b shows the first and fourth location of the scan window. Figure 3 shows sequential scan pattern examples, and figure 9c a compound (parallel) scan pattern example.

Data Sequencer. The hardwired data sequencer features a rich and flexible repertory of scan patterns [4] for moving scan windows along scan paths within memory space. Address sequences needed are generated by hardwired address generators having a powerful repertory of generic address sequences [2], [13]. After having received a scan pattern code a data sequencer runs in parallel to the rest of the hardware without stealing memory cycles. This accelerates Xputer operation, since it avoids performance degradation by addressing overhead.

Reconfigurable ALU. Xputers have a reconfigurable ALU (rALU), which usually consists of global field-programmable interconnect (for reconfiguration), hardwired logic (a repertory of arithmetic, relational operators), and field-programmable logic (for additional problem-specific operators) [3]. Figure 1a shows an example: the rALU of the MoM-3 Xputer architecture: 4 smart register files provide 4 scan windows. A rALU has a hidden RAM (hidden inside the field-programmable integrated circuits used) to store the configuration code.

rALU Configuration is no Microprogramming. Also microprogrammable von Neumann processors have a kind of reconfigurable ALU which, however, is highly bus-oriented. Buses are a major source of overhead [7], especially in microprogram execution, where buses reach extremely high switching rates at run time. The intension of rALU use in Xputers, however, is to push overhead-driven switching activities away from run time, over to loading time as much as possible, in order to save the much more precious run time.

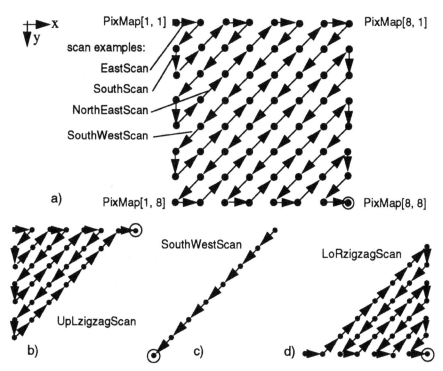

Fig. 3. JPEG Zig-Zag scan pattern for array PixMap [1:8,1:8], a) and its subpatterns: b) upper left triangle UpLzigzagScan, d) lower right LoRzigzagScan, c) full SouthWest-Scan

Compound Operators. An Xputer may execute expressions (which we call *compound operators*) within a single machine step, whereas computers can execute only a single operation at a time. The rALU may be configured in such a way, that one or more sets of parallel data paths form powerful compound operators connected to one or more scan windows (example in figure 7).

Execution triggering. A compound operator may be activated (sensitized) by setting a flag bit (and passivated by resetting this flag bit). Each operator currently being active is automatically executed whenever the scan windows connected to it are moved to a new location. E.g. during stepping through a scan pattern of length n this operator is executed n times.

Summary of Xputer Principles. The fundamental operational principles of Xputers are based on *data auto sequencing* mechanisms with only sparse control, so that Xputers are deterministically data-driven (in contrast to data flow machines, which are indeterministically data-driven by arbitration and thus are not debuggable). Xputer hardware supports some fine granularity parallelism (below instruction set level: at data path or gate level) in such a way that internal communication mechanisms are more simple than known from parallel computer systems (figure 9d and e, for more details about Xputer see [4], [5]).

Source code example (see figure 5)	Action described
ScanPattern (see line (2)) EastScan is 1 step [1,0]	single step scan pattern **EastScan**: 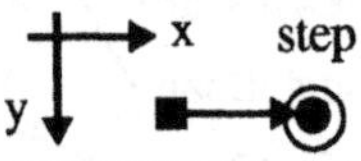 x step vector: [1,0] y
ScanPattern (see line (3)) SouthScan is 1 step [0,1]	single step scan pattern **SouthScan**: 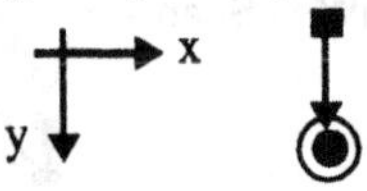 x step vector: [0,1] y
ScanPattern (see line (4)) SouthWestScan is 7 steps [-1,1]	multiple step scan pattern: 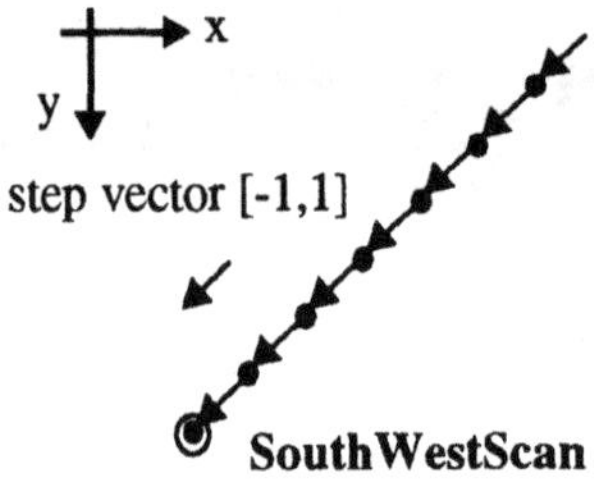 x y step vector [-1,1] **SouthWestScan**
ScanPattern (see line (5)) NorthEastScan is 7 steps [1,-1]	like SouthWestScan (see above), but reversed order sequence

Fig. 4.   Scan patterns declared for the JPEG example (see also figure 5)

3 A Programming Language for Xputers

This section introduces the high level Xputer programming language MoPL-3 (Map-oriented Programming Language) which is easy enough to learn, but which also is sufficiently powerful to explicitly exploit the hardware resources offered by the Xputer. For an earlier version of this language we have developed a compiler [16]. MoPL-3 is a C extension, including primitives for data sequencing and hardware reconfiguration.

3.1 MoPL-3: A Data-procedural Programming Language

This section introduces the essential parts of the language MoPL-3 and illustrates its semantics by means of two program text examples (see figure 5 and figure 8): the constant geometry FFT algorithm, and the data sequencing part for the JPEG zig-zag scan being part of a proposed picture data compression standard. MoPL-3 is an improved version of MoPL-2 having been implemented at Kaiserslautern as a syntax-directed editor [16].

From the von Neumann paradigm we are familiar with the concept of the *control state* (current location of control), where control statements at source program level are translated into program counter manipulations at hardware machine level. The main extension issue in MoPL compared to other programming languages is the additional concept of *data location* or *data state* in such a way, that we now have simultaneously two different kinds of state sequences: a single sequential control state sequence and one (or more concurrent) data state sequence(s). The control flow notation does not model the underlying Xputer hardware very well, since it has been adopted from C to give priority to acceptance by programmers. The purpose of this extension is the easy programming of sequences of data addresses (scan patterns) to prepare code generation for the data sequencer. The familiar notation of these MoPL-3 constructs is easy to learn by the programmer.

Function Name	Corresponding Operation
rotl	turn left
rotr	turn right
rotu	turn 180°
mirx	flip x
miry	flip y
reverse	reversed order sequence

Table 1. Transformation functions for scan patterns

3.2 Declarations and Statements

The following Xputer-specific items have to be predeclared: scan windows (by <u>window</u> declarations), rALU configurations (by <u>rALUsubnet</u> declarations), and scan patterns (by <u>ScanPattern</u> declarations). Later a rALU subnet (a compound operator) or a scan pattern may be called by their names having been assigned at declaration. Scan windows may be referenced within a rALU subnet declaration.

Scan Window Declarations. They have the form: <u>window</u> <group_name> <u>is</u> <window_specs>';'. Each window specification has the form: <window_name(s)> <window_size> <u>handle</u> <point>. Figure 6 shows an example, where a 2-dimensional window named 'SW1' with a size of 2 by 2 64-bit-words, and two windows named 'SW2' and 'SW3' with the size of a single 64-bit-word each, are declared. The <point> behind <u>handle</u> specifies the word location inside the window, which is referenced by scan patterns. The order of windows within a group refers to physical *window numbers* within the hardware platform. E.g. the above windows 'SW1' through 'SW3' are assigned to window numbers 1 through 3.

rALU Configuration. A compound operator is declared by a rALUsubnet declaration of the following form: <u>rALUsubnet</u> <group_name> <u>is</u> <expression_ assignment(s)>, where the compound operators are described by expressions. All operands referenced must be words within one or more scan windows. Figure 7 illustrates an example of a group 'FFT' which consists of two compound operators with destination window 'SW2', or 'SW3', respectively, and a common source window 'SW1'.

Scan Pattern Declarations. Scan patterns may be declared hierarchically (*nested scan patterns*), where a higher level scan pattern may call lower level scan patterns by their names. Parallel scan patterns *(compound scan patterns)* may be declared, where several scan patterns are to be executed synchronously in parallel. Scan pattern declarations are relative to the current data state(s). A scan pattern declaration section has the form **ScanPattern** <declaration_item(s)> ';'. We distinguish two types of declaration items: simple scan pattern specifications <simple_spec> (linear scan patterns only: examples in figure 4) and procedural scan pattern specifications <proc_spec>. More details will be given within the explanation of the following two algorithm examples.

Activations. Predeclared rALU subnets (compound operators) may be activated by **apply** statements (example in line (44) of figure 10, where group 'FFT' is activated), passivated by **passivate** statements, and removed by **remove** statements (to save programmable interconnect space within the rALU). Scan window group definitions can be activated by **adjust** statements (example in line (43) of figure 10). Such adjustments are effective until another **adjust** statement is encountered.

Parallel Scan Patterns. For parallel execution (compound) scan patterns are called by a name list within a **parbegin** block. See example in line (46) of figure 10, where the scan patterns 'SP1', 'SP23' and 'SP23' are executed in parallel (which implies, that three different data states are manipulated in parallel). Pattern 'SP23' is listed twice to indicate, that two different scan windows are moved by scan patterns having the same specification. The order of patterns within the **parbegin** list corresponds to the order of windows within the adjustment currently effective (ThreeW, see line (43) in figure 10). E.g. scan pattern 'SP1' moves window no. 1, and 'SP2' moves windows no. 2 and 3. Each scan pattern starts at current data state, evokes a sequence of data state transitions. The data state after termination of a scan pattern remains unchanged, until it is modified by a **moveto** instruction or another scan pattern.

Nested Scan Patterns. Predeclared scan patterns may be called by their names. A scan pattern may call another scan pattern. Such nested calls have the following form: <pattern_name> '(' <pattern_definition>')' ';'. An example is shown in line (46) of figure 10, where scan pattern 'HLScan' calls the compound scan pattern definition formed by the **parbegin** block explained above. The entire scan operation is described as follows (for illustration see figure 9). Window group ThreeW is moved to starting points [0,0], [2,0], and [2,8] within array CGFFT by line (45) - see initial locations in figure 9c. Then the (inner loop) compound scan pattern (parbegin group in line (46)) is executed once. Then the (outer loop) scan pattern 'HLScan' executes a single step, where its step vectors move the window group ThreeW to new starting points. Now again the entire inner loop is executed. Finally the inner loop is executed from starting points being identical to the end points of the outer loop scan pattern. After last execution of the inner loop the windows have arrived at final locations shown in figure 9c.

3.3 JPEG ZIG-ZAG SCAN EXAMPLE

The MoPL-3 program in figure 5 illustrates programming the JPEG Zig-Zag scan pattern (named JPEGzigzagScan, see figure 3) being part of the JPEG data compression algorithm [10], [12], [15]. The problem is to program a scan pattern for scanning 64 locations of the array PixMap declared in line (1) of figure 5 according to the sequence shown in figure 3a. Note the performance benefit from generating the 64 addresses needed by the hardwired address generator such, that no time consuming memory

/* assuming, that rALU configuration has been declared and set-up */

```
Array       PixMap [1:8,1:8,15:0];                        (1)
ScanPattern EastScan       is  1 step  [ 1, 0],           (2)
            SouthScan      is  1 step  [ 0, 1],           (3)
            SouthWestScan is  7 steps [-1, 1],            (4)
            NorthEastScan is  7 steps [ 1,-1],            (5)
                                                          (6)
            UpLzigzagScan is                              (7)
            begin                                         (8)
                while (@[<8,])                            (9)
                begin  Eastscan;                          (10)
                    SouthWestScan until @[≤1,];           (11)
                    SouthScan;                            (12)
                    NorthEastScan until @[,≤1];           (13)
                end                                       (14)
            end UpLzigzagScan;                            (15)
                                                          (16)
                                                          (17)
            JPEGzigzagScan is                             (18)
            begin                                         (19)
                UpLzigzagScan;                            (20)
                SouthWestScan;                            (21)
                rotu (reverse(UpLzigzagScan));            (22)
            end JPEGzigzagScan;                           (23)
            /* end of declaration part*/                  (24)
                •                                         (25)
                •                                         (26)
                •                                         (27)
begin   /*statement part*/                                (27)
        moveto PixMap [1,1];                              (28)
        JPEGzigzagScan;                                   (29)
end                                                       (30)
```

Fig. 5. MoPL program of the JPEG scan pattern shown in figure 3

access cycles are needed for address computation. Figure 3 shows, that the JPEG scan pattern may be partitioned into the three subsequences shown by figure 3b through d. Lines (2) thru (5) in figure 5 declare four scan patterns used later to synthesize the scan patterns shown in figure 3 and explained in figure 4. Scan pattern declaration statements have the following form:

<name_of_scan_pattern> <maximum_length_of_loop> **STEPs** <step_vector>.

Scan Pattern Declaration. The step vector specifies the next data location relative to the current data location (data state) before executing a step of the scan sequence. A positive integer specifies the maximum length (maximum step count) of the scan pat-

Source code example:	Size of the scan windows:

window ThreeW is
 	
    SW1 [1:2,1:2,63:0]
    handle [1,1],
	handle [1,1] ⎯
	name of window: SW1
	word size [63:0] 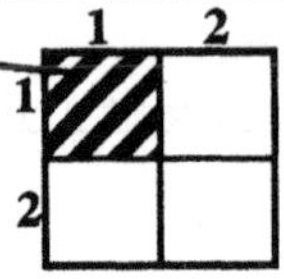

SW2, SW3
    [1:1,1:1,63:0]
    handle [1,1];
	names of windows: SW2, SW3
	word size [63:0] 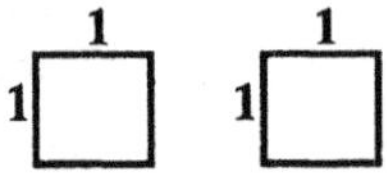

Fig. 6.   Scan window declaration of the FFT example (see also figure 8)

tern. A scan pattern may be terminated earlier than predeclared when an escape clause has become true (which will be explained later).

Calling Scan Patterns. Predeclared scan patterns may be called by statements. The 4 declared scan patterns (figure 4), which are needed for the JPEG zig-zag scan, are called in the two **while** loops at lines (9) thru (14) and at lines (18) thru (23) in figure 5, e.g. see line (10), where the scan pattern 'EastScan' is called (similar to a procedure call in C). By an escape a scan may also be terminated before <maximum_length_of_loop> is reached.

Escapes. In this case there will be an escape from the scan pattern, when the boundary of the data map is reached or exceeded. E.g. see the **until** clause (escape clause) in line (11) indicating an escape on having reached a leftmost word within the 'PixMap' array (see figure 3: the first execution of 'SouthWestScan' at top left corner of the array reaches only a loop length of 1). The condition @[≤1,] says: escape if within current array a data location with an x subscript ≤1 has been reached. The empty position behind the comma says: ignore the y subscript.

Data State Initialization. Before the execution of the first scan pattern, you have to specify the starting point in the data map. For this purpose we use another data state manipulation statement, the **moveto** statement. With this statement (a data goto) you are able to realize absolute jumps of the scan window inside the data map. E.g. see line (28) in figure 5, where the scan window is moved to the upper left corner of the array 'PixMap', which is the starting point of scan pattern 'JPEGzigzagScan' call at line (29).

Hardware-supported Escapes. To avoid overhead for efficiency the **until** clauses are directly supported by the MoM hardware features of escape execution [4]. To support the **until** @ clauses by off-limits escape the address generator provides for each dimension (x, y) two comparators, an upper limit register, and a lower limit register.

Structured Scan Pattern. The above MoPL-3 program (figure 5) covers the following strategy. The first **while** loop at lines (9) thru (14) iterates the sequence of the 4 scan calls 'EastScan' thru NorthEastScan for the upper left triangle of the JPEG scan, from PixMap [1,1] to PixMap [8,1] (figure 3). The second **while** loop at lines (19) - (23) covers the lower right triangle from PixMap [8,1] to PixMap [8,8]. The 'South-

Source code example	rALUsubnet FFT is SW2 = SW1 [1,1] + SW1 [2,1] * SW1 [1,2], SW3 = SW1 [1,1] - SW1 [2,1] * SW1 [1,2];
rALU configuration	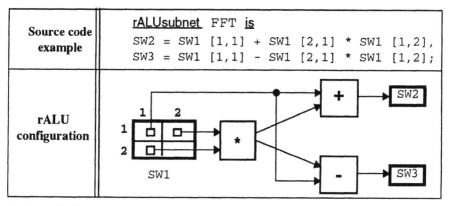

Fig. 7.   Declaration of the 2 compound operators of the FFT example (see figure 8)

WestScan' between both while loops at line (30) from PixMap [8,1] to PixMap [1,8] connects both triangular scans to obtain the total JPEG pattern. In line (22) two scan pattern transformation functions (rotu, reverse) are used. With these functions (see table 1) one can easily realize new scan patterns by using predeclared scan pattern, which structure is similar to the newer ones.

3.4 CONSTANT GEOMETRY FFT EXAMPLE

The second example illustrates parallelism by running several windows synchronously. It is the constant geometry Fast Fourier Transform (FFT) illustrated by figure 9, with a data map (CGFFT) size of 9 by 16 words (figure 9a). Figure 8 shows the MoPL-3 section declaring the scan patterns and the rALU configuration. The declaration of the scan pattern starts with the keyword ScanPattern. 'HLScan' is the outer loop, whereas 'SP1' and 'SP23' are used for inner loops running in parallel (compound scan pattern, see Figure 9c). 'SP1' is used for scan window 'SW1' and 'SP23' is used for two scan windows 'SW2' and 'SW3'. The configuration of the rALU is specified in two parts: the window declaration (size declaration, see Figure 6) and the declaration of the rALUsubnet referencing the window group (see Figure 7).

The declaration of the scan windows and their sizes starts with the keyword window. With handle you can specify the reference point of a window (see Figure 6). This point will be needed, when you move a scan window to a specific place in the data memory. The window group named 'ThreeW' (see above: including 3 scan windows with the names 'SW1', 'SW2' and 'SW3') is declared by the keyword window followed by the name of the group and the keyword is etc. (see line (36) in Figure 8).

Problem-specific Compound Operators. Their declaration starts with rALUsubnet followed by the name of the rALU subnet group (here: 'FFT') and the keyword is. Thereafter the compound operators are specified (see figure 7 and also lines (39) thru (41) in figure 8). Having been activated by adjust and apply (line (43) and (44) in figure 10) these operators are executed automatically in every single step of a scan pattern.

Execution of the Scan Pattern. Figure 9 shows an algorithm implementation example, a 16 point constant geometry FFT, where three scan windows run in parallel.

```
Array      CGFFT  [ 1:9, 1:16, 63:0];                        (31)
ScanPattern                                                  (32)
           SP1    is  7  steps  [0,2],                        (33)
           SP23   is  7  steps  [0,1],                        (34)
           HLScan is  3  steps  [2,0];                        (35)

window     ThreeW is                                         (36)
           SW1      [1:2, 1:2, 63:0] handle [1,1],           (37)
           SW2, SW3 [1:1, 1:1, 63:0] handle [1,1];           (38)

rALUsubnet FFT is                                            (39)
           SW2 = SW1[1,1] + SW1[2,1] * SW1[1,2],            (40)
           SW3 = SW1[1,1] - SW1[2,1] * SW1[1,2],            (41)
```

Fig. 8. Declaration part of the FFT example (operator definition omitted)

Figure 9a shows the signal flow graph and the storage scheme (the grid in the background). The 16 input data points are stored in the leftmost column. Figure 8 shows the MoPL-3 program section, which moves the scan windows to proper starting points and calls the nested compound scan pattern (such as illustrated in figure 9c). Line (45) in figure 10 makes the handles of windows number one through three move to the 3 starting points of scan patterns (line (46)) within the array CGFFT:

Weights w are stored in every second column, where each second memory location is empty (for regularity reasons). Figure 7 shows the window adjustments: the 2-by-2 window no. 1 is the input window reading the operands a and b, and the weight w. Windows no. 2 and 3 are single-word result windows. Figure 7 also shows the compound operator and its interconnect to the three windows.

This is an example of fine granularity parallelism, as modelled by figure 9e, where several windows communicate with each other through a common rALU. Figure 9c illustrates the nested compound scan patterns for this example. Note, that with respect to performance this parallelism of scan windows makes sense only, if interleaving memory access is used, which is supported by the regularity of the storage scheme and the scan patterns.

This section has introduced the essentials of the language MoPL-3 by means of two algorithm implementation examples. The main objective of this section has been the illustration of the language elements for data sequencing programs and the illustration of its comprehensibility and the ease of its use.

4 CONCLUSIONS

The paper has briefly summarized the new Xputer machine paradigm, has demonstrated its basic execution mechanisms, and, has discussed its high efficiency having been published earlier. The paper has introduced a new high level Xputer programming language MoPL-3 and has illustrated its conciseness, comprehensibility and the ease of its use in data-procedural programming for Xputers. An earlier version of the language (MoPL-2) has been implemented at Kaiserslautern on VAX station under

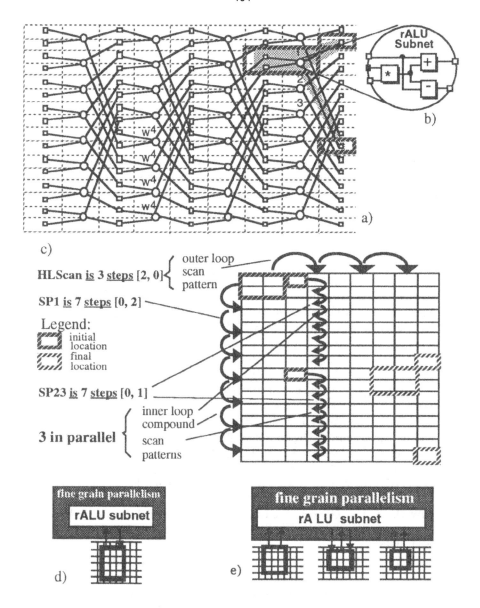

Fig. 9. Constant geometry FFT algorithm 16 point example using 3 scan windows synchronously in parallel: a) signal flow graph with data map grid and a scan window location snapshot example, b) rALU subnet, c) nested scan pattern illustration (also see figure 8), d) illustration of fine grain parallelism: single window use, e) multiple window use.

```
begin                                                    (42)
adjust ThreeW;                                           (43)
apply FFT ;                                              (44)
moveto CGFTT [0,0], [2,0], [2,8] ;                       (45)
HLScan ( parbegin SP1, SP23, SP23 parend ) ;            (46)
end                                                      (47)
```

Fig. 10.   Statement part of the FFT example

ULTRIX. It is an essential new aspect of this new computational methodology, that it is the consequence of the impact of field-programmable logic and features from DSP and image processing on basic computational paradigms. Xputers, their languages and compilers open up several promising new directions in research and development - academic and industrial.

5 REFERENCES

1. A. Ast, R.W. Hartenstein, H. Reinig, K. Schmidt, M. Weber: A General Purpose Xputer Architecture Derived from DSP & Image Proceesing; in: (ed.: M.A. Bayoumi) VLSI Design Methodologies for DSP Architectures; Kluwer, 1994.

2. M. Christ: Texas Instruments TMS 320C25; Signalprozessoren 3; Oldenbourg, 1988.

3. R. Freeman: User-Programmable Gate Arrays; IEEE Spectrum, December 1988

4. R.W. Hartenstein, A.G. Hirschbiel, K. Schmidt, M. Weber: A Novel Paradigm of Parallel Computation and its Use to Implement Simple High Performance Hardware; Int'l Conf. on Information Technology, Tokyo, Japan, Oct. 1990.

5. R.W. Hartenstein, A.G. Hirschbiel, M. Weber, The Machine Paradigm of Xputers and its Application to Digital Signal Processing Acceleration; Proc. of 1990 International Conference on Parallel Processing, St. Charles, Oct. 1990.

6. R.W. Hartenstein, A.G. Hirschbiel, M. Riedmüller, K. Schmidt, M. Weber: A Novel ASIC Design Approach Based on a New Machine Paradigm; IEEE Journal of Solid-State Circuits, Vol. 26, No. 7, July 1991.

7. R.W. Hartenstein, G. Koch: The Universal Bus Considered Harmful; in [8].

8. R.W. Hartenstein, R. Zaks: Microarchitecture of Computer Systems, North Holland, 1975.

9. A.G. Hirschbiel: A Novel Processor Architecture Based on Auto Data Sequencing and Low Level Parallelism; Ph.D. Thesis, Kaiserslautern University, 1991.

10. J. Hoffmann: Redundanz raus; Computer Time, Heft 6, 1991.

11. G. J. Lipovski: A Stack Organization for Microcomputers; in [8].

12. L. Matterne et al.: A Flexible High-performance 2-D Discrete Cosine Transform IC; Proc. Int'l Symp on Circuits and Systems, Vol. 2, IEEE New York, 1989.

13. N.N.: DSP 56000/56001 Digital Signal Processor User's Manual; Motorola, '89.

14. G.K. Wallace: The JPEG Still Picture Compression Standard; CACM 34,4, April 1991.

15. M. Weber: An Application Development Method for Xputers; Ph.D. Thesis, Kaiserslautern University, 1990.

Implementing On Line Arithmetic on PAM

Marc Daumas[1], Jean-Michel Muller[1] and Jean Vuillemin[2]

[1] Laboratoire de l'Informatique du Parallélisme - CNRS
École Normale Supérieure de Lyon
Lyon, France 69364
[2] Paris Research Laboratory - Digital Equipment Corporation
Rueil Malmaison, France 92563

Abstract. On line arithmetic is a computation tool able to adapt to the precision expected by the user. Developing a library of on line operators for FPGAs will lead in a near future to the spread of brick-assembled application-dedicated operators. In the implementation of the basic arithmetic operations (addition, multiplication, division and square root), we have met some new problems: our work has involved changes in the VLSI design methodology in order to achieve some effective performances. We shall present the modified on-line algorithms and their adaptation to the cell oriented FPGA architecture. The correct integration of some retiming barriers has proved to be critical as far as speed is concerned.

Introduction

With the advances in programmable logic we observe a strong demand from the users for a library that implements the arithmetic functions. The DEC PeRLe 1 board [3] is the perfect platform for developing, testing and prototyping such a library. The board is a configurable universal coprocessor built from 23 Xilinx XC 3090 chips [11].

On line arithmetic [5] operates on numbers flowing serially one digit at a time most significant digits first. Digit serial arithmetic is widely used in signal processing where the communication links cannot handle the parallel transmission of the signal and whenever hardware area is critical. Some recent work about on-line arithmetic on FPGAs can be found in [8].

We have described and thoroughly tested a fully optimized working prototype for the addition and multiplication. We have concluded the development of the division and the square root operation. The block architecture of the FPGAs has affected our design of an hardware efficient circuit; moreover, we have incorporated some retiming barriers in the algorithm to hide the circuit commuting time.

In Section 1, we present a set of basic on line arithmetic operations and the PeRLe board architecture with the example of an on line adder configuration. Section 2 describes the general architecture for the multiplication, the division and the square root operator. The Section 3 is dedicated to the modified algorithms that incorporate some of the needed retiming barriers.

1 Environment

1.1 On line Operation

The circuit uses a redundant number system such as Avizienis' signed digit systems [1], which include the radix 2 Borrow Save representation (BS) [4]. Any real number X can be written as follows with the BS notation. We define the value X_i from one decomposition of X truncated after the ith most significant digits.

$$X = \sum_{k=-\infty}^{\infty} x_k 2^{-k} \text{ with } x_k \in \{-1,0,1\}$$

$$X_i = \sum_{k=-\infty}^{i} x_k 2^{-k}$$

Each on line operator $S = X \oplus Y$ is characterized by its delay δ: the digit s_i of the result is computed just from the value of $X_{i+\delta}$ and $Y_{i+\delta}$. Since one new digit of the operands is available at each clock cycle, a new digit of the result is produced each cycle after δ initialization cycles (see Fig 1).

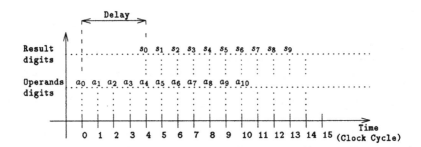

Fig. 1. Delay of an Operator

Used in an adapted environment, a group of on line operators induces some parallelism by the pipeline at the digit level as presented Figure 2: an operation is initiated before all the digits of its operands are available.

The on line basic operations have been defined in many references, including [2, 5, 7, 10]. We use the following sign function $S(W)$ adapted from [5].

$$s(W) = \begin{cases} 1 \text{ if } W \geq 1/2 \\ 0 \text{ if } -1/2 < W < 1/2 \\ -1 \text{ if } W \leq 1/2 \end{cases}$$

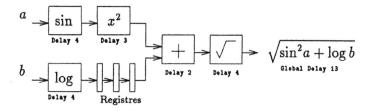

Fig. 2. Sequence of On Line Calculations

Addition The actual circuit is deduced from the carry free parallel BS adder. The addition has also been defined from the recurrence below. The size of $W^{(j)}$ is bounded and does not depend on the length of X and Y.

$$W^{(j)} = 2(W^{(j-1)} - s_{j-1}) + \tfrac{1}{4}(x_j + y_j)$$
$$s_j = S(W^{(j)})$$

Multiplication To compute the product of two numbers on-line, we have to store both numbers X_j and Y_j as their digits flow in. The multiplication algorithm is based on the shifting accumulator W.

$$W^{(j)} = 2(W^{(j-1)} - p_{j-1}) + y_j X_j + x_j Y_{j-1}$$
$$p_j = S(W_0^{(j)})$$

Division The *natural* division algorithm is $W^{(j)} = 2W^{(j-1)} - q_{j-1}D$. The on line algorithm compensates for the incoming digits x_{j+3} and d_{j+3}. Since the delay is 3, the difference between the on line algorithm and a natural division is small enough, the redundancy of the number representation allows some temporary errors. The divisor is prescaled $D \in [1, \tfrac{3}{2}[$.

$$W^{(j)} = 2W^{(j-1)} - q_{j-1}D_{j+2} - \tfrac{1}{8}d_{j+3}Q_j + \tfrac{1}{8}x_{j+3}$$
$$q_j = S(W_0^{(j)})$$

Square Root This operation is very close to the division process. A second full length adder is needed to accumulate q_j^2 at the correct position into the accumulator.

$$W^{(j)} = 2W^{(j-1)} - 2q_{j-1}Q_{j-1} + q_j^2 2^{-j} + \tfrac{1}{8}x_{j+2}$$
$$q_j = S(W_0^{(j)})$$

1.2 Board Architecture

The Xilinx XC 3090 chip [11] is composed of a core of 20×16 logic cells with some capabilities for signal routing and a ring of 72 input–output buffers (see Fig 3). The signals are conveyed to the logic cells through the communication routers and the programmable wire connections. Although the commuting time of a router is very small, a circuit involving nearest neighbor communication will mostly use the wire connections which are much faster. Whenever we had to broadcast an information to a row or a column of cells, we have used a vertical or horizontal long lines.

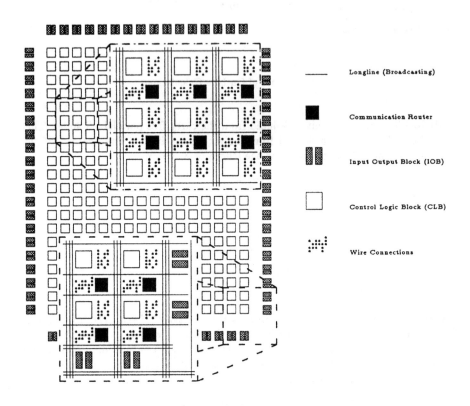

Fig. 3. Chip Internal Architecture (Logic Cell & IO Buffer)

A logic cell (CLB) implements two functions of its inputs from a truth table and contains two registers to store any binary result (see Fig 4). There are some restrictions configuring an FPGA compared to building a VLSI: a cell has only two outputs, whereas both the registers and the logic unit may produce two useful results; it is not possible to load both register with two different values and leave the two logical units for some other functions; the long lines are assigned a limited number of input and output of each logic cell reducing the possibilities for routing and placement.

The DEC PeRLe 1 board gathers 23 Xilinx XC 3090 chips. In this work, we will focus on the 4 × 4 2D computation matrix (see Fig 5). Seven more programmable chips are involved in the data path for the user control and the communications with the host.

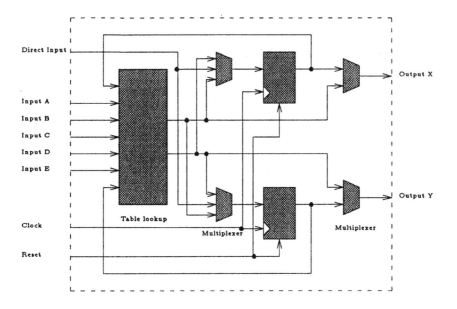

Fig. 4. Logic Cell Functionalities

1.3 Addition

We have deduced the on line adder presented Figure 9 from the parallel adder described in [1] (see Fig 8). This architecture uses two rows of *Plus Plus Minus* cells (see Fig 6 and 7). In order to implement the on-line adder on the PAM we have used only 2 CLBs (see Fig 10). Four registers are available in the two CLBs of the adder, only three of them are used.

2 General Architecture

2.1 Basic Organization

The multiplication, the division and the square root operator share the same general organization: the intermediate result is accumulated and shifted with two other terms obtained from the product of a number by a digit (see Fig 11). The circuit computes the sum of the three operands, and possibly some local transformation $\tilde{f}$ on $W^{(j)}$.

$$W^{(j+1)} = 2\tilde{f}(W^{(j)}) + a'_j \times A^{(j)} + b'_j \times B^{(j)}$$

We have implemented two of the adders presented Figure 8 and the operators for the digit products $a'_j \times A^{(j)}$ and $b'_j \times B^{(j)}$. The function $\tilde{f}$ only involves a few of the most significant digits and is obtained by adding some logic to the data path for

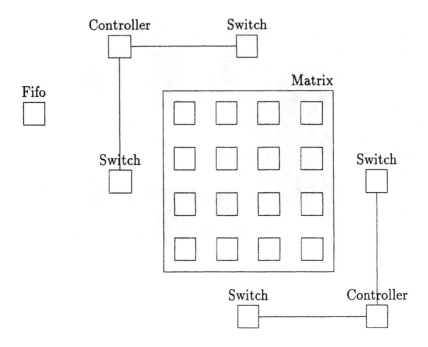

Fig. 5. Computation Matrix and Surrounding Logic

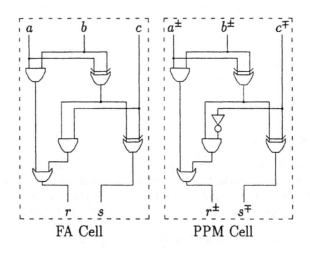

Fig. 6. Addition Cell

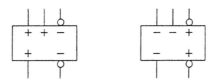

Fig. 7. Representing the PPM Cell

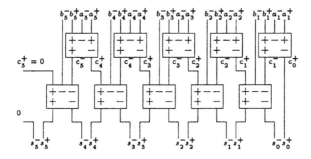

Fig. 8. Borrow Save Adder

these specific digits. The words $A^{(j)}$ and $B^{(j)}$ are constructed as follows for most of the operations: the circuit places the value of a^j at the position $j + d$ in $A^{(j)}$.

$$a_k^{(j+1)} = \begin{cases} a_j & \text{if } k = j + d \\ a_k^{(j)} & \text{otherwise} \end{cases}$$

The circuit counts the iteration number j with a cursor: each segment k of the multiplier stores one bit St_k. At iteration j, if St_k is set, it means that $k = j$,

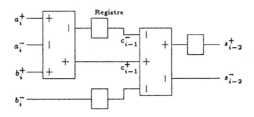

Fig. 9. On Line Adder

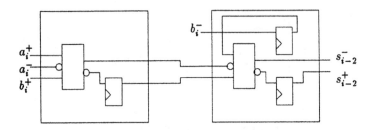

Fig. 10. FPGA On Line Adder

hence $a_{j+d}^{(j+1)}$ is set to the value of a_j. The cursor controls the behavior of the cell $a_{j+d}^{(j+1)}$ regarding a_j. For the square root operation, the algorithm only stores one operand. The other term is used to accumulate the value $q_j 2^j$. The definition of $B^{(j)}$ is presented below, no register is needed.

$$B^{(j)} = q_j 2^j$$

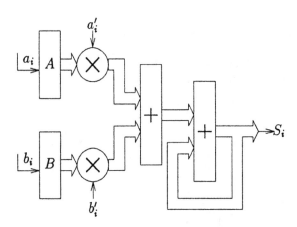

Fig. 11. General Architecture of the Implemented On-line Operators

2.2 Implementation Details

We consider one segment of the operator: it works on one digit taking care of the incoming carry signals and propagating the outgoing carries. The length l of the registers needed is fixed by the length $n = 2 \times l$ of the inputs as detailed in [7]. The final circuit is obtained by repeating l times the segment. Using as much as possible the functionalities of each logic cell, we have optimized the segment down to 8 cells.

The partial products $a'_j \times a_k^{(j)}$ and $b'_j \times b_k^{(j)}$ can be computed on one cell each; one BS adder uses 2 PPMs, the two adders are implemented with a total of 4 cells. This leads to the scheme proposed Fig 12.

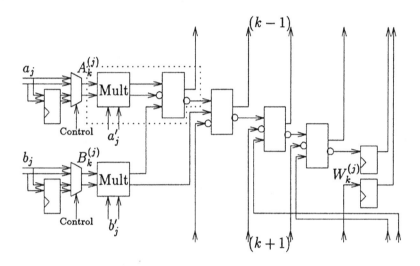

Fig. 12. Low Level Map of a Segment

All the segments of one chip are aligned on two columns to share the same signals on the vertical long lines. The cells of the first line of the chip generates the signals to be broadcasted as presented Figure 13.

To incorporate the state counter in the segment, we have grouped the digit multiplication of $a'_j \times a_k^{(j)}$ and the first PPM cell (see the dotted box in Fig 12). The two output functions of the grouped cell have exactly 5 inputs: each one can be implemented with one logic cell generating only one output signal. The first of the two cells obtained stores and propagates the state counter. The reset signal is used by many cells across the chip and is not assigned any vertical long line. To broadcasts the reset signal to all the circuit, we have used the second cell obtained to feed one of the horizontal long line on each segment. The reset signal is sent on the second vertical long line of this cell and repeated on one of the horizontal long lines of the segment. The arithmetic function of the second cell uses only the sign of a'_i: the signal sent on the first long line of the cell is the sign of a'_i in the Signed Digit representation $d = (-1)^s \times m$, the second long line is available for the fast transmission of the pre-reset signal to all the segments.

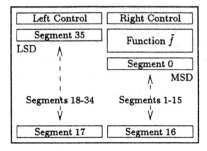

Fig. 13. Segments Organization in a Chip

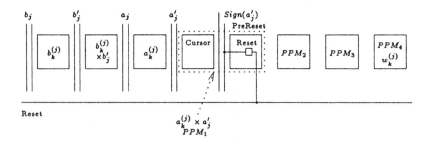

Fig. 14. Longlines usage of one Segment

3 Enhancements

3.1 Multiplication

Pipelining — The information in the multiplier moves from the left side to the right side of a segment (see Fig 12) and from the LSD segment to the MSD segment (see Fig 13). No information is sent in the opposite direction. Introducing a retiming barrier of latches between two parts of the circuit only involves a time change of one cycles between two areas. It is possible to introduce two retiming barriers in the segments and one just before the evaluation of p_j.

Scaling — A chip totally occupied by the multiplier produces 74 digits. A multiplier extended to the other chips of the board is able to generate a very large product. Going from one chip to another in the circuit adds two retiming barriers; this is possible because the information flows from the least significant digit segment to the most significant digit segment. The state signal propagates from the MSD segment to the LSD segment: the predicted signal is sent 4 cycles in advance from one chip to the next one to cross the retiming barrier. To use a linear interconnection in the PeRLe board mesh architecture, we have routed the signals from the rightmost chip of each row to the leftmost chip of the row and then to the chips of the lower row (see Fig 15).

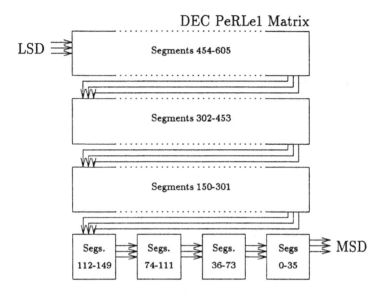

Fig. 15. Data Circulation on the PeRLe 1 Board

3.2 Division and Square Root

In the division and the square root operation, the digit q_j is produced by the head of the operator and used by the partial product $q_j \times D_{j+2}$. It is not possible to incorporate a retiming barrier due to the communications between the different parts of the circuit. The circuit *predicts* a digit of the quotient one cycle ahead with the following recurrence, adding one unity to the delay. The *predicted* digit of the quotient is send backward through the time barrier.

$$W^{(j)} = 2W^{(j-1)} - q_{j-1}D_{j+3} - \tfrac{1}{16}d_{j+4}Q_j + \tfrac{1}{16}x_{j+4}$$
$$q_{j+1} = S(W_1^{(j)} - q_j D_1)$$

Conclusion

The scheme used for our implementation of the BS on-line basic operators has been studied thoroughly. This work has lead us to a fast space-optimized scalable device.

The cell oriented architecture of the Xilinx XC 3090 chips has forced us to adopt some strategies in the design of the circuit.

The basic operations, with no pipeline, run with a slow clock (50 ns); the longest signal typically involves 10 cells. The circuit obtained for the multiplication with the 3 barriers has a correct behavior with a clock cycle of 33 ns (30 MHz). Two more retiming barriers could have been included in the multiplier with no change in the size of the segment. With a clock cycle smaller than 33 ns, the placement and the routing are so critical that most of it must be exactly specified by the user. Yet the tools available from Xilinx on the XC 3090 chips do not present a user interface comparable to the PeRLe1DC library.

The operator for the multiplication computes the result up to 1210 correct digits. The reasonnable implementation of the division and the square root produce results up to 72 digits.

References

1. A. Avizienis, "Signed digit number representation for fast parallel arithmetic," *IRE Transaction on Electronic Computers*, Volume EC-10, 1961.
2. J.C. Bajard, J. Duprat, S. Kla & J.M. Muller, "Some operators for on-line radix 2 computation," to appear in *Journal of Parallel and Distributed Computing*, also available from *Laboratoire de l'Informatique du Parallélisme* RR 92-42, October 1992.
3. P. Bertin, D. Roncin & J. Vuillemin, "Introduction to programmable active memories," *Systolic Array Processors*, Prentice Hall, also available from *Paris Research Laboratory*, PRL-RR 24, March 1993.
4. C.Y. Chow & J.E. Robertson, "Logical design of a redundant binary adder," *4th IEEE Symposium on Computer Arithmetic*, October 1978.
5. M.D. Ercegovac, "On line arithmetic: an overview," *Real Time Signal Processing VII*, SPIE, Volume 495.
6. —, "A general hardware oriented method for evaluation of functions and computations in a digital computer," *IEEE Transactions on Computers*, Volume C-26, N. 7, July 1977.
7. S. Kla Koué, "Calcul parallèle et en ligne des fonctions arithmétiques," *Laboratoire de l'Informatique du Parallélisme*, PhD Dissertation 31-93, February 1993.
8. M.E. Louie & M.D. Ercegovac, "On digit recurrence division implementations for field programmable gate arrays," *11th IEEE Symposium on Computer Arithmetic*, June 1993.
9. J.M. Muller, "Some characterization of functions computable in on-line arithmetic," to appear in *IEEE Transactions on Computers*, also available from *Laboratoire de l'Informatique du Parallélisme* RR 91-15, 1991.
10. K.S. Trivedi & M.D. Ercegovac, "On line algorithm for division and multiplication," *IEEE Transactions on Computers*, Volume C-26 (7), July 1977.
11. Xilinx Inc., "The programmable gate array data book," *Product Briefs, Xilinx*, 1987.

Software Environment for WASMII: a Data Driven Machine with a Virtual Hardware

Xiao-yu Chen[1] Xiao-ping Ling[2] Hideharu Amano[1]

[1] Department of Computer Science, KEIO University, Japan.
[2] Department of Computer Science and Engineering,
Kanagawa Institute of Technology, Japan.
{chen,ling,hunga}@aa.cs.keio.ac.jp

Abstract. A data driven computer WASMII which exploits dynamically reconfigurable FPGAs based on a virtual hardware has been developed. This paper presents a software system which automatically generates a configuration data for FPGAs used in the WASMII. In this system, an application program is edited as a dataflow graph with a user interface, and divided into a set of subgraphs each of them is corresponding to the configuration data of an FPGA chip. These subgraphs are translated into program modules described in a hardware description language called the SFL. From the SFL programs, a logic synthesis tool PARTHENON generates a net-list of logic circuits for the subgraphs. Finally, the net-list is translated again for the Xilinx's CAD system, and the configuration data is generated. Here, the ordinary differential equation solver is presented as an example, and the number of gates is evaluated.

1 Introduction

Technologies around the FPGA (*Field Programmable Gate Array*) have been rapidly established in these several years. Now, the FPGA which works at 250MHz (toggle frequency) including more than 10000 gates is available, and a small scale microprocessor can be implemented on one FPGA chip.

Reconfigurable FPGAs represented with Xilinx's XC3000/4000 family[1] have been giving a large impact to computer architectures because of their flexibility. In these FPGAs, logic circuits are configured according to the configuration information stored in the configuration RAM inside the chip. By inserting the configuration data again, the logic circuits can be easily changed.

With the best use of its flexibility, there are many researches on "flexible computer architectures"[2][3][4]. However, since it takes a long time (*μsecs* or *msecs*) to change the hardware function on an FPGA for inserting the configuration data, application is limited. In order to cope with this problem, an extended FPGA chip was proposed[5][6]. As shown in Figure 1, several SRAM sets each of which is corresponding to the configuration RAM for an FPGA chip are added inside the chip. They are switched by a multiplexor for changing the connection between the SRAM and the logic circuits.

Using this structure, multiple functions can be realized in a single FPGA chip, and changed quickly. This method is called a *Multifunction Programmable Logic Device(MPLD)*. The configuration data in a SRAM set is called *a configuration information page*. Such a system can be extensively used in the area of image/signal processing, robot control, neural network simulation, CAD engines, and other applications which require hardware engines.

The MPLD, however, has two major problems. First, if required configuration data is larger than that of the configuration RAMs on the MPLD, the problem cannot be solved. This is similar to a computer without a secondary memory.

Next, the state of a sequential logic circuit realized on an FPGA is disappeared when a configuration RAM is replaced. It means that data in registers on an FPGA is also disappeared. As a result, it is difficult to control the time when a configuration RAM can be replaced.

To cope with the former problem, techniques of virtual memory are applied. A backup RAM unit is attached outside the MPLD, and the configuration data can be carried into the unused configuration page inside the chip.

For the latter problem, a data driven control mechanism is introduced into the MPLD. Each configuration RAM is replaced when all tokens are flushed out of the circuit.

The chip structure with the data driven control and the virtual hardware mechanism is called the *Single-chip WASMII*. A parallel system using the single-chip WASMII chips is called the *Multi-chip WASMII*[7][8].

In this system, an application program is edited as a dataflow graph with a user interface, and divided into a set of subgraphes each of them is corresponding to the configuration data of an FPGA chip. Then, these subgraphs must be translated into a configuration data for an FPGA chip.

In this paper, a software system for generating a configuration data is described. First, the WASMII system is introduced in the Section 2. The divided subgraphs are translated into program modules described by a hardware description language called the SFL. From the SFL programs, a logic synthesis tool PARTHENON generates a net-list of logic circuits for the subgraphs. In the Section 3, design and implementation of these translators are described. Finally, the number of gates which are generated from an example application with the system is evaluated.

2 WASMII

2.1 The virtual hardware

First, the internal configuration information pages (RAMs)of the MPLD are connected with the off-chip backup RAM through a bus. When an internal page is not used (not connected with elements of the FPGA), a new configuration information can be carried from the backup RAM. By replacing and preloading the configuration data from the backup RAM, a large scale hardware can be realized with a single FPGA chip.

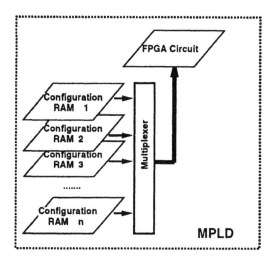

Fig. 1. The concept of MPLD.

We call this mechanism the *virtual hardware*, and use the terms according to the terminology in the virtual memory. An internal configuration data RAM is just called an internal *page*, and the behavior of loading configuration information from the backup RAM to the internal pages is called the *preloading*. When a page is connected to elements in the FPGA and forms the real circuit, the page is called *activated page*.

2.2 Introducing data driven mechanism

The next problem of the MPLD is that it is difficult to manage this mechanism because all states of sequential logic circuits realized on an FPGA are disappeared when a page is replaced. In order to solve this problem, a data driven control mechanism is introduced for the activation and preloading of the page in the virtual hardware[7].

The target application is represented with a data-flow graph consisting of nodes which can be any function like an adder, multiplier, comparator, or more complex functions. Although the node has a register for storing the data during its computation, it must not store any information after its computation. Each node starts its computation when tokens arrive at all input. That is, this mechanism is a pure data-flow machine.

In order to realize the data-flow mechanism with the virtual hardware, the input token register with firing mechanism and token router are introduced as shown in Figure 2.

- *The token-router* is a packet switching system for transferring tokens between pages. It receives tokens from the activated page and sends them to the input token registers. A high speed multistage packet switching network[?] is utilized.

– *Input-token-registers* store tokens outside pages. A set of registers is required to every page of the WASMII chip.

Outside the WASMII chip, a scheduler, which is the only intelligent part of the system, is prepared. A small microprocessor system is used, and it will be connected with a host workstation. It carries configuration data from the backup RAM to internal pages inside the chip according to the order decided by a static scheduling algorithm described later.

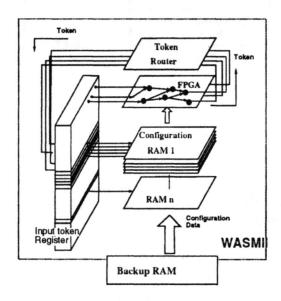

Fig. 2. A single-chip WASMII.

We call this chip architecture a single chip WASMII. A target dataflow graph is divided into subgraphs, each of which is mapped to a page of the virtual hardware. Using the current state of technology, only one or two floating calculation units can be implemented in an FPGA chip. However, several units will be implemented in the near future. Here, we assume that several nodes of a dataflow graph can be executed with an FPGA chip.

When all required input-tokens arrive at the input-token-registers, the corresponding page is ready to be activated. After all tokens are flushed out of the current activated page, one of ready pages is activated by the order assigned in advance. In an activated page, all nodes and wires are realized with a real hardware on the FPGA. Each node starts its computation completely in the data driven manner. Tokens transferred out of the activated page are sent to the input registers through the token-router, and they enable the other pages to be ready.

2.3 Multi-chip WASMII

WASMII chips can be easily connected together to form a highly parallel system. The token-router is extended so as to transfer tokens between pages in different WASMII chips.

Each WASMII chip has its own backup RAM, and subgraphs are statically allocated to each chip. Here, we adopt a simple nearest neighbor mesh connection topology. In the multi-chip WASMII, relatively large latency for the token routing in the mesh structure can be hidden by the data driven operation. In order to avoid causing a bottleneck, schedulers must be distributed. Since the scheduling is fixed when the dataflow graph is generated, each scheduler can execute its job independently.

2.4 The WASMII emulator

In order to demonstrate the efficiency of the WASMII system, WASMII emulator is under-developing. 4 × 4 array of WASMII chip emulators are connected as a mesh structure. Unfortunately, there is no FPGA chip with the virtual hardware mechanism. Therefore, all pages which are not activated are stored in the backup RAMs, and transferred to FPGA chips when the page is activated.

Each WASMII chip emulator consists of a main FPGA chip (XC3090), backup RAM, an input-token-registers and a token-router chip. Input-token-registers are realized with a small scale FPGA chip(XC3042). A router chip is a banyan type switch which works at 50MHz. Each chip has 16 input/output lines. A single microprocessor board (the main CPU is 68040) connected with workstations via Ethernet works as a scheduler.

3 The software system for WASMII

3.1 Overview of the software system

WASMII is expected to be utilized as a special purpose engine for signal/image processing, image recognition, voice recognition, LSI CAD, robot control, and other application fields. In these systems, a loaded program is repeatedly (or continuously) executed many times, and the time for preprocessing is not an important problem. Now, three applications (ordinary differential equation solver (ODESSA), production system MANJI[9] and neural network simulator NEURO generate a dataflow graph for the WASMII system.

Figure 3 shows an overview of the software system. The configuration data for each page is generated as follows:

− The target dataflow graph is divided into subgraphs corresponding to a page. The cyclic structures which may cause the deadlock are also eliminated. This procedure is called the *graph decomposition*.

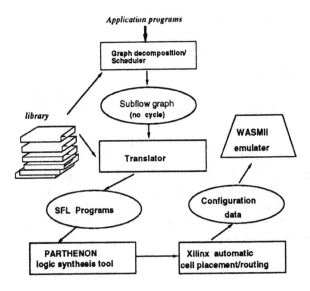

Fig. 3. Software system of the WASMII.

- The execution priority of decomposed subgraphs is calculated as well as the execution order of nodes in a subgraph. This procedure is called the *scheduling*.
- Program text for each page written in a hardware description language SFL is generated automatically with the translator. The SFL node library consisting of various types of nodes are used for the generation.
- The net-list of gates corresponding to each page is obtained by the logic synthesis tool PARTHENON, and then it is translated into the Xilinx's net-list format for generating configuration data.

3.2 Graph decomposition and scheduling

From the name of the node in the dataflow graph, the number of gates required for the node is obtained with checking the SFL node library. The graph is divided into subgraphs so that the total number of gates which required for the node are smaller than that of the target FPGA (XC3090).

Next, the execution priority of decomposed subgraphs is calculated as well as the execution order of nodes in a subgraph. Although the execution of WASMII is basically done in the data driven manner, the following two operations are managed according to the order or priority assigned in advance:

- *Preloading:* A page must be loaded from the backup RAM according to the order decided in advance.

— *Page activation:* A page must be selected when there are multiple pages whose input tokens are ready. This selection is done by the priority assigned in advance.

In the preprocessing stage, decomposed subgraphs are analyzed, and the activity priority and preloading order are decided. A simple level scheduling algorithm called LS/M[10] is utilized for the ordering.

3.3 Translator and node library

SFL and PARTHENON SFL/PARTHENON is used for hardware description and logic synthesis of the target subgraph. PARTHENON is a VLSI design system developed by NTT[11], which consisting of the logic simulator SEC-ONDS, logic synthesis tool SFLEXP, and optimizer OPTMAP. The SFL(Structured Function description Language) is a front-end hardware description language for PARTHENON. Although it is similar to the VHDL in a part, the pipeline operation is easy to be described in the SFL. SFL/PARTHENON is widely used in Japanese universities and industries for education and fabrication of VLSI chips.

In the WASMII software system, the data driven operation in the node library is described with the best use of pipeline description of the SFL.

3.4 Generating SFL description

The structure of the translator is shown as Figure 4.

Here, as an example, solving Van Der Pol's equation (a non-liner system simulation) is introduced. The equation is represented with a connection graph like an old analog computer as shown in Figure 5. "int", "add", "mul", "inv", "con" are integrator, adder, multiplier, sign reverse function, and constant value generator respectively. This connection graph can be almost directly used as a dataflow graph for the WASMII software system. By using the front end system called ODESSA[10][3], the graph is edited. After the graph decomposition, the dataflow graph shown in Figure 6 is generated. Here, this graph is divided into two subgraphs (responding to page_no1 and page_no2 shown in Figure 7), and cyclic paths which may cause the deadlock are eliminated in each page.

From the subgraph "page_no1", the following SFL descriptions are automatically generated by the translator:

```
%i ''/plasma/wasmii/library/mul.sfl''
%i ''/plasma/wasmii/library/add.sfl''
      ......
module page_no1 {
  submod_type    mul {
    input input1<17>; input input2<17>;
    output output1<17>; instrin mul_instrin; }
```

---

[3] ODESSA was developed for a front end system for a multiprocessor called $(SM)^2$.

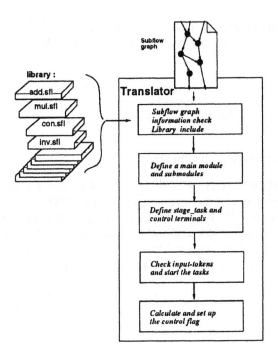

Fig. 4. Structure of the Translator

```
submod_type    add {
  input input1<17>; input input2<17>;
  output output1<17>; instrin add_instrin; }
...... }
```

Like the VHDL and other hardware description languages, hierarchical description is allowed in the SFL. The translator searches the SFL node library with the node name of the subgraph. Here, module "add" and "mul" corresponding to the 16bit adder and multiplier are included. Input/output interface is only described in this part.

Then, the following statements are generated for each node:

```
stage_name add_no1_stage { task       add_no1_task()  ; }
stage_name mul_no2_stage { task       mul_no2_task()  ; }
......
```

In the SFL, tasks performed in the pipelined manner are defined with "task" statement in the pipeline stage named by "stag_name" statements. Here, each node in the subgraph is generated in their own stages.

In WASMII, each node is activated when tokens arrive at all input of the node. The arrival of tokens are noticed with activation of the inside control

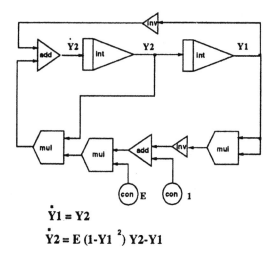

$$\dot{Y}1 = Y2$$
$$\dot{Y}2 = E (1-Y1^{2}) Y2-Y1$$

Fig. 5. Connection Graph of Van Der Pol's Equation.

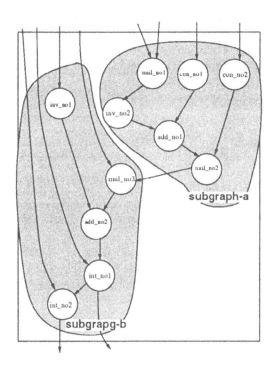

Fig. 6. Dataflow Graph of Van Der Pol's Equation.

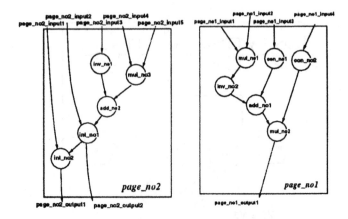

Fig. 7. Graph Decomposition for Pages

terminals defined with "instrself" statements (add_no1_flag and mul_no2_flag).

```
instrself add_no1_flag ; instrself mul_no2_flag ;
      ......
any { (inv_no2_flag)&(con_no1_flag):
  generate add_no1_stage.add_no1_task(); }
any { (add_no1_flag)&(con_no2_flag):
  generate mul_no2_stage.mul_no2_task(); }
      ......
```

Using the statement "any", each task starts its computation when the following conditions are satisfied. In this case, the task starts when both flags are set, that is, tokens arrive.

Each node receives tokens from both inputs, and starts its computation. After computation, it set appropriate flag for activation of the successor nodes. Thus, each node consists of two states as follows:

```
stage add_no1_stage {
  state_name   state1 ; state_name   state2 ;
  first_state  state1 ; state  state1 par {
    add_no1.add_instrin (inv_no2_tmp,con_no1_tmp);
    goto state2 ; }
  state  state2 par {
    add_no1_tmp=add_no1.add_instrin (inv_no2_tmp,con_no1_tmp).output1;
    add_no1_flag();
    finish ; } }
stage mul_no2_stage {
  state_name   state1 ; state_name   state2 ;
```

```
first_state  state1 ; state  state1 par {
  mul_no2.mul_instrin (add_no1_tmp,con_no2_tmp);
  goto state2 ; }
state  state2 par {
  mul_no2_tmp=mul_no2.mul_instrin (add_no1_tmp,con_no2_tmp).output1;
  page_no1_output1=mul_no2_tmp;
  mul_no2_flag();
  finish ; } }
......
```

In "state 1", a node receives data from two inputs, and in "state 2", computation is performed by calling the library modules. Then, flags are set (add_no1_flag(), mul_no2_flag()) for the successor nodes, and the task is finished.

4 Evaluation results

Here, the number of gates generated from the WASMII software system are evaluated. Table 1 shows the generation results of solving the Van Der Pol's equation shown in Figure 5.

Table 1. The required hardware for Van Der Pol's equation.

Type	Number of Gates	Number of CLBs
add	849	149
mul	2005	272
int	849	149
inv	1	0
con	4	0
page-no1	4860	700
page-no2	4553	724
total circuit for VanDerPol	9413	1429

In this example, 17 bits (16 bits for data and 1 bit for sign) fixed point number is utilized. While the simple look ahead adder "add" requires a small number of gate, the array-type high speed multiplier "mul" requires a lot of gates. In this example, a quick multiplier is required to avoid the being bottleneck of the system. The integrator is complex node including multiply and add. However, since the multiplicand is just a fixed number (dt), the required gates are the same as that of the adder. Only a few gates are required for the sign reverse function (inv) and constant generator (con). Since these gates are actually included in the other node, they consume no CLBs.

Table 1 also shows the required CLBs (Configuration Logic Blocks) which is a unit PLD for Xilinx's XC3000/4000 family LCA. Since a XC3090 chip supports 320 CLBs, this equation requires five configuration pages corresponding to XC3090s.

5 Conclusion

Now, the framework of the software system has been developed, and the ordinary equation solver is available. Other two applications, the neural network simulator and the production system are under development. These applications will be executed on the WASMII emulator, which is now under implementation, for establishment of basis for future development of the real WASMII chip.

References

1. XILINX Corp.: "Programmable gate array's data book", 1992.
2. M. Wazlowski, L. Agarwal, T. Lee, A. Smith, E. Lam, P. Athanas, H. Silverman and S. Ghosh, "PRISM-II Compiler and Architecture", Proceedings of IEEE Workshop on FPGAs for Custom Computing Machines, IEEE Computer Society Press, pp.9-16, 1993.
3. S. Casselman, "Virtual Computing and The Virtual Computer", Proceedings of IEEE Workshop on FPGAs for Custom Computing Machines, IEEE Computer Society Press, pp.43-59, 1993.
4. T. Sueyoshi, K. Hano, Togano and I. Arita, "An Approach to Realizing a Reconfigurable Interconnection Network Using Field Programmable Gate Arrays", Trans. IPS Japan, No.33, pp.260-269, 1992.
5. N. Suganuma, Y. Murata, S. Nakata, S. Nagata, M. Tomita and K. Hirano, "Reconfigurable Machine and Its Application to Logic Diagnosis", International Conf. on Computer Aided Design, pp.373-376, 1992.
6. S. Yoshimi(FUJITSU Inc.), "Multi-function programable logic device", Japan Patent(A) Hei2-130023, 1990.
7. X.-P. Ling and H. Amano, "WASMII: a Data Driven Computer on a Virtual Hardware", Proceedings of FPGAs for Custom Computing Machines, pp.33-42, 1993.
8. X.-P. Ling and H. Amano, "Performance Evaluation of WASMII: A Data Driven Computer on a Virtual Hardware", Proceedings of the 5th International PARLE Conference, LNCS 694, pp.610-621, 1993.
9. J. Miyazaki, H. Amano and H. Aiso, "MANJI: A parallel machine for production system", Proceedings of the 20th Annual HICSS, pp.236-245, 1987.
10. X.-P. Ling and H. Amano, "A static scheduling system for a parallel machine $(SM)^2 - II$", Proceedings of the 2nd International PARLE Conference, LNCS 365-I, pp.118-135, 1989.
11. NTT Data Communication Company, Japan.: "PARTHENON Reference Manual", 1989. "PARTHENON User's Manual", 1990.

Constraint-based Hierarchical Placement of Parallel Programs

Mat Newman, Wayne Luk and Ian Page

Programming Research Group, Oxford University Computing Laboratory,
Wolfson Building, Parks Road, Oxford, UK, OX1 3QD

Abstract. This paper continues our investigation into the feasibility of exploiting the structure of a parallel program to guide its hardware implementation. We review previous work, and present our new approach to the problem based upon placing netlists hierarchically. It is found that appropriate constraints can be derived from the source code in a straight-forward way, and this information can be used to guide the subsequent placement routines. Comparisons with traditional placement procedures based on simulated annealing are given.

1 Introduction

The ability to compile programs written in a language such as occam with well-defined semantics and transformation rules facilitates the development of provably-correct systems. By compiling such programs into hardware, we can extend this verifiability to cover systems containing both hardware and software.

We have been investigating methods for compiling parallel programs into hardware, using FPGAs as our target technology [1, 2]. One of our goals is the rapid compilation of programs into hardware yielding faster implementations than traditional software compilation. A major bottleneck in this approach is in the automatic placement and routing of the netlists generated by our hardware compilers. We are currently looking at several methods which will speed up this process to acceptable levels.

2 Compiling Handel into hardware

Several compilers have been developed for a variant of occam known as Handel, which includes the basic programming constructors such as:

SEQ P Q	(sequential composition – do P then do Q)
PAR P Q	(parallel composition – do P and Q concurrently)
WHILE E P	(while E is true, do P)
IF E P Q	(if E then P else Q)
v1,...,vn := E1,...,En	(assign expression Ei to variable vi, $1 \leq i \leq n$)
C!E	(output expression E on channel C)
C?v	(assign value from channel C to variable v)

Programs written in Handel are transformed into synchronous circuits using a token passing system. Each control block has signals called *request* (abbreviated to *req* or *r*) and *acknowledge* (abbreviated to *ack* or *a*). As an example, the control circuitry for SEQ *P Q* must ensure that the process *P* has finished before the process *Q* can begin. This is achieved by simply connecting the *ack* signal from *P* to the *req* signal of *Q*. Our convention states that the *req* line of a block is raised for one clock cycle, at which time that block may begin its computations. When the block has finished it in turn raises its *ack* line for one clock cycle, after which it remains dormant until it receives another *req* signal. Initially there is a special STARTER block which has no *req* line, and raises its *ack* line for one clock cycle when the circuit is started. Figure 1 illustrates the control circuitry for some of the constructors.

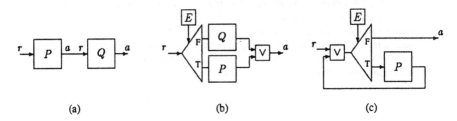

(a)	(b)	(c)

Fig. 1.   Hardware for (a) SEQ *P Q*, (b) IF *E P Q* (if *E* then *P* else *Q*) and (c) WHILE *E P* (while *E* do *P*). The ∨-labelled component is an or-gate. The triangular-shaped component is a demultiplexer, which steers its horizontal input to either the T (true) or F (false) output depending on the boolean expression *E*.

Traditionally one would take the flattened netlist, consisting of basic logic gates and latches, and feed it to an automatic place and route (APR) package for laying out into the FPGA. This results in generally acceptable layouts, but it often takes a large amount of time. Given that our netlist is not generated randomly, but rather from a piece of (hopefully!) structured code, we feel that there should be some way of exploiting the structure of the source program in guiding the placement phase of the layout process.

3 Syntax-directed layouts

We shall briefly review our first approach to this problem, details of which can be found in [2]. This approach adopts a deterministic procedure for laying out a circuit where the complete placement and routing information was derived from the structure of the source program. The layout scheme utilised parametrised Register, Constructor and Wiring blocks, as illustrated in Figure 2.

The Constructor Block holds the control circuitry and expression evaluation hardware, while the Register Block contains the variables. The control information flows orthogonally to the data, and we avoid wiring through control circuitry by utilising a specific Wiring Block. The adopted convention specifies that the control circuitry has its *req* on the left and the *ack* on the right. While this compiler produces functioning circuits, these tend to have an inefficient layout with a poor aspect ratio: programs with many statements tends to have long and thin rectangular layouts. While such layouts are amenable to compaction techniques, we have decided to look at less rigid interpretations of the source code structure in guiding the placement of circuits.

	Constructor Block
Register Block	Wiring Block

Fig. 2. Syntax-directed compiler layout template.

One can think of the two approaches outlined above as being at opposite ends in a spectrum of possibilities for circuit layout. At one end we use APR tools to generate a layout from a flattened netlist; this approach is suitable for designs containing mostly random logic. At the other end of the spectrum, we deterministically place and route the circuit using pre-developed macros, with relatively little flexibility. A complete picture of such a spectrum may include:

– Use APR tools on the flattened netlist.
– Use library modules for recognised pieces of code, use APR for the rest.
– Use the structure of the source code to guide the APR software.
– Use a fixed datapath architecture, and use APR for the control circuitry.
– Have fixed datapath and control paths derived from the source code syntax.

It is not clear that the above list is in any particular order in terms of time to layout, quality of the resulting circuit and so on. Also one can imagine using combinations of the techniques described. The approach studied here uses a combination of library modules and source code structure to guide an APR phase.

4 Modular compilation and constraints

Certain pieces of circuitry are used many times when parallel programs are compiled into hardware, and the components comprising this circuitry are invariably placed near to each other in the final layout. This leads one to use library modules of hand-placed components for often-used circuits. One such example concerns variables which we implement as registers. In our compiler, the circuitry generated to implement a variable depends on two factors: firstly the width of the variable, and secondly the number of different sources for assignments to the variable. The second factor is a result of the need to multiplex the assigned data into the data-in lines of the registers implementing the variable.

At present our compiler has a fixed library template for a multi-bit variable, and it produces individual gates for the multiplexing hardware whose layout has not yet been determined. This arrangement facilitates the optimisation of the multiplexing hardware at a later stage. We can assume however that for the majority of cases the multiplexing circuitry should be placed near to the variable in an efficient layout scheme. We can thus specify constraints which tell the APR software to try to keep the multiplexing circuitry near to the corresponding variable.

For our experiments we have devised a Constrained Hierarchical Netlist format known as Chopin, in which constraints between components can be expressed. As an example there is a *Near* relation, where *Near (P,Q,s)* specifies that the blocks *P* and *Q* are to be placed near to each other with a strength *s*. The weaker the strength, the more leeway the APR software is given in the relative placement; a default value can be used if the strength is not specified. This constrained netlist can be produced by our hardware compiler with almost no additional computing cost.

We have concentrated so far on generating constraints due to hardware being associated with a common component – a variable, a channel or a control block. "Communications" between components in our circuits can be utilised to provide an additional level of constraint information. Thus constraints are generated from the following types of program statements:

```
x := y
c?x
c!y
Op(x,y)
```

Here Op is a binary operation, such as addition or a comparison operation.

When we are compiling a program and we come across one of the above statements, a corresponding *Near* constraint is generated. For example the statement x:=y causes the constraint *Near (Var x, Var y, s)* to be generated. For statements of the form Op(x,y), we need constraints to specify the relationship between the two variables and the operator hardware; for instance x+y would result in the constraint *Near (Var x, Var y, Add, s)* where *Add* refers to a specific instantiation of an adder circuit.

One feature of Chopin is its ability to capture constraints hierarchically: for a statement such as z:=x+y, the constraint generated when x+y is compiled is referred to in the constraint associated with the assignment to z.

5   Using the constraints

Once we have generated our netlist of hardware and constraints, we need to be able to utilise these constraints in guiding our placement tools when laying out the hardware. For algorithms involving cost functions, such as simulated annealing and the majority of other well-known heuristic placement algorithms, the "obvious" place to incorporate these constraints is in either the cost function or in the move generation procedure (for more details on the simulated annealing algorithm see [3]).

It can easily be seen, however, that incorporating the constraints as additional features in the cost function will result in a slower algorithm than one without the constraints. This is because currently we assume that to obtain acceptable results, the constraint-based algorithm has to perform the same number of loop iterations, and to reduce the temperature parameter at the same rate, as the one without the constraints. Similarly, trying to replace a simple move generation procedure (such as "swap the contents of two random locations") with one which takes the constraints into account could result in longer run-times for the algorithm, but hopefully with better final results. In this paper, however, we shall focus on methods for producing better results than traditional ones in comparable times.

The approach we have taken is to split off segments of our total netlist into sub-netlists. These sub-netlists are then placed individually, using a standard simulated annealing approach. The final placement of the sub-netlist is then used to define a fixed macro, which performs the function of that sub-netlist. This macro contains all the port information necessary to connect it up to the rest of the original netlist. Once these macros have been combined with the remaining circuitry into a final netlist, this too is then placed using a simulated annealing approach. Experimental software has been developed to implement this method, and we outline the results below. We shall also comment on some of the ways in which it can be improved.

In our experiments we used a simple model consisting of a rectangular array of cells, each of which can either be a combinational gate or be a single-bit register. In order to compare our results with placements achieved using a simulated annealing algorithm over the flattened netlist, we allowed both methods the same amount of CPU time to produce a placement of the circuit on this array of cells. Both methods were given the same cooling schedule and the same design size in which to place the components. Components were selected to be moved with a probability inversely proportional to their area. This was done because in our current implementation, it is more costly to move large components than moving small ones. Hence the movement of a large component is likely to be rejected by the acceptance function especially at the later stages of the annealing procedure, and so we should not waste time examining such moves.

Our cost function for these experiments was a simple estimated wire length (EWL) calculation, based upon the Manhattan metric: for each net we added half the edge length for the bounding rectangle of the net. Since we were doing comparative experiments we did not attempt to route our final placements. The issue of the routability of designs is covered in the final section.

6   Some examples

To demonstrate our approach we now outline three example compilations, and compare the process of placing the derived circuit firstly by running a simulated annealing algorithm over the flattened netlist, and secondly by running the algorithm presented in the preceding section.

The first example is a very simple piece of code (see Figure 3) which illustrates the benefits of using macros for variables and operators.

```
VAR x,y : 4
SEQ
    x := 9
    y := 5
    x := x + y
```

Fig. 3.   A simple example.

225

While it is clear that using pre-placed macros for the variables and the adder will result
in nearly minimal wire lengths for the internal hardware of these components, the majority
of the wiring for such a circuit is in the data paths connecting the components together. Thus
one might expect to lose some quality in the final placement due to the constraints over how
the adder and variables are designed. In fact this is not the case, and as Table 1 shows there
are significant gains as a result of using such macros (all timings are in seconds).

Without macros		With macros	
EWL	Time	EWL	Time
121	74	74	73

Table 1.   Results with and without hardware macros.

An example as simple as that just given does not lend itself to a hierarchical style of
compilation, since there is not much structure to exploit. In order to investigate hierarchical
compilation we need a more complex example.

The code given in Figure 4 implements a bubble sort routine. To identify subsets of
components that can benefit from pre-placement, a data-flow analysis of the source code is
performed to find out how components implementing variables and channels communicate
with one another, and the result is shown in Table 2. An entry is placed in this table when
an expression or a statement in the source program indicates a connection between two
resources; for instance, the expression

```
Tmp_crnt <= Tmp_next
```

and the statement

```
Tmp_crnt := Tmp_next
```

indicate that the variables Tmp_crnt and Tmp_next have to communicate twice, hence
the (Tmp_next,Tmp_crnt) entry is 2. Using this table, subset of components with greater
connectivity can be dealt with first in our hierarchical method. Note that the communications
inside the inner WHILE loop will be executed more frequently than those outside, and we
could weight them accordingly. At present all constraints are given an equal weighting.

Three different derivative netlists were generated by our compiler from this program.
Firstly, we generated a flattened gate-level netlist. Secondly, a netlist using macros for the
variables and adders was produced; as expected, we were able to obtain better placements
from the second netlist than from the first in the allotted time. Thirdly, we generated a
sub-netlist comprising just those expressions involving the variable Crnt and the channel
Addr, the pair which scores the highest value in the data-flow analysis. We then ran our
simulated annealing routine on this netlist to produce a fixed-placement macro. Next, this
macro was inserted into the netlist of the remaining components, and the resulting netlist

```
VAR Crnt, Last, Tmp_crnt, Tmp_next : 4
CHAN Addr, Din, Dout : 4
SEQ
  Last := 15
  WHILE (Last != 0)
    SEQ
      Crnt := 0
      PAR
        Addr ! Crnt
        Din  ? Tmp_crnt
      WHILE (Crnt != 15)
        SEQ
          PAR
            Addr ! (Crnt+1)
            Din  ? Tmp_next
          IF (Tmp_crnt <= Tmp_next)
            SEQ
              PAR
                Addr ! (Crnt+1)
                Dout ! Tmp_crnt
              PAR
                Addr ! Crnt
                Dout ! Tmp_next
            Tmp_crnt := Tmp_next
          Crnt := Crnt+1
  Last := Last - 1
```

Fig. 4.   A bubble sort routine.

	Last	Crnt	Tmp_crnt	Tmp_next	Addr	Din	Dout	Constants
Last	1	×	×	×	×	×	×	2
Crnt	0	1	×	×	×	×	×	2
Tmp_crnt	0	0	0	×	×	×	×	0
Tmp_next	0	0	2	0	×	×	×	0
Addr	0	4	0	0	0	×	×	0
Din	0	0	1	1	0	0	×	0
Dout	0	0	1	1	0	0	0	0
Total	1	5	4	2	0	0	0	0

Table 2.   Data-flow analysis for bubble sort routine.

was placed. The times for placing the sub-netlist and the composite netlist were added to produce the final timings (incorporating the sub-netlist macro into the composite netlist takes negligible time). The averaged results from many such trials is given in Table 3. The results show that the hierarchical method always returns better circuits with shorter EWL than the versions with macros and with flattened netlists, over a range of specified placement times.

Flattened		Macros		Hierarchical	
EWL	*Time*	*EWL*	*Time*	*EWL*	*Time*
1584	498	950	345	878	348
–	–	961	297	910	279
–	–	1016	232	938	235

Table 3.   Bubble sort results with and without hardware macros, and using hierarchical placement.

As another example, the run-length encoder code given in Figure 5 was examined. Here the routine repeatedly accepts values on the Din channel until a new value is encountered, it then outputs the value and the number of occurrences (modulo 16) of that value on Dout.

```
VAR Crnt, Count, Prev : 4
CHAN Din, Dout : 4
SEQ
  Count := 0
  Prev  := 0
  WHILE (TRUE)
    SEQ
      Din  ? Crnt
      IF (Crnt = Prev)
        Count := Count + 1
      SEQ
        Dout ! Prev
        Dout ! Count
        Count := 1
        Prev  := Crnt
```

Fig. 5.   A run-length encoder.

Here a data-flow analysis reveals that the variables Prev and Crnt are involved together in two statements: one assignment and one comparison. This is a weaker connection than the four times that the channel Addr and the variable Crnt were combined in the bubble

Flattened		Macros		Hierarchical	
EWL	*Time*	*EWL*	*Time*	*EWL*	*Time*
455	148	292	119	296	117
–	–	314	100	312	100

Table 4. Run-length encoder results with and without hardware macros, and using hierarchical placement.

sort program, so it is no surprise that hierarchically compiling this circuit by pre-placing a sub-netlist based upon Prev and Crnt does not do significantly better than placing the entire circuit in one go. Table 4 summarises the results for this circuit.

7 Further considerations

Our work indicates that a hardware compiler can generate placement information at little extra cost. This information can be exploited by a layout routine to produce good results more quickly than a direct placement of the flattened netlist. Similar ideas are being explored for compiling other languages, such as Ruby [4], into hardware.

It is clear that in order for our method to obtain good results, the program should contain appropriate subsets of components that can benefit from pre-placement. Identifying such subsets is at present achieved by simple data-flow analysis. It is open to future research to find other ways of discovering such subsets.

There are two obvious ways to improve our placement routines. First, the sub-netlists are currently placed without regard to the remaining parts of the circuit. As a result, gates connected to components outside of the sub-netlist may be placed in the interior of the created macro instead of at the periphery (this is similar to performing Min-Cut without terminal propagation). Making such cells more likely to be placed on the border of a macro is likely to improve the final estimated wire length.

Second, our current macro generator produces a macro as a collection of individual components with fixed relative positions, as opposed to a single large component. Hence the movement of a large macro is more costly than moving a small macro. Producing a homogeneous macro block should speed up the movement procedure in the simulated annealing algorithm.

Future work will verify that improved placements are still achievable when routing is taken into consideration. One major stumbling block to most placement routines is that high congestion areas occur, and this leaves some nets without valid routings. The density of the macros we create will play a crucial part in ensuring routability. It would also be interesting to study the impact of device-specific features on the performance of our compilation approach.

At present we use constraints derived from our circuits to guide placement in a hierarchical fashion. It may be possible to use these constraints in other contexts, for example as input constraints to the timing-based placement tool described in [5].

Clearly our approach can be used in developing any kind of hardware, from custom circuits to printed-circuit board designs. For implementation in partially-reprogrammable FPGAs, it would be desirable to constrain the layout to facilitate fast reprogramming of critical parts of the device. The size of FPGAs currently available means that only relatively small programs can be compiled into them. Our approach relies on the source program having discernible structures to exploit, and is thus more amenable to larger programs. As larger FPGA devices and boards populated with multiple FPGAs become available, our approach should prove useful in compiling programs of reasonable size automatically into hardware.

Acknowledgement

Thanks to members of the Oxford Hardware Compilation Research Group for discussions and suggestions. The support of U.K. Science and Engineering Research Council, ESPRIT OMI/HORN project and Oxford Parallel Applications Centre is gratefully acknowledged.

References

1. Page, I. and Luk, W.: Compiling occam into FPGAs, in *FPGAs*, W. Moore and W. Luk, Eds. Abingdon EE&CS Books, pp. 271–283, 1991.
2. Luk, W., Ferguson D. and Page, I.: Structured hardware compilation of parallel programs, in *More FPGAs*, W. Moore and W. Luk, Eds. Abingdon EE&CS Books, pp. 213–224, 1994.
3. van Laarhoven, P.J.M., Aarts, E.H.L. and Liu, C.L.: Simulated annealing in circuit layout. Nieuw Archief Voor Wiskunde, 9(1), pp. 13–39, 1990.
4. Luk, W.: Systematic serialisation of array-based architectures. *Integration, the VLSI Journal*, 14(3), pp. 333-360.
5. Raman S., Liu C.L. and Jones, L.G.: Timing-based placement for an FPGA design environment, in *More FPGAs*, W. Moore and W. Luk, Eds. Abingdon EE&CS Books, pp. 213–224, 1994.

ZAREPTA: A Zero Lead-Time, All Reconfigurable System for Emulation, Prototyping and Testing of ASICs

Tormod Njølstad, Johnny Pihl, and Jørn Hofstad

University of Trondheim
The Norwegian Institute of Technology
Faculty of Electrical Engineering & Computer Science
N-7034 Trondheim-NTH
NORWAY
(tormod.njoelstad@fysel.unit.no)

Abstract. Primarily, our ZAREPTA system addresses the need for a low-cost static ASIC tester. By utilizing reconfigurable FPGA technology, the ZAREPTA's total functionality (and 400 DUT pins) can be configured, controlled and monitored by PC software. The main principles of the tester, the block diagram, the software, and the FPGA designs will be explained. However, as a spin-off, the ZAREPTA system may also be used for emulation and fast prototyping of small ASICs. Recently, the ZAREPTA system has been extended with 4 FPIDs, to provide programmable interconnect between the 13 Xilinx XC4005 of ZAREPTA. Bit-serial ASIC architectures with at most 120 external signal pins and 20000 gates may be emulated.

1 Introduction

The digital tester made by one of our students 12 years ago, has been used for testing of many of our CMOS VLSI prototypes. However, after all these years, it is not reliable any more. In 1992, we therefore decided to build ZAREPTA, which primarily is a new static ASIC tester [1][2][3][4]. The ZAREPTA system is based on the XC4000-series FPGAs from Xilinx, due to their very attractive properties in this context: high source and sink current per output pin, local RAMs, programmable pull-up or pull-down resistors, Boundary Scan support, fast reprogramming etc. The tester is implemented using 13 Xilinx XC4005. Each of these RAM-based FPGAs can be reconfigured from the PC bus at any time. It should be noted, as explained below, that only three different Xilinx configurations are required to implement the tester. These three configurations are independent of the ASIC under test.

The main features of the tester are:

- Static testing of CMOS ASICs
- DUTs (Devices Under Test) with at most 400 pins
- Each pin selectable to be either IN,OUT,I/O,CLKn,VCC,GND,or NC

Fig. 1. Photograph of the Zarepta system (prior to the FPID expansion).

- No need for a test jig
- Flexible implementation due to use of field-programmable gate arrays
- PC based software to configure, control and monitor the tester's total functionality
- Interface to commercial automatic test pattern generators (ATPG)
- Low-cost system compared to commercial ASIC testers

2 System Components

As shown in Fig. 2, the ZAREPTA board comprises a 20x20 pin grid array ZIF socket, thirteen Xilinx XC4005 FPGAs, 5 octal registers and bus interface to a personal computer. The PC bus interface provides access to these control and status registers, to the byte-parallel configuration ports of the different FPGAs, and to the actual internal registers implemented in each FPGA. 18 out of the total 20 ICs on the ZAREPTA board comply with the IEEE1149.1 Boundary Scan standard. They are connected in a Boundary Scan chain, *to make the tester itself testable*. The scan test chain is accessible through the PC bus. Every pin in the 20x20 socket are connected to a Xilinx I/O pad and to a jumper to

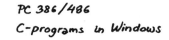

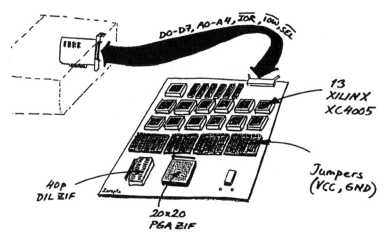

Fig. 2. System components.

select whether this pin should be VCC, GND or a signal. In the latter case, the corresponding Xilinx pin will be set up by PC commands as an input, output or a specified clock, as explained below.

3 The Three Xilinx Configurations for the Static ASIC Tester Mode

The 400 jumpers facilitate connection of power and ground to the DUT. Unless using a lot of relays, there is no way to automate this. However, with PC commands we *check* for wrong GND or VCC jumper settings, using two Xilinx configurations with input pull-up and pull-down options, respectively, as shown in Fig. 3.

When the power supply is set up correctly and checked in the first two configurations as explained above, the third configuration is now downloaded in the FPGAs. Each pin may be a data pin or a specified clock, depending on the PC commands given to the different registers associated with each pin. When a pin is defined to be a data pin, the pin logic is as shown in Fig.4 (simplified for clarity). For each test clock cycle, the PC can write to FF1 and FF2 to define next data value and direction, and get current pin data value by reading TBUF. When the PC starts each new test clock cycle by a specified command, a pre-stored sequence of 16 bits specifies the time event for stimulus data setup and the response sampling time.

When a pin is defined to be a clock pin, the clock pin logic (simplified for

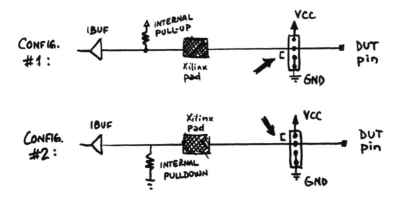

Fig. 3. Input pull-up and pull-down are used to check GND and VCC.

clarity) is as shown in Fig. 5. Here the 16 bits pre-stored sequence defines the time events for rising and falling edge of this particular clock.

Since the appropriate 16 bits data and clock patterns may be downloaded for each pin, many different clocking schemes and clock/data relationships are possible.

4 Interface to Commercial Test Software

We have developed a C library of high level functions for controlling the tester, to make it easy to build test programs. The functions are accessible in the Lab-Windows/CVI [12] environment under Microsoft Windows, making application software development a simple task. We are also working with an interface to the TDS ASCII format, which is a widely accepted test pattern standard [9] (e.g. optional output from the TestCompiler from Synopsys [10]).

5 Fast Prototyping with ZAREPTA

As a spin-off, we believe ZAREPTA may be a suitable tool for so-called fast prototyping and in-circuit emulation. In order to efficiently utilize the 13 FPGAs, they need to be interconnected. Recently, the ZAREPTA system has therefore been augmented with 4 IQ-160 field-programmable interconnect devices (FPIDs) from I-Cube. Fig. 6 shows an overview on how ZAREPTA is used for fast prototyping.

The 13 FPGAs may be configured individually from the PC to implement different parts of the functionality of the ASIC device under emulation (DUE),

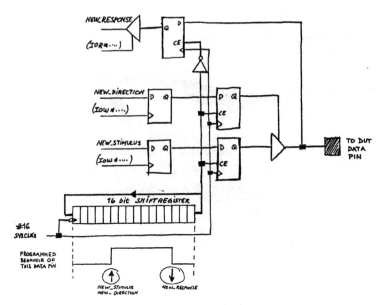

Fig. 4. Data pin logic.

and the FPIDs provide the required internal interconnections between the modules of the DUE as well as connections to the external world. As can be observed in Fig. 7, all 4 FPIDs are connected to the 13 Xilinx FPGAs and to the I/O connector representing the external signal pins of the DUE (max. 120). Since only 32 I/O pins per Xilinx part are used, the system is primarily suited for emulating ASIC designs with bit-serial data paths. To be conservative, we expect that designs with up to 20.000 gates may be emulated by the ZAREPTA system (assuming 30% utilization of the CLB resources in Xilinx).

Fast prototyping may be an alternative and a supplement to simulation for studying the functional behavior of a single module or a complete ASIC. By in-circuit emulation in the system environment, it is possible to detect and correct design errors and specification inconsistencies, which otherwise might not have been discovered before receiving the ASIC from the foundry. In system design, cooperating software and hardware may be developed and tested concurrently, even though no physical ASIC exists. Specifically, prior to layout and processing of an ASIC, its *real-time* behavior may be investigated. In adaptive signal processing, simulation may be too unrealistic, whereas in-circuit emulation can be invaluable. Also, when exhaustive simulation is required (e.g. for evaluation of subjective quality in audio and video codecs), emulation will speed up the computations.

We have used Synopsys' *Design Compiler* and *FPGA Compiler* to provide synthesis from a technology independent level (VHDL) to the Xilinx target technology. Exactly the same design descriptions may be used for the final standard cell or gate array synthesis as for the emulation. Obviously, the design descriptions need not adapt to a specific Xilinx-style, and we can always utilize the highest level of abstraction accepted by the synthesis tools.

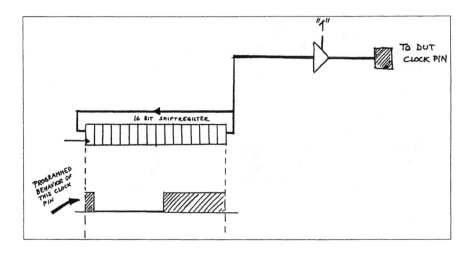

Fig. 5. Clock pin logic.

Fig. 6. Fast prototyping with ZAREPTA.

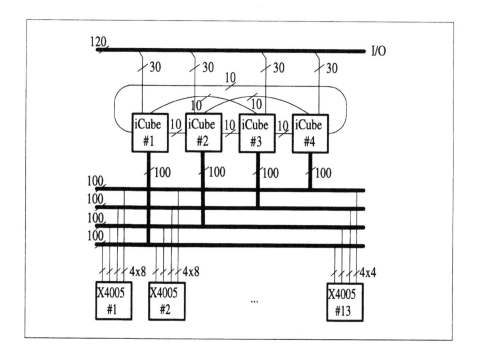

Fig. 7. Interconnection topology of FPIDs and FPGAs in ZAREPTA.

Four different UNIX-cshell/Synopsys scripts have been developed to facilitate synthesis from VHDL to Xilinx configurations, which automate most of the procedure. These scripts are:

- **zest:** Using Synopsys' *VHDL Compiler* the design is compiled from VHDL (IEEE1164) into Synopsys internal db-format. The Xilinx logic (CLBs) and I/O (IOBs) requirements, and the hierarchy are extracted from the design. The results from this script form the basis for further decisions on partitioning.
- **zcon:** For a given design, this script facilitates a query on connectivity. A report is generated on the connectivity between the subinstances of an instance in the design.
- **zpart:**A specified set of sub-modules (instances) are extracted from the design, and partitioned into one Xilinx part. A report on Xilinx logic and I/O requirements is generated. Depending on the results of this script, the partition may be accepted or rejected.
- **zbuild:** When the original design has been partitioned into between 1 and 13 sub-designs, this script uses Synopsys' *FPGA Compiler* to generate Xilinx netlists (XNF files without pin number assignments) for each of the sub-designs. The first connectivity program, **zcollect** [8], is then run on the XNF

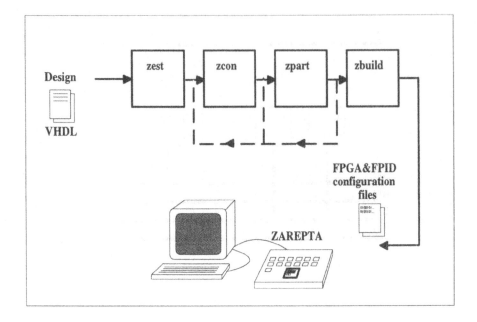

Fig. 8. Design flow for generating ZAREPTA configuration files from a VHDL design.

files. **zcollect** inspects the signal names of in- and out-signals of each FPGA and identifies common names. The XNF files are annotated with feasible pin assignments which are given by the connectivity restrictions imposed by ZAREPTA's hardware structure. This way, the Xilinx place and route tools are given largest possible freedom ensuring a good hardware utilization. An I/O pin definition file may be included to steer external ports (max. 120) of the design to pre-determined I/O pins. The Xilinx software (**ppr** and **makebits**) then places and routes each of the sub-designs, resulting in up to 13 loadable FPGA configurations (EXO files). Finally the second connectivity program, **zconnect** is run to create the iCube configuration files and convert them to Boundary Scan bit streams which can be loaded into the FPIDs.

The design flow using the scripts, is depicted in Fig. 8. The dashed lines indicate that the partitioning is semi-automatic, so that queries on connectivity and partitioning attempts is repeated until a satisfactory partition in terms of CLB and IOB requirements, is achieved. Thus, using the first three of these scripts, we may semi-automatically partition the VHDL design into the 13 Xilinx parts of ZAREPTA. Later on, we plan to automate the partioning process using a suitable algorithm. The two key ideas to be employed are: 1) Traverse the hierarchy to find the largest modules which fit into each FPGA without exceeding the I/O restrictions. 2) The optimal solution is not the one which gives the smallest CLB count, but the one which gives good enough resource utilization without spending too much effort on partitioning and building configuration files, since low development time is important.

After the loadable configuration files for the FPGAs and FPIDs have been generated, the configuration files are transferred to the PC controlling ZAREPTA, and the files are down-loaded. The design is now ready for emulation on the ZAREPTA system.

For DSP applications, the design can be specified at an even more abstract level with signal flow graphs (SFG), using Mentor's DSP Station. The synthesis tools of DSP Station produce bit-serial or bit-parallel solutions (VHDL netlists), suitable for further synthesis and mapping.

6   Examples on Use of ZAREPTA for Fast Prototyping

As an example of the fast prototyping capability of ZAREPTA, the system was used to emulate a 48-tap FIR filter design. The filter is an analysis bandpass filter for a filter bank. The filter was specified at the SFG level and bit-serial synthesis was performed in DSP Station. To simplify the partitioning, prior to bit-serial synthesis the design was entered as 10 sub-designs, each implementing 5 taps except the last one which implemented 3 taps of the FIR filter. A 5-tap FIR filter link is shown in Fig. 9. Output signal y connects to the acc input of the following link. The filter has an internal word length of 20 bits and an input/output word length of 16 bits. It was estimated that each of these 10 sub-designs would fit into one FPGA. A structural VHDL design was then manually created to reflect the connections between the 10 sub-designs as well as connections to external ports. Note that since the implementation is bit-serial the interconnection requirements between the 10 sub-designs are very low. The partitioning specification was then prepared and a master script calling the four scripts above was written. The master script first calls **zest** on the design at the top level, then performs partitioning with **zpart** using the partitioning specification (one call for each Xilinx part), and finally calls **zbuild** to create the FPGA and FPID configuration files. Approximately 3 man-hours was spent on developing the VHDL design description. Running the master script and down-loading the configuration files into ZAREPTA took approximately 3 hours of CPU time on a HP735 workstation. Note that once the VHDL design description and the partitioning specification has been created, the rest of the process is fully automatic.

7   Conclusion

With the ZAREPTA system, a static ASIC tester has been implemented. By introducing programmable interconnection devices, the ZAREPTA system now has been extended to be a simple, but versatile tool for fast prototyping and in-circuit emulation of small bit-serial ASICs.

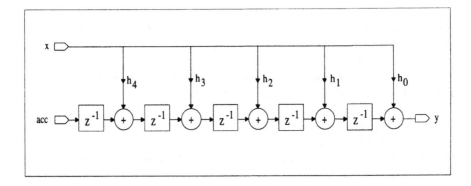

Fig. 9. 5-tap FIR filter link

Just like the old story from Sarepta about the widow's jar, which could never be emptied, the possibilities of our ZAREPTA are almost bottomless. The ambitious name is inspired by this old legend, but is of course also an acronym for what our ZAREPTA is: "A Zero lead-time, All Reconfigurable system for Emulating, Prototyping, and Testing of ASICs"!

References

1. T.Njølstad & H.Dale: "ZAREPTA. An ASIC emulator and tester", Technical report, Norwegian Institute of Technology, Trondheim, Norway, 1992 (Norwegian).
2. H.Dale: "Test station for integrated circuits", M.Sc.thesis, Norwegian Institute of Technology, Trondheim, Norway, 1992 (Norwegian).
3. S.Møien: "ZAREPTA- emulator and tester for ASICs", Term project report, Norwegian Institute of Technology, Trondheim, Norway, 1992 (Norwegian).
4. T.Njølstad, J.E.Øye, H.Dale and S.Møien: "ZAREPTA: A low-cost system for fast prototyping and testing of ASICs". Proceedings on the fourth Eurochip Workshop on VLSI Design Training, Toledo, Spain, 1993, pp. 150-155.
5. S.Møien: "The interface between the ZAREPTA system and the PC", Technical report, Norwegian Institute of Technology, Trondheim, Norway, 1993 (Norwegian).
6. S.Møien: "On digital design, fast prototyping and ASIC emulation of a digital filterbank", M.Sc.Thesis, Norwegian Institute of Technology, Trondheim, Norway, 1993 (Norwegian).
7. J.Hofstad: "Test of Altera EPS448 using ZAREPTA" Term project report, Norwegian Institute of Technology, Trondheim, Norway, 1993 (Norwegian).
8. J.Hofstad: "Fast prototyping of ASICs using the ZAREPTA system", M.Sc.Thesis, Norwegian Institute of Technology, Trondheim, Norway, 1994 (Norwegian).
9. TSSI: *TDS Options Guide*, August 1992.
10. Synopsys Inc: *TestCompiler and TestCompiler Plus Reference Manual*, Version 3.0, December 1992.
11. Xilinx Inc: *The Programmable Logic Data Book*, 1993.
12. National Instruments: LabWindows/CVI, 1994.

Simulating Static and Dynamic Faults in BIST Structures with a FPGA Based Emulator

Richard W. Wieler, Zaifu Zhang and Robert D. McLeod

Department of Electrical and Computer Engineering
University Of Manitoba, Wpg, MB, Canada, R3T-2N2

Abstract. Circuit emulation, using dynamically reconfigurable hardware is a high speed alternative to circuit simulation, especially for large and complex designs. Dynamic reconfiguration enhances the ability to efficiently analyse the test of combinational and sequential circuits by providing statistical information on fault grading, detectability, and signature analysis. In this paper we examine hardware accelleration of static and delay fault simulation, and the accelleration in simulating new BIST techniques.

1 Introduction

Emulation of circuits which are large or contain feedback loops, is desirable for several reasons. Emulation can considerably reduce the time taken in the analysis of fault grading and signature analysis for combinational and sequential circuits of small or large scale. A dynamically reconfigurable emulator, allows the user, to directly inject faults into a circuit for the purpose of test analysis. Towards this end we have been investigating the use of a field programmable gate array (FPGA) based platform. By using a FPGA platform, we allow for both rapid prototyping of the end product as well as rapid prototyping of test schemes.

Algotronix[1][3] currently has a FPGA based computer using an array of CAL1024 chips. This computer allows the user to download designs and control functions occurring on the computer. The ability to download large designs and have full control of clocking functions, is an ideal concept on which to base a hardware emulator.

Due to the increasing density of Integrated Circuits (IC) and ever increasing demands for high product quality in manufacturing and throughout the life cycle of an IC product, Built-In Self-Test (BIST) is becoming more and more popular[4][5][6][7]. All BIST techniques require two basic elements; test pattern generation and some form of test response compaction. These two components are equally important for obtaining high quality test. Test patterns generated are required to activate each fault's behaviour to at least one primary output, and compaction is required to capture and retain the faulty behaviour[5][8]. In general, with BIST schemes that use pseudorandom test, test patterns are most com-

---

1. Algotronix was recently purchased by Xilinx. For the purposes of this discussion the implementation is sufficiently generic to apply to other FPGA environments and particularly well suited to future technologies with increased support for dynamic reconfiguration.

monly generated by the use of linear feedback shift registers (LFSR), or cellular automata (CA)[5][9], and compacted by LFSR or CA signature analyses. The problem with LFSRs, or CAs, is that they may introduce a hardware overhead which may be undesirable and degrade the performance. A proposed alternative to augmenting a circuit with LFSRs is Circular Self-Test Path (CSTP)[1][2]. CSTP is a similar idea to the simultaneous self-test(SST) approach, presented in [10]. With a CSTP technique, less silicon area overhead is incurred. In theory, on an ASIC the CSTP loop may span a significant portion of the chip reducing the control overhead otherwise needed by more conventional scan based methods. In addition, the quality of the test patterns appear very high with aliasing being similar to that of traditional LFSR based signature analysis.

CSTP is not being widely utilized at this time. One reason for this slow acceptance is a lack of a fast simulator to analyse the CSTP circuit under test. A fast simulator is needed to assess the quality of a test process[11] like SST or CSTP. It is very expensive and time consuming to simulate sequential circuits, SST, or CSTP circuits, and to analyse the detectabilities and aliasing probabilities of the faults with serial computers. A fast simulator would be one that emulated the actual hardware, so a logical solution would be based on the use of reprogrammable devices. With this scheme it is possible to efficiently emulate both combinational and sequential circuits, on a large or small scale.

The paper includes an overview of the Algotronix FPGA technology and issues such as, why dynamic reconfiguration makes it well suited for these types of problems. We also illustrate that FPGA technology can be used not only for rapid prototyping of systems, but also rapid prototyping and verification of test structures. We also look at results of fault grading, fault detectability and aliasing, for both the LFSR signature analysis as well as a CSTP scheme. As well, the results of [1][2] have been verified for the state coverage, and the probability of a 1 occurring in a CSTP register.

2 Advantages of Control Store RAM Architecture for FPGAs

The Algotronix type FPGA is ideally suited towards emulation type applications. The Algotronix chips are known as CALs (Configurable Array Logic). These FPGAs consist of an array of programmable cells. Each cell may be connected (input and output) to its nearest neighbours.

The FPGA uses a static RAM control store. The control RAM controls multiplexers which in turn control both the functional blocks as well as routing. Only one mux in each functional block is controlled by data, instead of the control RAM. Because of this complete control over all aspects of the FPGA configuration, it is possible to change individual control RAM cells without affecting the remaining configuration of the chip. This enable the user to quickly and dynamically reprogram the FPGA. The dynamic reprogramability allows for efficient fault insertion.

Another very useful feature of this architecture is the bit of control store, in each cell which allows the user to read back the output of each function block. This allows for continuous monitoring of internal cells. Monitoring of cells internally, allows for I/O to be freed for other purposes, especially when cascading chips in an array configuration.

The architecture of arrayed logic cells, allows for transparent boundaries when chips are cascaded in an array fashion. This creates simplicity when partitioning a design over multiple chips. This architectural feature is exploited with the Algotronix FPGA computer. It also makes further expansion quite simple.

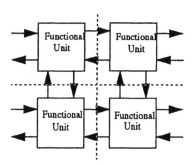

Fig. 1. Array Structure

3 Emulating Static Faults

The hardware emulator is derived from a FPGA based computer. The computer is made up of an array of Algotronix CAL1024 chips. The computer is on a board that fits into an AT compatible bus. It enables the user to download the design into the FPGA array and run the circuit in real time. The user may manipulate the control functions of the chip and the clocking, by way of the pc interface[3]. The current system we have, has a capability of emulating a circuit of several thousand gates. The actual hardware allows for 16,384, two input gate equivalents, although when implementing a CUT, the degree of resource utilization is reduced, due to routing overhead. Two points should be noted: i) this is considered a first generation FPGA technology, and ii) sub-circuits to be analysed for test will not likely be larger than several thousand gates, due to functional partitioning of large designs.

The fact that the emulation is actually occurring in hardware, reduces the time it takes to perform a simulation or generate a signature. Because of the reprogramability of the emulator, it is very easy to inject faults at crucial areas. The CAL1024 allows the user to change individual cells during write cycles. This feature has been manipulated to further expedite the fault injection process, and enhance test point insertion. More conventional hardware emulators or those based on chip level reprogramability have been and will continue to be important for hardware accelerated simulation. For applications such as BIST analysis, dynamic reconfiguration at the gate level is a definite asset. This is due to the fact that reconfiguration time is required to be minimal as the circuit is modified for each fault injected. To statistically evaluate the detectabilities and aliasing probabilities of the combinational circuit, the fault injection procedure could be further improved by simultaneously injecting faults which are in independent blocks. This is not possible on a serial machine simulation.

A variety of methods are under investigation in the use of the reconfigurable computer for emulation. The most straight forward involves the sequential emulation of a fault free circuit, storing the final signature for comparison with signatures of the circuit with the fault(s) injected. The basic procedure is illustrated in Figure 2 (a) and is denoted method 1. This method would allow for the largest of circuits to be investigated but requires more time than the following methods.

An improved method which still allows for maximum sized circuits consists of sequential emulation of the fault free and faulty circuit but involves the recording of multiple sig-

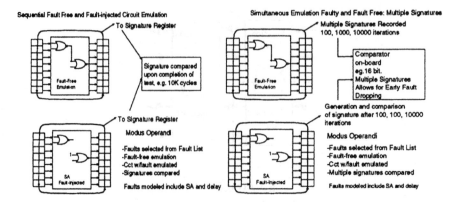

Figure 2 (a) Method 1: Sequential Emulation (left), (b) Method 2: Sequential Emulation with Multiple Signatures (right)

natures. These multiple signatures may be at 100, 1000, 10K, or longer iterations of the test cycle. Signatures generated during the emulation of the faulty circuit are compared to the multiple signatures stored. This allows for the fault to be dropped early in the test cycle and reduces overall test time. This method is illustrated in Figure 2 (b) and is denoted method 2.

The final method discussed denoted method 3 and illustrated in Figure 3 allows for immediate fault dropping upon detection of a difference in output between the fault free and faulty circuit. This is accomplished by simultaneously emulating the both the fault free circuit and fault injected circuit. The comparison circuit is a simple XOR gate. Upon the receipt of a difference, the fault in question can be removed from consideration.

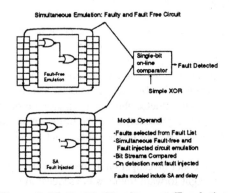

Figure 3 Method 3: Simultaneous Emulation

In each of these methods a fault list would be generated, likely based upon a reduced single stuck-at fault subset through gate level fault collapsing[12][11][13]. These stuck-at faults can be easily modelled (emulated). More sophisticate parametric fault models likely cannot be accommodated with current technology.

4 Circuit Under Test

The 74LS181, a four bit ALU[14] was chosen as the initial circuit used to test the emulator. The 74LS181 is a combinational circuit that has been widely used as a benchmark for various testing methodologies[15][16]. The procedures stated in this paper could easily be extended to larger and more complex circuits, however there is a physical limitations with the current emulator configuration,[1] so only slightly lager circuits can be emulated at this time.

Initially when testing the 74LS181, the simultaneous emulation method (method 3) was used. Two identical circuits were downloaded to the emulator. Each circuit used a 16 bit maximum length LFSR as an input. There are 14 inputs on the 74LS181 so a maximum of 2^{14} input patterns were possible, with this configuration. The outputs of the 74LS181 were fed through a comparator circuit.

The first task was to fault grade the ALU. The fault list consisted of 314 gate equivalent stuck-at faults. Since the Algotronix FPGA allows a maximum of two inputs per gate, the gates which are fed by more than three inputs were implemented with an appropriate type of several cascaded two input gates when modelled for the emulator. This did not effect fault coverage however, as the 74LS181 has 100% fault coverage. When testing a circuit with an undetermined, or less than 100% fault coverage, possible faults arising from the cascaded version of the multiple input gates could be ignored when fault grading, thus more closely emulating the original model. A one to one map between the original circuit and emulation implementation could be generated to ensure that only the faults which correspond to those of the original circuit are considered in the emulation procedure.

Fault grading of the 74LS181 resulted in all faults being detected within about 350 cycles for several different seeds. Once the fault had been detected, the circuit was reset and the next fault was injected. Using this process and running the computer, at a reduced bus speed of 8 MHz, the entire circuit was fault graded within one minute. This process would take longer if there were hard to detect faults or redundant lines in the circuit. However the procedure takes the same time per fault regardless of the complexity of the circuit, and increases slightly as the circuit size increases. This is a significant advantage of the emulator.

The detectability of the faults was the next test performed. The test was implemented in the following manner. Each single fault was injected into the CUT. The circuit was then clocked for 2^{14} cycles to insure the majority of possible input patterns were covered. Each time the comparator sensed the fault, it was recorded. The detectability profile of 74LS181 can be seen in Figure 4.

---

1. Although system size can be almost tripled, the hardware is no longer available from Algotronix.

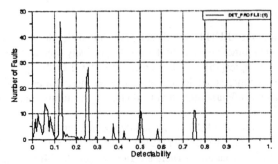

Fig. 4. Detectability Profile

5 Delay Fault Testing

We have explored several models for testing transition delay faults[14][18]. The first scheme is described as follows (Figure 5). The proposed scheme would consist of three similar circuits. One circuit would be the Golden circuit, another one would be the circuit, onto which the stuck-open faults would be implemented. The third circuit would also be a Golden circuit but would be delayed by one clock cycle. When a fault is detected from the test circuit, the delayed Golden circuit, is observed to find the delayed value of the line under test. In this way if the delayed circuit test point, and the test circuit have the same value, the delayed fault should be detectable in a normal testing scheme. We have attempted to implement this scheme with our current emulator configuration. Unfortunately the current software can not translate a design of this size to the bit steam needed to program the emulator.

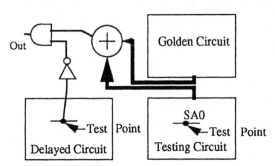

Fig. 5. Delayed Model Test Scheme for Slow-to-Rise Transition Delay Faults

Because of the constraints of our current system, we have implemented a second, less hardware intensive scheme for testing delay faults. The scheme involves adding flip-flops to the golden circuit. The flop-fops act as a shadow register of the previous state of the line connected to it. Because of the routing constraints of the circuit not all of the delay faults have been tested. Instead we took the 10 least detectable stuck-at-zero and the ten least detectable stuck-at-one faults, so that, transition faults corresponding to these faults will be hard to detect and have an impact on the necessary test lengths.

For the delay fault to be detected, it must be a slow-to-fall fault, when using the stuck-at-one fault to test. Likewise a slow-to-rise fault uses a stuck-at-zero fault to test for observability. If the stuck-at-zero fault of line l is dectected there is a probability (determined by the zero controllability of line l) that the slow-to-rise fault is also detected, as the slow-to-rise fault must be in a low state, the clock cycle before, if it is to be detected.

Our current method of testing shows that all of the faults we tested were observable. This is as expected. We did not at the time this paper was written test for the actual detectability of the faults. To test the detectability of the delay faults involves some minor modifications to compensate for routing inversions which occur on the Algotronix FPGAs.

The second scheme should be efficient enough for testing most delay faults, as it should only be necessary to test delay faults where the stuck-at fault has a low detectability, since the transition faults corresponding to those stuck-at faults, have a lower detectability [19].

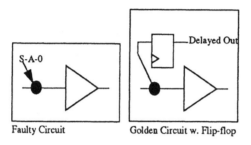

Faulty Circuit   Golden Circuit w. Flip-flop

Fig. 6. Delay Fault Scheme #2

6 Testing of the LFSR Signature

Signature analysis on the 74LS181 ALU has been implemented, with both different cycle lengths and different seeds. Given that the CUT was 100% fault testable, there could be 100% fault detection with a suitable LFSR signature analyses. That is, aliasing was not anticipated to be problem.

The testing methodology was changed to incorporate sequential emulation with multiple signatures (method 2). The golden circuit was clocked at multiples of 1000 cycles. After each clocking cycle of the golden circuit the 16 bit LFSR signature was recorded. Following that, each fault was injected and the circuit cycled again. Each faulty signature was then recorded and checked against the good signature for aliasing. After this test, no instances of aliasing occurred. This same procedure was then also extended to testing a CSTP scheme.

7 Circular Self-Test Path

CSTP links together registers into one circular path. Not all the registers in the circuit have to be utilized, but at least all the primary input and primary output registers must be used. The register acts like a simple D-latch or flip-flop when the circuit is in a normal mode of operation. When the circuit is under test, the bit from the previous latch is XOR'd with the

incoming bit[1]. This process acts as a modulo 2 sum of the previous state. It is this procedure that generates the test vectors as well as performing the compaction for signature
analysis. A schematic representation of the basic scheme is shown in Figure 7.

The most significant advantage of CSTP is the minimal amount of additional hardware
that is required. The additional hardware is illustrated in Figure 8.The ALU, using the configuration as shown in Figure 9, has been slightly changed to facilitate a CSTP test. Four of
the primary outputs are fed back to four of the primary inputs (F0-F3 and B0-B3 respectively) as "active inputs". The other primary inputs were set to a "0" value as "inactive
inputs" to form an ALU CSTP configuration.

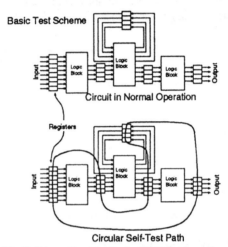

Fig. 7. Schematic of the Basic CSTP Configuration

The same emulation method that was used to test for aliasing (method 3) was used to
find the LFSR signature. The bits which are as long as the number of all the CSTP registers
(22 total, 14 input, and 8 output), were used for representing the signature contents. No
aliasing was found to be occurring.

Other tests on the CSTP design included testing for state-space coverage on the inputs,
and testing for the probability of any bit being set to one, on an input. Krasniewski and
Pilarski[1] have stated that after a short periods, the probability of any bit being set to one
on the input should approach 0.5. This was verified on our circuit as the probability of one
on all the inputs was very close to 0.5 (0.49657) at 10,000 cycles. This figure remained
close to 0.5 for increased cycle lengths.

The state space coverage was implemented for three randomly chosen different 8 bit
input taps, in which the chosen 8 bits include a hybrid of "inactive input" and "active
input" [1]. The state coverage at different cycle times are given, as shown in Table 1. The
three randomly selected 8 bit words for this experiment are denoted Tap1, Tap2 and Tap3.
We noted that after 1000 cycles, nearly 100.00% of states are generated on all three taps.
This verifies that the CSTP registers are a very good technique for pattern generation.

Circular Self-Test Hardware Overhead

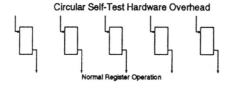

Normal Register Operation

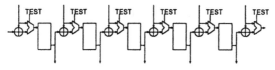

Test Mode Register Operation

Fig. 8. Hardware Required for CSTP

Table 1: State Coverage of Inputs

Cycles	Tap 1	Tap 2	Tap3	Average
100	37.81%	33.20%	32.03%	34.35%
1000	97.27%	96.88%	96.48%	96.88%
10,000	100.00%	99.61%	99.61%	99.74%

It is important to note that the test performed on the CSTP scheme took the same amount of time as the tests performed on the LFSR scheme. This would not be the case if a simulator was used. The ability to develop and verify different BIST strategies such as CSTP, will not only increase their acceptance, but will also decrease the time for a system to be validated. Although the CUT being used was a very basic design, this methodology can easily be extended to larger and more complex designs.

8 Recommendations for Future FPGA Development

We have taken our work in emulation technology as far as our current system limitations will allow. It may be possible to port some of this work to other FPGA technologies, however there are certain aspects of the current Algotronix technology that would greatly enhance future FPGA architecture. The following paragraphs outline some of the architectural aspects that could improve future emulation, based on FPGA technology.

Emulation, as previously stated, is greatly enhanced by dynamic in circuit reprogramability. By this we mean, not chip level, but functional block level and routing level reconfiguration. Advancements in this area with FPGAs currently commercially available could be a great resource.

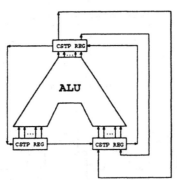

Fig. 9. 4 bit ALU CSTP Configuration

Reprogramability on the scale stated above, usually comes with a reduction in clock speed (due to the increased delays through control multiplexers). However any FPGA technology should try to optimize the architecture for fast clocking of global signals. As well the feature of a controllable on board clocking scheme would also be a great asset. If the clocking must be controlled by software, there will be a great loss in the performance of the emulator, due to the communication bandwidth limitations between board I/O and host I/O. There is no reason for this to occur if software can control an on board clock.

If a current FPGA vendor could meet the architectural requirements above, it would also be fortuitous to develop a multi-chip board and control software. It would be advantageous if a board, as described above, could be a stand alone unit with I/O access to a number of hardware platforms.

Acknowledgments

Support provided by the Natural Sciences and Engineering Research Council of Canada, the Federal Government's Centres of Excellence Micronet Program and the Canadian Microelectronics Corporation is highly appreciated.

References

[1]   A. Krasniewski, and S. Pilarski, *Circular Self-Test Path: A Low-Cost BIST Technique for VLSI Circuits*, IEEE Transactions on Computer-Aided Design, Vol. 8, pp.425-428, Jan. 1989.

[2]   S. Pilarski, A. Krasniewski, and T. Kameda, *Estimating Testing Effectiveness of the Circular Self-Test Path Technique*, IEEE Transaction on Computer-Aided Design of Integrated Circuits and Systems, Vol. 10, pp.1301-1317, Oct. 1992.

[3]   Algotronix Ltd., Configurable Array Logic User Manual, Edinburgh UK, 1991.

[4]   E.J. McClauskey, *Built-In Self-Test Techniques*, IEEE Design and Test of Computers, pp.21-36, April 1985.

[5]   P.H. Bardell, W.H. McAnney, and J. Savir, *Built-In Test for VLSI: Pseudorandom Techniques*, John Wiley & Sons, 1987.

[6]   V.D. Agrawal, C.R.Kime, and K.K. Saluja, *A Tutorial on Built-In Self-Test*, Part I, IEEE Design and Test of Computers, pp.73-82, March 1993.

[7]   V.D. Agrawal, C.R.Kime, and K.K. Saluja, *A Tutorial on Built-In Self-Test*, Part II, IEEE Design and Test of Computers, pp.69-77, June 1993.

[8]   Y. Zorian and A. Ivanov, *Programmable Space Compaction BIST*, Proc. of Int'l Symp. on FTC, pp.340-349, 1993.

[9] P.D. Hortensius, R.D. McLeod, W. Pries, D.M. Miller, and H.C. Card, *Cellular Automata-Based Pseudorandom Number Generators for Built-In Self-Test*, IEEE Transaction on Computer-Aided Design of Integrated Circuits and Systems, Vol. 8, No.8, pp.842-859, Aug. 1989.

[10] P.H. Bardell and W.H. McAnney, *Self-Testing of Multichip Logic Modules*, Proc. of Int'l Test Conference, pp.200-204, 1982.

[11] M. Abramovici, M.A. Breuer, and A.D. Friedman, *Digital Systems Testing and Testable Design*, Computer Science Press, New York, 1990.

[12] D.R. Schertz and G. Metze, *A New Representation for Faults in Combinational Digital Circuits*, IEEE Transaction on Computers, Vol.21, No.8, Aug. 1972.

[13] J.E Chen, C.L. Lee, and W.Z. Shen, *Single-Fault Fault-Collapsing Analysis in Sequential Logic Circuits*, IEEE Transaction on Computer-Aided Design of Integrated Circuits and Systems, Vol. 10, No.12, pp.1559-1568 Dec. 1991.

[14] *The TTL Data Book for Design Engineers*, Texas Instruments, Inc. Dallas, TX, 1981.

[15] S.B. Aker and B. Krishnamurty, *On the Application of Test Counting to VLSI Testing*, pp. 343-359, Chapel Hill Conference on VLSI 1985.

[16] J.L.A. Hughes and E.J. McClauskey, *Multiple Stuck-at Fault Coverage of Single Stuck-at Fault Test Sets*, Proc. of Int'l Test Conference, pp.368-374, 1986.

[17] Y. Ievendel and P.R. Menon, *Transition Faults in Combinational Circuits: Input Transition Test Generation and Fault Simulation*, 16th Int'l Symposium on Fault Tolerant Computing Systems, pp. 278-291, 1986.

[18] R. L. Wadsack, *Fault Modelling and Logic Simulation of CMOS and MOS Integrated Circuits*, Bell System Technical Journal 57(5), pp. 1449-1474, 1978.

[19] J.A. Waicukauski, E. Lindbloom, B.R. Rosen, and V. S. Igengar, *Transition Fault Simulation*, IEEE Design & Test of Computers, pp. 32-38, April 1987.

FPGA Based Prototyping for Verification and Evaluation in Hardware-Software Cosynthesis

Th.Benner, R.Ernst, I.Könenkamp, U.Holtmann,
P.Schüler, H.-C.Schaub and N.Serafimov

Technische Universität Braunschweig
Institut fuer Datenverarbeitungsanlagen
D–38106 Braunschweig, Germany

Abstract. COSYMA is a HW/SW-cosynthesis system for small embedded controllers. The final simulation of the COSYMA output leads to impractical computation time. Therefore, we decided to employ a hardware prototyping system. The HW/SW prototyping system consists of a SPARC processor, an FPGA-based coprocessor with HW/SW debugging features realized with a high speed microcontroller.

1 Introduction

COSYMA (CoSYnthesis of embedded Micro Architectures) is one of the first systems for hardware-software cosynthesis [1]. It is targeted to the design of small embedded controllers. Given an input description in a superset of C, C^x, consisting of one or more tasks with time constraints, and given a fixed core processor, COSYMA tries to map as much of the system as possible to software. When the time constraints cannot be met, it automatically partitions parts of the system description to an application specific coprocessor such that the timing constraints are met with minimum hardware overhead. The partitioning process regards and minimizes communication overhead. Significant speedups have been observed for real examples with more than 1000 lines of C code and a 33MHz SPARC as core processor. Figure 1 outlines the design flow. Partitioning is done on the basic block level using simulated annealing based on speedup, communication time and hardware overhead estimation. The software part is enhanced by a communication protocol, translated to C and compiled to object code. The hardware part is generated by high-level synthesis. The approach is flexible enough to permit the use of different high-level synthesis systems; currently these are OLYMPUS [2] from Stanford taking a HardwareC description as input and our own synthesis system BSS [3] taking a CDFG as input. A run time analysis is able to accurately estimate the execution time of the resulting hardware-software system on a clock cycle basis [4] and thus show if the timing constraints are met. Currently, we are working on an approach to adapt the estimation in hardware-software partitioning to the actual results. Once the timing constraints are met, the high-level synthesis system generates a hierarchical netlist on the logic level consisting of the controller and the data path in SLIF format (Stanford Intermediate Logic Format). In order to verify this netlist, we currently use HW/SW cosimulation. It turned out that logic level simulation of the coprocessor and the communication hardware is only possible for small examples in acceptable computation time, even though the processor is simulated at the register transfer level. So, we decided that for hardware-software co-verification a hardware prototyping system would be very helpful.

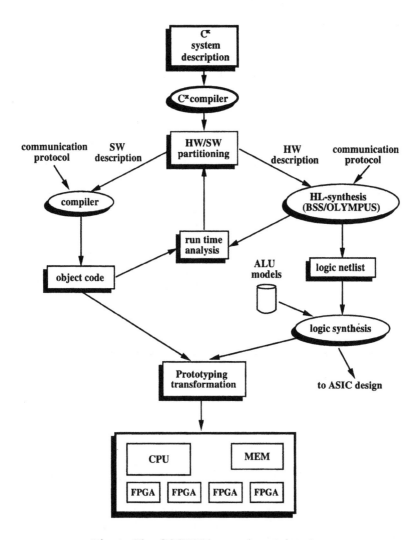

Fig. 1. The COSYMA experimental system

2 Prototyping System Requirements

Because the generated hardware-software system uses memory coupling, it seemed reasonable not to use a very expensive commercial prototyping system, which would have to be extended by a SPARC core, anyway, but to develop a specialized system. Because the coprocessor typically is rather small, we concluded that it should be possible to design the hardware-software prototyping system to run close to real time (33MHz SPARC clock) and so even use it for system evaluation in the real environment. Furthermore, symbolic debugging features with software breakpoints were considered essential for COSYMA verification. A major requirement was that the whole design flow from the high-level synthesis output should work without manual interaction.

3 The Architecture of the Prototyping System

The prototyping system consists of three different boards. Figure 2 gives an overview of the system. The LSI SPARC processor ([5]) and the coprocessor board are memory coupled by RAM A and RAM B of the coprocessor board. The arbiter separates the processor bus from the system bus, which enables the processor to fetch an instruction from its instruction RAM, while the coprocessor is fetching one or two 32 bit words in the same clock cycle.

The FPGA board contains four XILINX XC4010 ([6]). We chose the FPGA family, because of its sizes. For multiplications, we added an AMD 29C323.

For the purpose of time measurement, debugging and the connection to the host computer, a Motorola MC 68332 ([7]) is used.

3.1 The FPGA Board

The major function of the prototyping system is the emulation of an application specific coprocessor. A typical design generated by COSYMA could include up to five ALUs and 20 registers. This exceeds the capacity of one single FPGA. So, the design was partitioned into clusters. With up to 5 arbitrarily connected 32 bit ALUs and 10 more more registers, we would quickly run out of pins when allocating complete ALUs and registers to FPGAs. Instead we chose a bit slice architecture (fig. 3). The FPGA board consists of four XILINX XC4010, each handling an eight bit slice of all ALUs and registers. So, every FPGA holds up to five 8-Bit-ALU slices and an application dependent number of 8 bit register slices.

In order to handle the large number of control signals, a copy of the whole controller is included in each slice. Therefore it is ensured that all important control signals are generated on each slice. The carry signal of the ALUs is handled by a carry look ahead logic on one of the slices (no. 3). Additional flags for comparison decrease the controller overhead.

On every slice there are two RAM ports. Each of the ports is able to read or write 8 bits of the 32 bit word. Therefore, the control signals are generated on

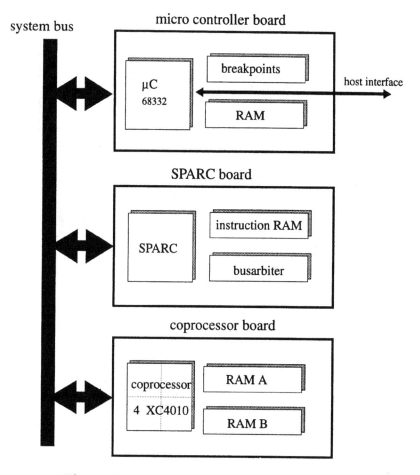

Fig. 2. The architecture of the prototyping system

one of the slices (no. 2). The memory consists of two banks, whose address spaces can be mapped in two different ways. The address space of port B can be mapped following port A. Another possibility is to address port A and port B alternately in steps of four bytes for interleaving. The address modes are configured by the micro controller.

3.2 Communication Between the SPARC and the FPGA Boards

Currently, processor and coprocessor communicate through shared memory with mutual exclusive access. In order to increase the parallelism in the future, every slice uses two separate memory ports. The coprocessor is able to fetch two 32

Coprocessor-Board

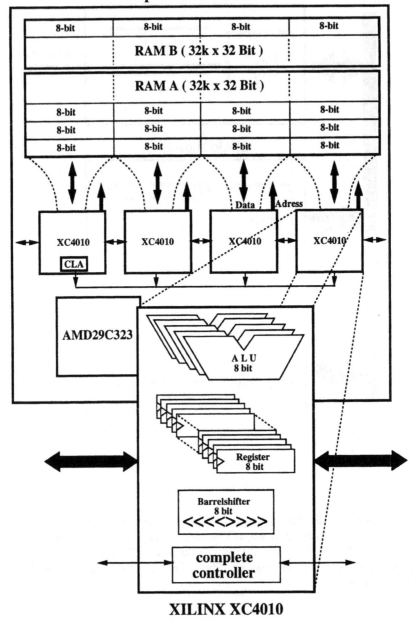

8-bit	8-bit	8-bit	8-bit
RAM B (32k x 32 Bit)			

RAM A (32k x 32 Bit)			
8-bit	8-bit	8-bit	8-bit
8-bit	8-bit	8-bit	8-bit
8-bit	8-bit	8-bit	8-bit

XILINX XC4010

Fig. 3. Architecture of the FPGA board

bit values concurrently. The two memory banks are mapped into the SPARC address space.

Up to now, the SPARC receives a HOLD signal when the coprocessor is active. We are going to increase the performance of the design by allowing parallel execution in the SPARC and the coprocessor. Therefore, a separate instruction memory is inserted, allowing the SPARC to fetch an instruction during a data fetch of the coprocessor.

3.3 Time Measurement and Debugging

As mentioned above, a microcontroller is used for time measurement and debugging. One interrupt channel of the MC68332 is triggered by a tag RAM, which contains the addresses of up to eight breakpoints. Whenever one of eight maskable tag breakpoint appears on the address bus, the microcontroller receives an interrupt while processor and coprocessor are stopped in the same cycle. Then, user is able to read the memory map.

The hardware breakpoint are also used for the time measurement. One of the MC68332 timers is triggered by the tag RAM. Each can be used either for time measurement or for debugging. The state of the system bus during a breakpoint event is stored in a register which can be read by the micro controller.

4 Design Flow

The result of the partitioning process in COSYMA consists of software for a processor core, and a coprocessor. For the software parts, COSYMA generates ANSI-C code which is then compiled.

As an output of high-level synthesis ([2],[3]), the coprocessor is described as a logic level netlist in SLIF format. This netlist is a technology independent description which has to be partitioned into the four bit-slices (fig. 4). Each slice uses a local copy of the hierarchical netlist description where all bits belonging to different slices are removed. This is currently done on text level by a script written in "perl". While this works well for the data path, special care has to be taken not to remove signals entering or leaving the controller or crossing bit-slices.

The description of modules like ALUs, RAMs, ... is extended by those signals necessary for the interconnection of the bit-slices. This happens already before partitioning. Because of a predefined set of interconnections, only a fixed set of signals (5) from data path is available in the controller.

After a translation to Verilog, logic optimization is done by the Synopsys system on each of the FPGA-netlists separately. At this point VHDL models of the ALUs and the barrel shifter are included. The synthesized slices are mapped to the FPGAs. The output of the Xilinx tools is a bitstream, which is then downloaded to the FPGAs controlled by the MC68332.

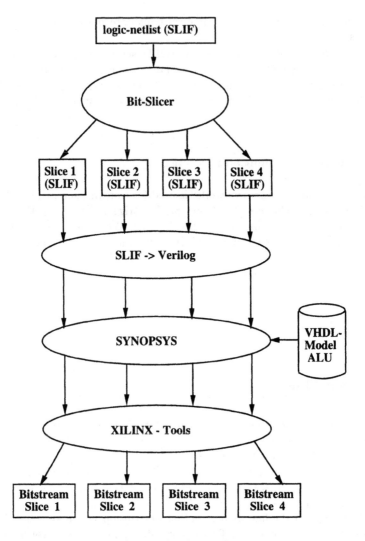

Fig. 4. The design flow

5 Future Directions

Now, the system is completed as wire-wrap prototype, which is clocked at 10 MHz. We are going to build a 30 MHz version using multilayer PCBs. One problem of the system is the capacity of the XILINX 4010. Using four XC 4025 instead of XC 4010, an internal multiplier could be realized on the chips and a large no. of registers.

We are going to extend COSYMA to small heterogeneous multiprocessor systems. As a first extension, the Motorola DSP 96002 will be added to the prototyping system. The DSP-Board will be prototyped using Aptix interconnection matrixes.

6 Acknowledgement

J. Herrmann and P. Meier were responsible for the mechanical construction of the prototyping system and H.-O. Leilich and F. Rabe gave very helpful hints when we debugged the hardware.

References

[1] R. Ernst, J. Henkel and Th. Benner, Hardware/Software Co-Synthesis for Microcontrollers, IEEE Design & Test of Computers, pp. 64–75, Dec. 1993.

[2] G. De Micheli et al., The Olympus Synthesis System, IEEE Design & Test of Computers, pp. 37–53, Oct. 1990.

[3] U. Holtmann, Hierarchical Behavioural Representation in the Braunschweig Synthesis System BSS, IFIP Workshop on Application of Synthesis and Simulation, Lenggries, 25-28.8.1993.

[4] W. Ye, R. Ernst, Th. Benner, J. Henkel, Fast Timing Analysis for Hardware-Software Co-Synthesis, Proc. of ICCD 1993, Cambridge, pp. 452–457 , 1993.

[5] LSI Logic Corporation, L64831 - SPARC Integrated IU/FPU Technical Manual, Milpitas, Calif. 1992.

[6] Xilinx, Inc., The Programmable Logic Data Book, San Jose, Calif., 1993.

[7] Motorola Inc., MC68332 User's Manual, Phoenix, Arizona, 1990.

FPGA Based Low Cost Generic Reusable Module for the Rapid Prototyping of Subsystems

Apostolos Dollas* and Brent Ward and John Daniel Sterling Babcock

Department of Electrical Engineering
Duke University
Durham, NC 27708-0291
USA

Abstract. The development of a model for sub-system reuse and the evaluation of currently available rapid prototyping platforms has led to the development of a GEneric Reusable Module (GERM). The GERM is a low-cost, stand-alone, reprogrammable development tool designed for prototyping digital subsystems. The GERM, and associated templates, aid the designer in rapidly prototyping and reusing subsystem designs. The GERM addresses also the introduction of students to FPGA technology in an environment which they can continue to use for more complex designs. Extensions of the GERM include combining multiple GERMs together to prototype larger subsystems and systems. The system was used successfully in computer engineering courses at Duke University.

1 Introduction

Decomposition of a system into functional blocks, or subsystems, is one of the more common approaches used to manage designs of complex systems. By dividing a system into small functional blocks, multiple designers and/or groups can focus on developing and testing specific subsystems [5, 2]. Tested, verified, and documented subsystems are ultimately integrated into a complete system for fabrication and delivery.

The subsystems created for a design can, in many cases, be reused in future systems and by other designers. Subsystems are often reusable in their original form, but in some instances, modifications to an existing design yield a newer and more applicable subsystem to meet a designer's requirements. The reuse of a subsystem will effectively save the designer time by eliminating the time needed to create the subsystem, and will ultimately lead to more rapid prototyping of systems. In the context of well parametrized design spaces, knowledge based CAD tools for rapid system prototyping have been successfully demonstrated [3, 4]. In an effort to establish the role of reusable subsystems in the more general case of rapid microelectronic system prototyping, a model of the design process was developed, shown in Figure 1.

---

* Now at the Technical University of Crete, Greece, dollas@ced.tuc.gr

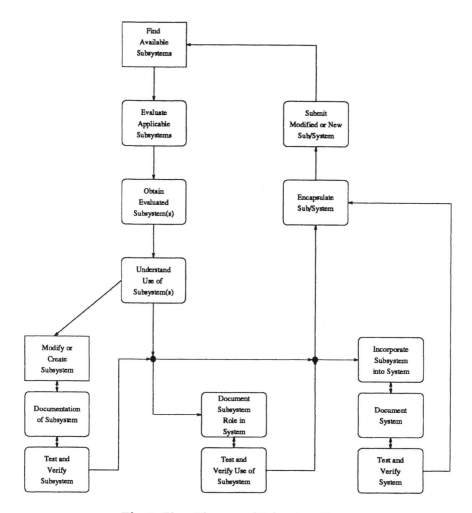

Fig. 1. Flow Diagram of Subsystem Reuse

At Duke University, we have studied issues of rapid system prototyping in academia [11, 10]. This paper presents results from an on-going research effort on rapid system prototyping through increased subsystem reusability. After developing a model for subsystem reuse [16], we determined that we needed an environment and hardware platform to rapidly prototype subsystems. We desired the option of designing subsystems using schematic capture, logic description, or behavioral code (such as VHDL [13, 6]), and then mapping the design into an FPGA based reprogrammable prototyping hardware system [15]. To determine the best platform to use in our research, we investigated and evaluated three FPGA prototyping environments.

Of the available prototyping environments, we evaluated the Anyboard [8,

9], the BORG [7], and the Protozone [12]. These boards represent prototyping systems which provide a low to medium level of reprogrammable resources, as opposed to Quickturn [17] and Splash [1] which have a high level of resources. The results of the evaluations are shown in Table 1. The results are reported in terms of High, Medium, and Low. The Anyboard and BORG boards are both multi-FPGA prototyping systems allowing for larger designs to be developed whereas the Protozone is a single-FPGA prototyping system.

Criterion and Platforms			
Criterion	Anyboard	BORG	Protozone
Hardware Resources	H	M to H	L
Experimentation Tool	H to M	H to M	H
Expected Ease of Use	M	H	H
Use of Standard Software	M	H	H
Cost	M	M	L
Availability	L	M	H
HW(SW) Support	L(M)	H(H)	M(H)
Documentation	H	H	M

H:High M:Medium L:Low

Table 1. Evaluation Results

After evaluating available prototyping systems, we determined that a hardware tool targeted specifically at subsystem development was needed to further study subsystem reuse. We desired a low cost, easy to use, and stand alone prototyping tool which would double as an educational aid in teaching undergraduates about FPGA technology and issues of reprogrammability and system prototyping. The ease of use of the Intel FLEXlogic family and its development environment [14] at the entry level was coupled with a slow learning curve to transition to Xilinx FPGA technology, and we determined that we needed a vehicle for easy introduction to FPGA technology in the Xilinx environment, which we use in more advanced applications (and more advanced classes). Thus, the GEneric Reusable Module (GERM) was designed with minimal components, connectors, and complexity as an experimental platform to study subsystem reusability and as an educational platform to introduce students to FPGA's at the sophomore level.

2 Curriculum Usage

The GERM has been used as an instructional tool for introductory digital design courses. We use it as a lab kit to introduce students to FPGA technology, reprogrammable hardware, subsystem design, and rapid system prototyping. The

GERM is also used to prototype and test subsystem designs which may then be connected together to prototype larger subsystems. The GERM board was developed in prototype, wirewrapped form in 1993, and in small scale production with printed circuit boards in early 1994.

GERMs have been used in undergraduate and graduate courses at Duke University, starting in the spring semester of 1994. The board supplements the current lab kit used in the sophomore level EE151, Introduction to Switching Theory and Logic Design, and it was also used in the senior/graduate level EE254, Fault-Tolerant and Testable Systems [2]. There are plans to use it also in EE251, Advanced Digital System Design, and EE261, Introduction to VLSI Design. The GERM can aid research into reusability because it is simple enough to use at the entry level and it will facilitate understanding of reusability issues in a realistic environment on realistic projects. We expect the GERM, its tools, and its design procedures to further aid the undergraduate and graduate students in the upper level courses where complex hardware design projects are expected to be produced in a semester. More advanced tools and facilities available to these students include BORG II boards and a dedicated FPGA laboratory.

3 GERM Design

A layout diagram of the GERM is provided in Figure 2, a list of the components in Table 2, and the header-pin designations in Figure 3. Although not shown in Figure 2, the board also has a header for serial downloading, which is very useful during design development and debugging. It turns out that the serial download cable is also very useful for class usage due to the larger number of available development systems (workstations, PC's) than EPROM programmers.

GERM Components		
1	XC3030PC-50	Xilinx FPGA
1	AMD2764	Advanced Microdevices EPROM
4	26 Pin Male Header	Connector
1	Red LED	Power Indicator
1	Green LED	GO Indicator
3	5 kohm Resistor	Pull-Up
2	200 ohm Resistor	Voltage Drop
1	SPDT Switch	Power Switch
1	10uF Capacitor	Bypass Capacitor for Power/GND Pins

Table 2. Components of the GERM

---

[2] For simplicity, the initial classroom use of the GERM was with the serial download cable and not the EPROM

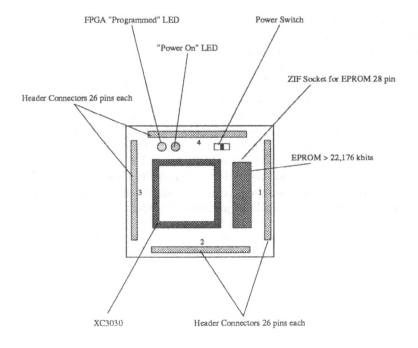

Fig. 2. Block Diagram of GERM

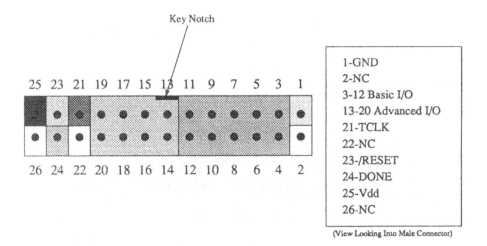

Fig. 3. Profile of Standard GERM Header Connector

Through the use of these standardized headers, connecting an array of GERMs or a GERM to another subsystem module is quick and well defined. A user may use 26 conductor ribbon cables with keyed header connectors, or single jumper wires. The Red LED indicates that the power is ON and the Green LED indicates that the FPGA has completed loading the configuration information from the EPROM. The GERM is approximately 3"x3.5"(7.62cm x 8.89cm).

Some attributes of the GERM include the following:

1. It has been designed primarily with the introductory level digital designer (the student) in mind. The GERM is to supplement current lab kits consisting of discrete TTL and GAL chips and wires.
2. It provides a simple development platform intended to prototype and emulate subsystems.
3. It is portable and low-cost, enabling students to potentially develop designs and circuitry in their dorm rooms as well as in the laboratory.
4. Many of the GERMs (each with a different subsystem configuration) can be connected together to develop more complex systems.
5. It provides a means of gradual transition from lower level subsystem design to higher level system design in the computer engineering curriculum.
6. The use of a regular EPROM rather than the once-programmable serial PROM allows for standalone operation (without a download cable) and reusability of the board itself without the expense associated with the PROM's.

4   GERM Design Process

Some experimental designs, to be discussed in Section 6, have been developed to demonstrate the capabilities and applicability in teaching undergraduates reusability and digital design with FPGA's.

Prototyping a digital design with a GERM board requires that a user have access to a schematic capture package (e.g. DATA I/O™ FutureNet or Viewlogic™ Viewdraw) or a behavioral description language (i.e. VHDL). The user designs the subsystem and then compiles the schematic into a bit file through the Xilinx Logic Cell Array suite of tools. The bit file is then programmed into the EPROM. Once the EPROM is programmed, it is placed in the ZIF socket on the GERM and power is turned on. The FPGA programs itself with the user design and the Green LED turns on to indicate that the design is loaded and ready to be tested. The steps in using the GERM are summarized below.

1. Schematic Capture or Logic Description of Circuit
2. Define Input and Output Pins
3. Compile Design Using Xilinx Tools
4. Program EPROM with .bit file
5. Insert EPROM in ZIF on GERM
6. Turn on Power
7. Wait for Green LED "GO"
8. Exercise Circuit Through Connectors

Alternatively to the last steps, the serial download cable option can be used on the GERM for downloading in a similar fashion to existing boards (e.g. the Xilinx demonstration boards).

5 Templates

We have found, in working with students and with other researchers, that templates (or prototype files) and tutorials are extremely valuable to the novice designer. Templates and basic tutorials greatly reduce the amount of time typically required to "get up to speed" when using such tools as Powerview, ABEL^{TM}, VHDL, writing code in C, documenting systems or subsystems in LATEX, and when reusing or creating new subsystems. Templates also provide a means for standardizing code design and documentation.

The templates we provide designers are typically "generic" in nature and supply the user with basic information and examples that the user may quickly modify to match his/her needs. The templates include those for documentation (LATEX), logic description (ABEL), behavioral description (VHDL), code development (C), and directory structure usage (tempdir). These files and directory structure can all be copied into the user's directory and modified, or they can be used as a reference for determining the proper usage of commands and structures. Thus, consistency with other designers' documentation and coding formats is maintained with little effort from the designer, and the time required to document and encapsulate designs is reduced as well. We have found that reusability at the subsystem level is hampered not only by the lack of proper tools, but also the mentality of the designers. The use of the GERM starting at the sophomore level and continuing through the senior level will facilitate reusability as designs will become of increasing complexity and thus require the use of previous ones (e.g. see experiment 2 in Section 6.2, which uses the circuit of experiment 1 in Section 6.1).

Templates describing a subsystem can be quickly modified and mapped into hardware for testing. Using established and supported facilities from both Viewlogic and Xilinx, we are able to quickly define a subsystem, simulate it, and then map it into the GERM for evaluation and connection to other subsystems.

6 Example Designs

Two of the primary designs used as examples for introductory level digital designers are described below. The first is an entry-level laboratory exercise that is designed to introduce the student to the schematic capture tools and the overall design process. The second is a more advanced laboratory exercise, designed to introduce the student to more complex circuitry and subsystem reusability.

6.1 Experiment 1: Simple Circuit

A simple circuit with a binary to seven segment converter is designed, with a debounced switch to be used as a manual clock input, a counter, and combinational circuitry to generate 7-segment LED display information. The circuit counts backwards through the sequence 3-2-1-0-3 etc. The binary data from the counter is converted into segment display information and sent off chip to a 7-segment LED display. The basic circuit schematic for the FPGA is shown in Figure 4.

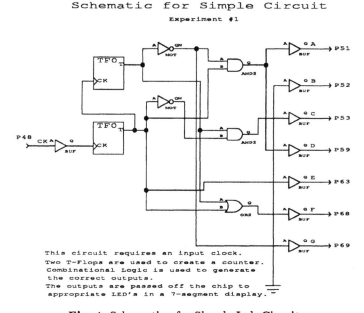

Fig. 4. Schematic of a Simple Lab Circuit

The circuit introduces a beginning student to the concepts of switch debouncing, counters, combinational logic, seven segment LEDs, FPGAs, and rapid prototyping without requiring the student to spend an hour or so breadboarding the circuit using discrete chips.

6.2 Experiment 2: Advanced Circuit

This more advanced circuit combines data registers, LED multiplexing, BCD to seven segment decoders, timers, clocking, and switch debouncing. The circuit also requires finite state machines, combinational logic, and requires power considerations when driving LEDs.

The student is required to develop a display driver, taking as input (sequentially) four 4-bit hex numbers, and driving four 7-segment displays in a multiplexed fashion, as is typically done in calculators. The numbers are entered in a four element deep, 4-bit wide FIFO. The student has access to a BCD to seven segment decoder from the parts library. The inputs to the chip include 4 bits of register data, 1 clock for clocking the registers (debounced), and one clock (jumpered) for either the user to manually clock across the seven-segment LED's or a 555 timer to clock across them. The outputs consist of seven bits to indicate values for the segments of the seven-segment displays, and four bits to provide power to the common anodes of each of the displays. The advanced circuit schematic for the FPGA is shown in Figure 5.

A deliberate goal of these exercises is to introduce reusability as a concept in rapid system prototyping. Reusability is more than the existence of libraries of designs, and includes standardization of documentation and the methodology that new designs require the use of previous ones. Indeed, the design component in this exercise is "how" to use registers, 7-segment display drivers, and other designs, together with a newly designed finite state machine in order to complete the design.

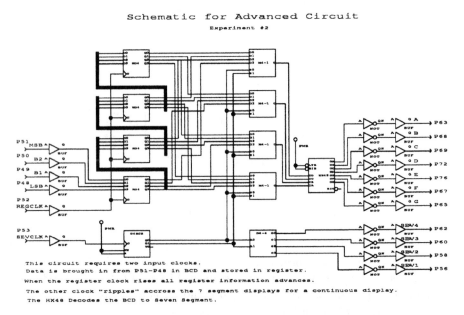

Schematic for Advanced Circuit

Experiment #2

This circuit requires two input clocks.
Data is brought in from P51-P48 in BCD and stored in register.
When the register clock rises all register information advances.
The other clock "ripples" accross the 7 segment displays for a continuous display.
The HX48 Decodes the BCD to Seven Segment.

Fig. 5. Schematic of an Advanced Lab Circuit

7 Evaluation

Due to its minimal configuration, the GERM lacks many of the advanced features of the larger systems. Some of the limitations and capabilities of the GERM are discussed below.

GERM	
Criterion	Evaluation
Hardware Resources	Low
Potential as an Experimentation Tool	High
Expected Ease of Use	High
Use of Standard Software	High
Cost	Low
Availability	Low
Hardware	Low
Software Support	High
Documentation	High

Table 3. GERM Evaluation

7.1 Limitations

- Users must manually partition large designs into multiple GERMs and then manually interconnect the GERMs to form larger networks.
- The interconnection of the modules can be done using GERMs as routers, however, this process requires a separate design for each of the GERMs.
- There is a small number of input/output signals on each cable (18), which may require multiple cables to be used between GERMs to fully implement a large design. The pin limitation is more evident in datapath and bus designs.
- The GERMs do not have protective circuitry on the programmable input/output pads, and therefore the user must be careful when interconnecting modules.

7.2 Capabilities

- The GERM can be programmed using either schematic or behavioral descriptions.
- It is easy to use, well-characterized, and needs little documentation to describe its functionality and use.
- The design of subsystems can be kept within the Xilinx design environment if desired or developed using Viewlogic tools and then converted through Xilinx utilities.
- GERMs can be programmed to act as many different subsystems and then connected together to form a larger subsystem or system.
- In addition to the Viewlogic front end tools, the Synopsys synthesis tools can be used as an alternative design path with language (VHDL) rather than schematic design entry.

A summary evaluation of the GERM with the same criteria as other small scale FPGA based boards is in Table 3.

8   Conclusions

We have found the reuse of subsystems useful in rapidly prototyping new subsystems and systems. The development and fabrication of the GERM board has facilitated research on reusability of subsystems.

The GERM is portable, small, and reprogrammable, and it serves an instructor as a simple, low-cost introductory teaching tool. The behavior of the GERM can be quickly changed by swapping EPROMs and attached modules. Thus, the GERM acts as a generic building block for subsystem prototyping and development.

Although the boards are newly developed, they have already been used in two Duke University classes at the Department of Electrical Engineering during the spring semester of 1994. Initial usage was limited to the serial downloading of student designs, but in the future the EPROM downloading will also be used.

In the future, GERMs using XC4000 series chips will be constructed to provide more user I/O and more internal programming resources. The GERM may eventually be equipped with an on board selectable clock, manual clock button, and Electrically Erasable PROMs. Sample GERM boards are available for distribution and evaluation with tutorials, demonstration designs, design files, templates, and user manuals from Duke University Electrical Engineering Department.

9   Acknowledgements

We would like to acknowledge the National Science Foundation for supporting our research in reusability and rapid system prototyping through grant MIP-9209866 and for funding a Field Programmable Gate Array Laboratory through grant DUE-9351480. The printed circuit boards were also funded by the National Science Foundation through the MOSIS service. Mr. Larry Calhoun of AMD Corporation and Mr. David Lam of Xilinx Corporation helped us greatly with generous part donations from their companies. We appreciate the interest and efforts of Professors John Board and Pete Marinos from the Electrical Engineering Department at Duke University who used the GERM in their classes and provided us with valuable feedback, and Professor David Overhauser from the same Department who contributed substantially to the project with many suggestions for improvements and assistance with CAD tools.

References

1. J. Arnold, D. Buell, and E. Davis. Splash 2. *Proceedings, 4th Annual ACM Symposium on Parallel Algorithms and Architectures*, pages 316–322, 1992.

2. A. Asur and S. Hufnagel. Taxonomy of Rapid-Prototyping Methods and Tools. In *Proceedings, The Fourth IEEE International Workshop on Rapid System Prototyping RSP-93*, pages 42–56. Computer Society Press, 1993.

3. W. P. Birmingham, A. P. Gupta, and D. P. Siewiorek. The MICON System for Computer Design. *IEEE Micro*, 9(5):61–67, October 1989.

4. W. P. Birmingham, A. P. Gupta, and D. P. Siewiorek. *Automating the Design of Computer Systems: The MICON Project.* Jones-Bartlett, 1992.

5. F. P. Brooks, Jr. *The Mythical Man-Month, Essays in Software Engineering.* Addison-Wesley, 1972.

6. R. Camposano, L. F. Saunders, and R. M. Tabet. VHDL as Input for High-Level Synthesis. *IEEE Design and Test of Computers*, 8(1):43–49, March 1991.

7. Pak K. Chan. BORG: A Field-Programmable Prototyping Board: User's Gukde. Technical report, University of California at Santa Cruz, 1992.

8. D. E. Van den Bout, O. Kahn, and D. Thomae. The 1993 AnyBoard Rapid-Prototyping Environment. In *Proceedings, The Fourth IEEE International Workshop on Rapid System Prototyping RSP-93*, pages 31–39. Computer Society Press, 1993.

9. D. Van den Bout, J. Morris, and D. Thomae et al. AnyBoard: An FPGA-Based, Reconfigurable System. *IEEE Design and Test of Computers*, pages 21–29, September 1992.

10. A. Dollas. Experimental Results in Rapid System Prototyping with Incomplete CAD Tools and Inexperienced Designers. In *Proceedings, The Second IEEE International Workshop on Rapid System Prototyping RSP-91*, pages 9–16. Computer Society Press, 1992.

11. A. Dollas and V. Chi. Rapid System Prototyping in Academic Laboratories of the 1990's. In *Proceedings, The First IEEE International Workshop on Rapid System Prototyping RSP-90*, pages 38–45. Computer Society Press, 1991.

12. Abbas El Gamal. Protozone: The PC-Based ASIC Design Frame - User's Guide. Technical report, Stanford University, Stanford, California, 1992.

13. IEEE Computer Society Press. *IEEE Standard VHDL Language Reference Manual, IEEE Std. 1076-1987*, 1987.

14. Intel Corporation. *PLDshell Plus/PLDasm User's Guide*, 1993.

15. S. H. Kelem and J. Seidel. Shortening the Design Cycle for Programmable Logic Devices. *IEEE Design and Test of Computers*, pages 40–50, December 1992.

16. B. T. Ward. An Experimental Study in Subsystem Reusability Toward Rapid Prototyping of Microelectronic Systems. Master's thesis, Department of Electrical Engineering, Duke University, 1993.

17. Howard Wolff. How Quickturn is Filling a Gap. *Electronics*, page 70, April 1990.

FPGA Development Tools:
Keeping Pace with Design Complexity

Bradly K. Fawcett and Steven H. Kelem

Xilinx Inc.
San Jose, CA, USA

Abstract. As the density and complexity of FPGA-based designs increases beyond 10,000 gates, highly-integrated and automated development tools are required. Several recent trends in development system capabilities are helping designers keep pace with growing design complexity, including FPGA-specific logic synthesis, increased design portability, improved design implementation tools, support for system-level simulation, and framework integration.

FPGAs have created a unique requirement for CAE software; their tools must deliver the ease-of-design and fast time-to-market benefits that have popularized FPGA technology, must be capable of implementing high density logic designs on an engineer's desktop system, and, in order to service a broad market, must be easy-to-use and compatible with the user's existing design environment. Several recent trends in FPGA development system capabilities are helping designers meet the twin challenges of growing design complexity and increasing time-to-market pressures.

As with other logic technologies, the basic methodology for FPGA design consists of three inter-related steps: entry, implementation, and validation (Figure 1). The design process is iterative, returning to the design entry phase for correction and optimization. Typically, generic tools are used for entry and simulation, but architecture-specific tools are needed for implementation.

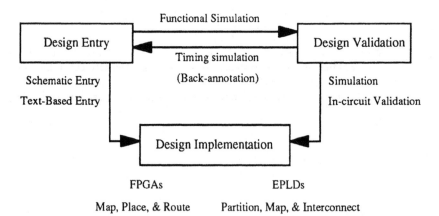

Figure 1. The basic FPGA/EPLD design methodology consists of three steps: entry, implementation, and validation.

1 Design Entry

Entry methods for FPGA design include schematics (using graphics-based schematic editors) and behavioral entry (requiring FPGA "fitters" - device-specific tools that optimize the logic to fit the target FPGA architecture).

For high-density FPGA designs, gate-level entry tools often are cumbersome, and the use of logic synthesis and high-level description languages (HDLs), such as VHDL or Verilog-HDL, can raise designer productivity. However, for a top-down, HDL-based design methodology to be useful, the synthesis tools must be effective in producing a gate-level design optimized for the target technology. Optimization algorithms for fan-in limited, lookup-table based architectures such as the Xilinx FPGAs are dramatically different than the algebra-based algorithms used for gate arrays. In this respect, logic synthesis for FPGAs is still an emerging technology.

Most FPGA development systems support hierarchical design entry; these development systems can combine hierarchical elements that are specified with multiple design entry tools, allowing the most convenient entry method for each portion of the design.

The ability to easily port a design to different device architectures provides several advantages to the system designer: the technology choice can be postponed until later in the development cycle when requirements are better defined, design migrations to reduce cost during the life of the product (such as migrating from an FPGA to a gate array) are simplified, and portions of the design can be easily re-used in future products, even if those products use different technologies. Ideally, new product development should be able to take advantage of the latest devices and technologies without having to duplicate earlier development efforts to re-use proven portions of previous designs.

In the past, users often had to make the technology decision (for example, choosing between an EPLD and an FPGA architecture) as the first step in the design process at the beginning of the design entry phase. Two recent developments have changed this scenario: the advent of design synthesis tools optimized for programmable logic architectures, and the development of 'universal' schematic libraries that support multiple device architectures. A design described in an HDL can be 'technology-transparent', relying on synthesis compilers to map the logic into the targeted technology automatically. The "Unified Library" of the new XACT^{TM} 5.0 development system from Xilinx typifies the advances being made in the development of 'portable' schematic libraries. All primitives and macros common to two or more Xilinx device families are consistent in name and appearance. Thus, migration of a design from one family to another requires a change of only the compilation target and, if needed, the editing of any family-specific symbols used in the design.

2 Design Implementation

After the design is entered, implementation tools map the logic into the resources of the target FPGA's architecture, determine an optimal placement of the logic, and select the routing channels that connect the logic and I/O blocks. Design implementation tools apply a high degree of automation to these tasks. These tools tend to be unique to each FPGA architecture, but should have a smooth interface to their supporting cast of entry and validation tools.

Xilinx's automated implementation tools typify the advances being made in this field. An automatic design compilation utility, XMAKE, retrieves the design's input files and performs all the necessary steps to create the FPGA configuration program: translating the input files to the Xilinx Netlist Format (XNF), merging together the elements of a hierarchical design, deleting unused logic, mapping the design into the FPGA's logic resources, placing and routing the logic and I/O blocks, and generating the configuration program.

The automated partition, place, and route algorithms are timing-driven; that is, timing analysis of the signal paths within the application is performed during the placement and routing of the design. Users can specify performance requirements along entire paths in an FPGA design (as opposed to the traditional method of assigning "net criticality" to individual nets), and the implementation programs use this information to guide the placement and routing process.

Optionally, user-designated partitioning, placement, and routing information can be specified as part of the design entry process (typically, right on the schematic). The implementation of highly-structured designs can greatly benefit from the basic floorplanning techniques familiar to designers of large gate arrays.

3 Design Validation

Validation (testing) of FPGA designs typically is accomplished through a combination of in-circuit testing, simulation, and static timing analysis. The user-programmable nature of FPGAs allows designs to be tested immediately in the target application. However, as designs increase in density and complexity, the number of circuit paths that may have timing problems increases, and timing simulation becomes an invaluable tool. To support timing simulation, FPGA implementation tools include timing calculators to determine the post-layout timing of implemented designs, including the actual delays of routing paths. This information is annotated into gate level libraries for full timing simulation. To manage increasing design complexity better, a growing number of users employ board-level and system-level simulation spanning multiple device types, in addition to simulating each FPGA on its own. Alternatively, static timing analyzers examine a design's logic and timing to calculate the performance along signal paths, identify possible race conditions, and detect set-up and hold-time violations, without requiring user-generated input stimulus patterns or test vectors. However, most users limit the use of static timing analysis to fully-synchronous designs only; the technique is difficult to apply accurately to asynchronous circuits.

4 Frameworks And Tool Integration

The typical FPGA development environment includes a mix of generic design tools and architecture-specific implementation tools. Ideally, these tools are molded into an integrated, easy-to-use development environment. Most FPGA vendors provide their own design management software. However, more-tightly integrated tool sets are now available. For example, Viewlogic Systems, Cadence Design Systems, and Data I/O are among the CAE vendors that package, sell, and support design kits that provide full front-to-back design capabilities by melding FPGA vendors' implementation tools into their own frameworks.

Meaningful Benchmarks for Logic Optimization of Table-Lookup FPGAs

Steven H. Kelem

Xilinx, Inc.

Abstract. This paper discusses benchmarks for optimization to table-lookup FPGAs. We discuss a scientific method for systematically generating a set of benchmarks for measuring the effectiveness of a synthesis tool/algorithm for a particular FPGA architecture. The benchmarks have the useful properties of being generated easily, having an *a priori*, known best result, covering all the possible configurations of a lookup table, and yielding a simple metric. This metric can be used to compare different synthesis tools/algorithms for their efficiency in mapping to a given FPGA architecture. This is in contrast to the *ad hoc* sets of benchmarks, for which it is difficult to compare results of different optimization tools/algorithms.

1 Introduction

This paper discusses benchmarks for optimization to table-lookup FPGAs. We discuss a scientific method for systematically generating a set of benchmarks for measuring the effectiveness of a synthesis tool/algorithm for a particular FPGA architecture. Logic optimization is a technique (set of algorithms) that was originally developed for reducing the number of product terms in a PLA. These algorithms have been modified or re-designed to reduce the number of resources required to implement a circuit or set of equations in architectures other than PLAs. These include multi-level PLAs[1], fine-grained architectures such as gate arrays, and medium-grained architectures such as table-lookup FPGAs[2]. A minimal set, or "kernal", of 4-input functions have been discovered for technology mapping of FPGAs[3], but the results of this mapping has not yet been compared to existing techniques.

To measure the effectiveness of an optimization tool/algorithm designs should be run through the tool/algorithm, once for delay optimization, and once for area optimization. Then the resulting circuit size and delay can be measured for each optimization criterion. Then a metric derived from these measurements can be used to compare different optimization tools or algorithms, or to tune a given algorithm.

The approach we take is a statistical one, where a large number of samples from the space of meaningful configurations for the tables in the table-lookup FPGA. This has the effect of not being biased towards any particular benchmarks, and thus more accurately predicts how an optimization tool will perform on real designs.

2 Table-Lookup FPGA Architectures

The CLB in Xilinx' 4000-series FPGA contains two, independent, 4-input table-lookup Function Generators, F and G. The CLB can be configured so that the outputs of F and G are inputs to the 3-input Function Generator, H. (The third input to H may be a primary input to the CLB.) In this mode any function of five variables can be implemented. Because the eight inputs to the two 4-input Function Generators are independent of one another, some Boolean functions of up to nine variables can be implemented in one CLB. These include many useful functions, including the 9-input exclusive OR, which, without the TLU architecture, would be expensive to implement.

2.1 Number of Unique Functions

The number of functions that can be realized in the table-lookup architecture can be determined as follows. The n-input table-lookup function generator is a 2^n-bit RAM. The number of possible values, and therefore the number of possible functions, in the RAM is 2^{2^n}. The 4-input function generators can implement 65,536 functions, the 5-input function generators 4,294,967,296. The 4K CLB has 40 bits in the RAM CLB, and can implement 1,099,511,627,776 functions.

The number of functions that the CLBs can implement is large, however, many of the functions are degenerate—some of the inputs to many of the n-input functions are not necessary. For example, the 5-input CLB can implement all Boolean functions of 1, 2, 3, and 4 variables. In particular, a 5-input CLB with one unused input can implement a 4-input function f as $f(a,b,c,d)$, $f(a,b,c,e)$, $f(a,b,d,e)$, $f(a,c,d,e)$, or $f(b,c,d,e)$, depending on which of the five inputs to the CLB is not taking part in the function.

We can calculate the number of degenerate and non-degenerate n-input functions for arbitrary n[1]. Table 1 shows the number of degenerate and non-degenerate n-input functions for n from 0 to 6.

Table 1. Numbers of Non-Degenerate n-Input Functions

n	# degenerate	# non-degenerate	Total	% degenerate
0	0	2	2	0%
1	2	2	4	50%
2	6	10	16	37.5%
3	38	218	256	14.8%
4	942	64,594	65,536	1.44%
5	325,262	4,294,642,034	4,294,967,296	0.00769%
6	25,768,825,638	18,446,744,047,940,725,978	18,446,744,073,709,551,616	1.4×10^{-7}

[1]. There is an elegant proof of this, but these margins are insufficient to hold it. Contact the author for details.

3 Generating Test Cases

Because the number of possible functions realizable in a Table-Lookup Architecture is large (2^{2^n}), enumerating all of them is impractical. Our goal is to generate a large number of test cases by statistical means. We want to generate test cases for which we know the desired result *a priori*. First we generate all possible configurations of a single CLB in a TLU, then we see how to generate predictable multi-CLB configurations. From the analysis alluded to in the previous section, we know how many n-input functions are realizable, and how many of the functions are degenerate. Further, we wish to generate only i-input functions for an n-input CLB, for $0 \le i \le n$. So it is important to be able to detect when a randomly-generated test case does not meet the desired conditions.

Given a truth table for a n-input function, we wish to be able to detect which, if any, input variables are not used in the function. The method for doing this can be determined by examining a truth table for symmetry. Symmetry can be detected in time proportional to size of the truth table. If T is the size of the truth table, this time is $o(T \log T)$. Because $T = 2^n$, the size of the truth table is exponential in the number of inputs, n, to the CLB. This time is $o\left(n2^n\right)$.

3.1   Generating Test Cases for a n-Input Function Generator

The procedure for generating test cases for a n-input Function Generator is to generate n-input truth tables randomly. A truth table is represented as a 2^n-bit configuration of a CLB. The first bit in this configuration corresponds to the truth table entry for all inputs=0, the last to the entry for all inputs=1. When an n-input function is desired, we use the symmetry-detecting procedure to check whether any of the input variables are unused. If any are unused, then the configuration is rejected and another is chosen randomly and subjected to the same test. Once a configuration is chosen which meets the desired conditions, the test case can be written in the desired format.

4 Conclusion

We have discussed a method for evaluating the complexity of a randomly-generated logic minimization test cases with an *a priori* optimal result. This provides a statistical basis for test cases that can be used to compare logic optimization programs. The results observed with these test cases correlate well with real logic designs.

References

1. Karen A. Bartlett, *et al*: Multilevel Logic Minimization Using Implicit Don't Cares. *IEEE Transactions on Computer-Aided Design of Integrated Circuits*, 7(6): 723–740, June 1988.
2. Stephen D. Brown, *et al*: Field-Programmable Gate Arrays. Kluwer Academic Publishers, Engineering and Computer Science Series. Boston, 1992.
3. Steve Trimberger: A Small, Complete Mapping Library for Lookup-Table-Based FPGAs. In the *2nd International Workshop on Field-Programmable Logic and Applications*. IFIP Working Groups 10.2 and 10.5, August 1992

Educational Use of Field Programmable Gate Arrays

David Lam
University Programs
Xilinx, Inc.
2100 Logic Drive,
San Jose, CA 95124

Abstract. *Traditionally, digital logic designs in undergraduate courses are described on paper and implemented with TTL SSI/MSI components. These standard logic devices have proven to be an inexpensive approach, but require "wirewrapping" and other similar means of circuit board assembly; as a result, a significant portion of the students' effort is focused on completing and debugging the physical connections between devices. This paper describes an alternative approach using Field Programmable Gate Arrays that avoids these assembly issues, allowing the students to focus on the logic design process, with the added benefit of exposing the students to the use of modern CAE tools.*

Since its introduction by Xilinx in 1985, the FPGA has played a major role in revolutionizing digital system design. The flexibility of the architecture and the ease of use of the software make FPGA devices an ideal choice for a wide range of applications. In undergraduate education, FPGAs are used in courses ranging from first-year introductory levels to senior design projects. Multiple TTL components can be integrated into a single FPGA, eliminating the need for wirewrapping and allowing the students to focus on their designs.

In the past, digital courses with labs required students to purchase kits that contained standard logic devices The purpose of the lab assignments was to reinforce the concepts learned during lecture and to actually build what was described on paper. For small design projects using less than a handful of components, the objective was normally met. The problem surfaced when building large designs using multiple components. The task of wirewrapping was excessively tedious and time consuming. Time spent on debugging the wirewrapped connections increased enormously. The emphasis of the project was no longer on digital design, but on wirewrap debugging techniques.

Design Flow

Traditionally, logic design was done with paper and pencil and implemented with discrete components and wires that connected the various devices in the system. The design flow required much attention to detail and was subject to a wide margin of error. Documentation was limited to the design on paper and any associated notes written by the designer. The FPGA allows designers to view their project from a system-level perspective. Designing with FPGAs eliminates the arduous task of debugging large wirewrapped circuits. Designs are either entered using a text-based language or a schematic editor, or a combination of both.

The FPGA design cycle consists of three basic steps: entry, implementation, and verification. The open architecture of the FPGA implementation software allows the use of various third-party design entry and simulation packages. The Xilinx Netlist Format is a standard format used to interface to these packages. The simplicity of the interface has prompted many universities to develop translation programs for their "homebrewed" CAE tools to take advantage of the technology. Entering the design using a commercial or public domain CAE tool set provides the designer with better documentation, a platform to functionally simulate prior to technology mapping, greater control over the entire design process, a short development cycle, and the realization of the final design in a standard, off-the-shelf device.

Functional and timing verification is typically performed using third-party simulation packages and in-circuit testing. Report files and a static timing analyzer also are available in the implementation software. Designing an FPGA can be performed easily and quickly on a desktop computer, resulting in more complex projects done in a shorter period of time.

Prototyping Boards

A prototyping board is a valuable tool for measuring the success of a laboratory assignment. Student designs can be realized on a prefabricated board without having to breadboard their circuit. The same board can be used by various design groups to implemented projects without having to rewire the components. The enthusiasm of using Xilinx FPGAs has led to the development of many different custom prototyping boards.

FPGAs used in Academia

There are many ways to use the FPGA technology in education. It is a perfect fit as a learning tool for computer science and electrical engineering courses that emphasize digital design.

Hundreds of VLSI designs are sent to MOSIS each year from universities across the United States. MOSIS is a wafer fabrication facility funded by the National Science Foundation. Chips are fabricated, packaged, and returned to the respective university within eight week period. Unfortunately, a majority of the designs are not functional due to poor design methodologies. Often, student designs are not properly tested before they are sent to MOSIS. Turnaround time is also a problem; students may lose interest or the school term may end before the finished product is received. Using FPGAs to prototype designs can significantly improve yields. Every design can be fully tested in-system to ensure that it is functionally correct. The end result is a motivated student who understands the importance of good design practices.

At a recent NSF "Summer Workshop on Microelectronic Systems Education," a behavioral description of a "Craps" game was described in VHDL and targeted for a Xilinx FPGA for rapid prototyping prior to MOSIS fabrication. The design was then compiled in ViewSynthesis with a target technology in mind. The first path was Xilinx FPGAs for rapid prototyping. The resulting netlist from the synthesis compiler was translated into the XNF format ready to be used by the Xilinx. The resulting binary representation of the design was downloaded to a demonstration board for in-circuit testing. Using the same source after verification, the design was retargeted to a standard cell library and processed through the Lager/Octtools set. The Lager/Octtools public domain tool set contains programs that translate the output file from the synthesis compiler to a format accepted by MOSIS.

Tufts University uses commercial CAE tools to enter FPGA designs in their Introductory Digital Logic Design course. The goal of the course is to teach students proper techniques encompassing the specification, analysis, design, and implementation of digital logic using industrial grade tools. Students first gained familiarity with the CAE tools and the top-level design approach by designing a 4-bit adder/subtractor accumulator. The second lab experiment required the students to design a traffic light controller and download it into an XC4000 demo board. The final experiment was a simple 4-bit microprocessor with a 4-bit data and address bus that could perform 16 instructions. Code was generated and included in the microprocessor's ROM. The program and the design's operation were tested in-circuit.

The University of California at Davis is offering an upper level course entitled "Synthesis Approach to Digital System Design" for graduate and senior level students. The entire course is devoted to the challenges of design complexity and fast turnaround time. Their choice of tools was Synopsys for VHDL synthesis and simulation and the Xilinx FPGA devices for rapid prototyping. The emphasis of this course is to gain "hands-on" experience with logic synthesis tools using circuits of reasonable complexity.

In a Georgia Tech computer architecture course, the MIPS RISC microprocessor from Patterson and Hennesy's "Computer Organization and Design" text was successfully synthesized using VHDL tools from VIEWlogic. There are plans to implement the design in the XC4000-based BORG prototyping board in the next computer architecture course.

Summary

The flexibility of the architecture, and the simplicity of the design tools, make FPGAs the ideal technology for prototyping logic designs in the university environment. Design implementation software in the past several years has become highly automated. Third party tools are virtually integrated into user interfaces. A design may be prototyped by simply invoking a command that will take a third party netlist and create a configuration bitstream ready for downloading. Design changes can be implemented rapidly and efficiently, and students can get immediate feedback on those changes. Little knowledge of the FPGA architecture and implementation programs are needed to successfully implement designs. An entire digital design project can be completed within a term. The FPGA technology is significantly changing the way traditional logic design is taught.

HardWire:
A Risk-Free FPGA-to-ASIC Migration Path

Bradly K. Fawcett, Nick Sawyer, and Tony Williams

Xilinx Inc.
San Jose, CA, USA

Abstract. HardWire LCAs are architecturally-equivalent, mask-programmed versions of Xilinx FPGA devices, where the programming elements have been replaced with fixed metal connections. Built-in scan test logic used in conjunction with automatic test vector generation software results in 100% fault coverage. Completed FPGA database files are used to generate HardWire LCA masks and test programs, ensuring compatibility with the corresponding programmable device and minimizing the engineering resources required for the conversion.

Field Programmable Gate Arrays (FPGAs) combine the density and flexibility of mask-programmed gate arrays with the convenience and time-to-market benefits of a user-programmable device. However, the 'overhead' of on-chip programming elements and the circuitry to support them results in FPGA die sizes that are significantly larger than the equivalent-density gate arrays. As a result, FPGA component prices tend to range anywhere from three to ten times the cost of equivalent mask-programmed gate arrays. In high-volume applications with a stable design, FPGA users often consider migrating the design to a gate array as a cost reduction path.

However, such FPGA-to-gate array migrations are not without risk and cost. Converting designs to a different technology will change the timing of signal paths, and the gate array version of the design must be exhaustively simulated. Special features utilized within the FPGA, such as three-state buffers or dedicated carry logic, must be re-designed in the gate array implementation. Test vectors must be created for the gate array, a task often as time-consuming as the original circuit design. In short, the FPGA-to-gate array conversion process can be resource intensive, and exposes the designers to all the risks that were avoided by using an FPGA initially.

These considerations led to the development of the HardWire[TM] Logic Cell Array (LCA[TM]) families. HardWire devices are mask-programmed versions of the popular XC2000, XC3000, and XC4000 FPGAs. In the standard FPGAs, the logic functions and interconnections are determined by configuration data stored in static memory cells. In the HardWire components, the memory cells and the logic they control are replaced by metal connections. All other circuitry in the HardWire devices is identical to the corresponding FPGA's internal circuitry. Thus, a HardWire LCA is a semicustom device manufactured to provide a specific functionality, yet is completely compatible with the FPGA it replaces.

1 HardWire LCA Architecture

The underlying architecture of the HardWire LCA devices is identical to that of their FPGA counterparts, with a matrix of Configurable Logic Blocks (CLBs) surrounded with a perimeter of Input/Output Blocks (IOBs). All other architectural features of the FPGAs, such as global clock buffers, internal three-state buffers, carry logic circuitry, and boundary scan test logic, are also replicated in the HardWire LCAs. The interconnect topology is preserved; however, the programmable interconnect of the FPGAs is replaced by metal connections implemented in a single mask layer.

Unlike the FPGAs, configuration data does not need to be supplied to HardWire parts. However, several 'configuration modes' are available to ensure compatibility in the target system. If the customer chooses the "instant-on" option, the device wakes up functioning like a programmed FPGA. Optionally, a HardWire device can emulate any configuration mode of the corresponding FPGA; this capability allows HardWire LCAs to function in configuration daisy-chains with standard FPGAs; the HardWire LCA is an identical replacement for the programmable device.

2 HardWire LCA Test Architecture

Test programs for HardWire devices are generated automatically with 100% guaranteed fault coverage; customer-generated test vectors are not required. Testing of HardWire devices is facilitated by special on-chip "scan test" logic. Dedicated test latches, called TBLKs, are included in each logic block and I/O block. For example, Figures 1 shows the CLB TBLK locations for the HardWire versions of the XC3000 FPGA family. The placement of these test latches is critical, since signals exiting the CLBs and IOBs can fanout to multiple destinations. The TBLKs are completely transparent to the normal operation of the circuit.

Scan testing allows the contents of all internal flip-flops to be serially shifted off-chip, and for automatically-generated test vectors to be shifted into the device, thus enabling all flip-flops to be initialized to any desired state. The TBLKs are connected

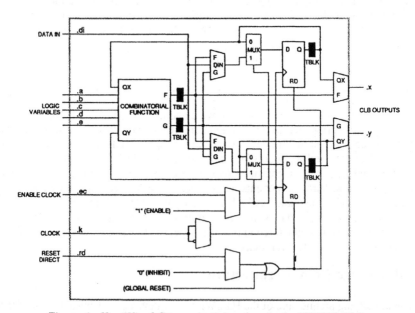

Figure 1. HardWire LCA test latch locations in the XC3000 CLB

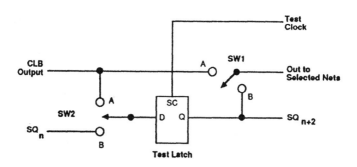

Figure 2. TBLK block diagram

together in a serial chain. The path begins at the Scan In pin, sequences through each CLB, then through each IOB, and exits at the Scan Out pin. Automatically-generated test vectors can be shifted into the device, initializing all internal flip-flops to known states. Similarly, the contents of all internal flip-flops can be shifted out of the device.

The internal architecture of a TBLK is shown in Figure 2. In normal operation, switch SW1 is in the A position and all test latches are bypassed. The device is placed in Test Mode (SW1 is in position B) when unique conditions are present on several configuration pins and a "password" is serially shifted into the device. Depending on the position of switch SW2, a test latch can receive data from either an IOB or CLB output or the previous TBLK in the chain. Synchronized by a special test clock, the latches operate in two phases. The first phase serially loads all test latches to place a specified vector at the inputs to all blocks to be tested (SW2 = B). Next, all latches are stored in parallel with the expected output data from the CLBs and IOBs (SW2 = A). Then, phase one is repeated, serially clocking out the results while simultaneously loading the next test vector.

3 The Conversion Process

The HardWire device is manufactured using the information from the FPGA design file, ensuring compatibility with the programmable device. Customers provide the routed and verified LCA file from the completed programmable design, Xilinx engineers perform a semi-automated design rule check, test vectors are generated using an ATGP (automatic test generation program), and the custom mask layer is created. 100% test coverage is guaranteed. Prototypes are fabricated using the same manufacturing lines as production devices, and can be supplied in four weeks time.

In contrast, converting an FPGA design to a conventional gate array involves converting the netlist to the gate array vendor's format, possible design changes to add test logic or insure pin compatibility, a design rule check, functional simulation, placement and routing, back-annotation of timing parameters, full timing simulation, test vector generation, and the creation of two to four mask layers.

HardWire devices typically cost 50% to 70% less than their FPGA counterparts. The combination of FPGA and HardWire devices offer a fast and easy way to get electronic systems to market, while ensuring a subsequent low-risk, high-volume cost reduction path.

RECONFIGURABLE HARDWARE FROM PROGRAMMABLE LOGIC DEVICES

Nigel Toon
European Marketing Manager, Altera Corporation

Reconfigurable hardware is an emerging technology that utilises SRAM based field programmable logic devices to implement functions in hardware to accelerate processing functions. Using the reconfigurable aspect of the programmable logic device this hardware function can then be changed allowing alternative functions to be implemented. This article will describe a hardware platform - the Altera Re configurable Interconnect Peripheral Processor (RIPP 10) that is now available to support research in this area.

1. INTRODUCTION

Imagine the leading edge personal computer that you would be able to buy in five years. High resolution graphics with 24bit 'True' colour for video reproduction; real-time image compression /decompression; voice recognition; handwriting recognition and with sufficient processing power to perform complex processing functions such as real-time video manipulation; high speed database transaction enquiries; or advanced scientific functions. It is possible to consider that some types of processing functions currently performed by super computers could be performed on your desktop or notebook computer. The question is how?

2. HARDWARE CO-PROCESSORS

The concept of off loading specific processing task's to peripheral devices is commonly understood. A graphics co-processor can perform Bit-Blt functions for windowing graphic environments faster then the central processor; or a maths co-processor can perform arithmetic functions. This concept works by having specific hardware functions masked into custom silicon devices. These devices manipulate the data and perform an algorithm in hardware that would otherwise be performed by the processor executing a software program. In this way the data can be manipulated and processed in nano-seconds as opposed to milli-seconds or longer for the processor executing a software program.

2.1 RECONFIGURABLE HARDWARE
Now consider instead of having a custom silicon device, that performs a fixed function to accelerate a processing task, that you have a Re configurable Hardware resource that can be programmed to perform a particular algorithm and then at some later point be reprogrammed to perform another function. A software program could be partitioned into

some functions that would be processed by the central processor and into functions that would be performed in hardware and as a result could either be many times more complex or could be performed many times more quickly. In a microprocessor there is an Arithmetic Logic Unit (ALU) which is 'programmed' by the instruction set to perform a simple task i.e. add two values, perform a bit shift, compare two values etc. In the same way you can consider having a much larger area of re configurable hardware that can be 'programmed' to perform a task but where the task is a graphics rending function, or an MPEG image decompression, or voice recognition, or analysis of meteorological data.

3. IN-CIRCUIT RECONFIGURABLE LOGIC DEVICES

The concept of Re configurable Hardware has only become realisable with the emergence of high density In-Circuit Re configurable (ICR) Logic devices such as the SRAM based Altera FLEX devices. These devices use static RAM memory cells to store logic functions and interconnect connectivity information. The configuration data is typically held in an associated serial EPROM from which the device re-configures itself at power-up. Alternatively the configuration data can be held in any non-volatile memory source and can be down-loaded to the device. In this way different configuration files can be down-loaded to the device dynamically changing the function of the device. In mainstream applications , programmable device designers have held back from using the re configurable aspect and typically the device is configured once with a fixed function. However some people have recognised how this dynamically re configurable property could be used to investigate the concepts of re configurable hardware.

4. RECONFIGURABLE PROCESSORS

With-in the academic community and at some leading edge R&D centres a number of research projects are being undertaken to investigate and utilise this re configurable hardware concept. These projects tend to be focused on limited-purpose re configurable processing applications that are focused on specific compute intensive functions such as DNA pattern matching, seismic data analysis, database manipulation , and simulation projects. These re configurable hardware processing based solutions are easily outperforming super computers on the same tasks. These promising results have spawned other investigations and the whole concept of re configurable hardware seems on the verge of rapid growth.

4.1 ALTERA RIPP 10 BOARD
To support these projects Altera Corporation has developed a PC Compatible ISA bus add-in card called the Re configurable Interconnect Peripheral Processor (RIPP 10) board. This board provides 100,000 usable gates of re configurable programmable logic, populated with Altera FLEX8000 programmable logic devices and I-Cube Programmable Interconnect devices. Onto the board it is also possible to add SRAM devices allowing complete processing functions to be implemented. This board is being used by a number of universities and research companies to investigate Re configurable Processing applications.

4.2 ALTERA RIPP 10 APPLICATIONS

Currently Ceram Inc,USA a computer acceleration company together with a computer aided natural resources engineering company is utilising the RIPP10 board in the analysis of Seismic data. Today analysis is constrained by processing resources, and re configurable hardware offers the opportunity to accelerate certain tasks. The project begins with an analysis of the software functions to identify the processing bottlenecks. The functions that significantly slow the overall processing task are mapped into the programmable logic off-loading the central processor from the task. These functions include post-stack seismic processing, mapping functions and database server functions for well and seismic access. Computer architecture simulation is another area with a number of projects using re configurable hardware to simulate advanced computer structures this work includes investigations into stochastic processing (see insert) and also work being carried out at the University of South Carolina on simulating new processing structures. The major problem that needs to be overcome is the issue of implementing algorithms in the programmable logic. It will be necessary to develop compilers that can take a high level software description and partition the code into target code for the microprocessor and into configuration data for the programmable devices so that they are able to implement the correct hardware algorithms.

5. SUMMARY

As the compiler problems are worked on and solved and specialised programmable logic architecture's evolve targeted at these Re configurable Processing applications it is possible to envisage a PCMCIA add-in Re configurable Processing card for your notebook computer that will provide a general purpose acceleration for a wide range of processing functions - enabling your notebook computer to perform processing tasks faster then a super computer. The technologies that will evolve have the potential to accelerate the processing capability of a desktop or notebook PC into new application areas and to add functionality that would otherwise not be possible .

On some Limits of XILINX Based Control Logic Implementations

Attila Katona and Péter Szolgay*
katona@miat0.vein.hu , szolgay@mars.sztaki.hu

Department of Information Technology and Automation, University of Veszprém
*Computer and Automation Institute of HAS, Hungary

Abstract. In this paper we gave some methods how the complexity of a design description can be quantified and what is the largest complexity that can be implemented on a given type of XILINX chip using the standard XACT design system. These methods can be used to partition a large design task, given by either a circuit schematic or an algorithm, to smaller ones which can be implemented in FPGA chips.

1.Introduction

In digital systems built up from VLSI parts the component list follows the traditional computer architecture: Processor - Memory - Control unit. A typical design task uses standard processor and memory chips/blocks while the control logic part is specific to a given problem. Using PLDs (Programmable Logical Devices) or FPGAs (Field Programmable Gate Arrays) it is possible to integrate all the control logic functions in a few chips [4]. There is a wide design-software support to provide continuous help to the designer in the whole process. As an example, the XILINX FPGA chip and the XACT design package are considered here. We identify two basic problems concerning design methodology [5]:

(i) partitioning a large design description, given by either a circuit schematic or an algorithm on a hardware description language, to smaller ones which can be implemented in FPGA chips;

(ii) how the complexity of a design description can be quantified and further what is the largest complexity that can be implemented on a given type of XILINX chip using the standard XACT design support.

Here we are going to present an approach to the second problem which, of course, may help to solve the first one, as well.

2.   On the Limits of Design Methods

2.1.   The Limits Originating from the Structure

A design task may be given by a TTL level circuit diagram in which case it is automatically converted to logic-block-level circuit diagram. In the XC3020PC68 chip there are 58 user programmable I/O blocks and an array of 8x8=64 CLBs (Configurable Logic Blocks). Each CLB in LCA (Logic Cell Array) has five combinatorial logic variable inputs and four other inputs: clock, enable clock , direct data in, and asynchronous reset. Each CLB has two outputs [2]. Table 1. shows the routing resources of a $k \times l$ array of logic blocks inside the structure. A $k \times l$ array has

k rows and l columns of CLBs. The following constraints were considered:
(i) a synchronous network design is assumed using the global clock net.
(ii) the interconnections of a $k \times l$ size CLB array are realized through the border of the array.
(iii) let us suppose that only the border problem can cause routing failure that means we do not examine now the routability inside and outside the $k \times l$ block. We suppose that there is no routing problem anywhere else.
(iv) synchronous I/O blocks are supposed.

Proposition 1. The 8x8 chip is routable under the above constraints if there are less than 144 nets in the design (nets are the interconnections which connect CLB outputs to CLB inputs).

k,l	the number of possible connections (see figure 1.) 9*k+9*l	the number of inputs and outputs 10*k*l	
1	18	10	
2	36	40	
3	54	90	
4	72	160	
5	90	250	
6	108	360	
7	126	490	
8	144	640	

Table 1.

Figure 1.

The figure shows the routing resources around a CLB. Two horizontal long lines, three vertical long lines, four direct connections and five lines are running into each direction through the switching matrices. The proof of the proposition can be derived from the chip layout.

2.1.1. Complexity Limits Derived from the Circuit Schematic to Implement

There are some additional design requirements effecting the final results. The prescription of the bounding pads and the speed of the circuit are the most critical. There are certain limitations coming from the size of the used FPGA chip, namely the number of I/O pads, and flip-flops. These values of the circuit diagram can be obtained easily and may be compared to the target FPGA chip. There are not so simply computable limits characterising the complexity of a design task:
(i) wire density, the number of nets in the schematic
(ii) assigned cells (CLBs)/ total cells
(iii) the number of combinatorial inputs
(iv) The maximal wire density along the one-dimensional layout model of a task can be a measure of complexity.[1]

2.1.2. Complexity Limits Derived from the Algorithm to Implement

The other possibility to specify a design task is the algorithmic level description - a high level description. For hardware specification the VHDL is a de facto standard. The VHDL description can be transformed into the .XNF type internal file of the

XACT system. It is allowed to come from other PLD design systems as CUPL or ABEL. In all these systems the tasks are described by an algorithm. Based on the Halstead method in [3] a complexity measure, was calculated from the number of distinct operators and operandi and from the total number of operators and oprandi.

2.2. Design Experiences

We have solved some examples with the XACT system. Certain designs were not possible to be realized on a single chip because of the limited number of CLBs. The number of CLBs is one of the most important constraints. We can easily count the number of flip-flops and I/O pads but not the number of CLBs in the schematic. In some examples both the algorithmic and the circuit schematic level description were given. The algorithmic level and circuit level complexity measures were composed for these examples. A close correlation were found between the two complexity measures. Table 2. shows the routing limits for an artificially generated test design family in which all the CLBs were used. The basic building block of the test schematics were similar to the LCA architecture because here we wanted to study the effect of routing to the complexity. After the automatic translation all the 64 CLBs were used. In the Nets column the number of the nets in the schematic is given. We did not count the combinatorial output to flip-flop input nets because their connection was made within the CLBs. When 223 nets were in the design then there was no unrouted pin. The number of unrouted pins can be different even in the case of the same design due to the fact that XACT design process starts from a random placement.

Name	IOBs	Nets	Unrouted pins
CLBSN	0	223	0
CLBSO I	25	248	1
CLBSO II	25	248	9
CLBSP I	24	247	11
CLBSP II	24	247	8
CLBSQ	20	243	7
CLBSR	10	233	2
CLBSS	0	223	0

Table 2. Unrouted pins versus nets

3. Conclusion

Based on our experiences there is not a single parameter which can be selected to describe the complexity of a task given by a circuit schematic, but a multidimensional parameter space is required. Nevertheless we identified the wire density as the most important parameter [5].

References

[1] P.Szolgay," On algorithms in the parallel design of logic and layout of circuits with functional blocks", Int. J. of Circuit Theory and Applications, Vol.20, pp.411-429, (1992.)

[2] The Programmable Gate Array Data Book 1992, Xilinx, San Jose, California

[3] M.H.Halstead, Elements of Software Science, Elsevier North-Holland, Amsterdam, 1977.

[4] Special Section on Field Programmable Gate Arrays (ed. A. El Gammal), Proc. of the IEEE Vol. 81, No. 7, July 1993 pp. 1011-1083

[5] A.Katona and P.Szolgay, "Limits of XILINX based control logic implementations", IT-2-1993, University of Veszprém

Experiences of using XBLOX for Implementing a Digital Filter Algorithm

Gerhard R. Cadek[1], Peter C. Thorwartl[1], and Georg P. Westphal[2]

[1] Vienna University of Technology, IAEE, CAD-Division, A-1040 Vienna, AUSTRIA
[2] Atomic Institute of the Austrian Universities, A-1020 Vienna, AUSTRIA

Abstract. A preloaded digital filter algorithm for filtering the output signal of a Germanium gamma ray detector was implemented on a XC4000 device using the XBLOX software from Xilinx. A system frequency of 20 MHz was achieved due to multiple pipeline stages and extensive usage of the high speed carry paths. This paper describes author's experiences gained using XBLOX intensively. The strong and weak points of the tool will be discussed.

1 The Application

For high-rate gamma spectrometry using a pre-loaded filter which automatically adapts its noise filtering time to the actual pulse interval is a proved method. Up to now this pre-loaded filtering of the charge pulse signal at the output of the detector has been accomplished using conventional low noise analog circuitry [1]. A new concept was developed using a 20 Msamples/s 12 bit analog to digital converter in companion with a high speed customised digital signal processor. The resulting circuitry offers a better noise rejection ratio and drift performance than the analog circuitry. The digital filter consists of four blocks. The input

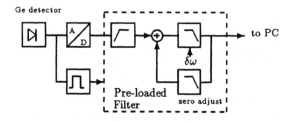

Fig. 1. Block diagram of the pre-loaded filter

unit differentiates the digital input from the ADC. After performing integration, low-pass filtering with variable cut-off frequency has to be done. The results are stored into an on-chip 32 word deep, 16 bit wide FIFO. A standard PC data acquisition card reads out the FIFO for further processing. Additional circuitry inhibits drifting due to noise by a loop-back correction. First prototypes had to be available within one month. The digital circuitry had to accomplish:

1. 20 MHz system clock
2. 12, 16, and 24 bit integer arithmetic
3. two's complement notation
4. 32x16 bit on-chip FIFO
5. about 250 flip flops

2 Device and Tool Selection

Following our experience we decided to use a device from the Xilinx XC4000 family. This was due to the on-chip RAM capability of the XC4000 series and due to our practical experience in implementing high speed pipelined arithmetic circuitry into these devices. We decided to use the XBLOX software to spare us boring gate-level data-path design. Design entry and the simulation task were done using Viewlogic's Viewdraw and Viewsim, respectively, on a PC. The XBLOX processing-, placement-, and routing-task was performed using the Xilinx software on a SUN SPARC10.

3 The Implementation

Design entry using XBLOX is very fast since you need not to worry about the implementation of a register, counter or adder. Some special adaptations of the circuit were made to meet the XC4000 architecture. The nine stage 12 bit wide shift register was implemented using a 9x12 bit on-chip SRAM instead of using conventional registers. This solution requires only 8,5 CLBs compared to 108 for the conventional one.

The first tool problems arose when we tried to connect busses of different widths. Since there is no XBLOX element available for this purpose, we had to convert the busses into Viewlogic ones which could easily be processed. The XBLOX bus conversion elements do not support the negative bus notation which was used for the right hand bits of the comma (fractional notation).

Although the XBLOX library offers a big number of data path elements, there is no element for defining a barrel shifter. Since we used two's complement notation our barrel shifter needed to perform sign extension too. The problem was solved by converting the signals to Viewlogic ones where the definition of the 24 bit 0-7 digit shift-right barrel shifter required only one symbol. Despite these difficulties the design was defined within three days. The backannotation of the Xilinx netlist after inserting pre-layout timing information into the Viewsim netlist format was tedious and took about after a quarter of an hour. Due to a bug of the Xilinx to Workview interface *xnf2wir* you need to copy your design from the network drive to a local disk because a file is opened twice.

During XBLOX processing another major bug was detected. If a hard macro like an adder is directly followed by a register stage, these registers will be implemented in the same CLBs. But if negative bus notation is used, the generated hard macro uses the register output FFX twice within a CLB for these negative bits instead of using FFX and FFY outputs, respectively. This bug leads to a fatal error during a PPR-run. So far this is only correctable by editing the *.hm-file after analysing the XBLOX output.

Automatic global clock insertion is another weak point of the XBLOX software. If the clock signal is fed only into one combinatorial input - this may be only an inverter - XBLOX does not use a global clock buffer automatically, which leads to enormous routing problems and a poor timing performance. Thus the global clock must be handled manually by using the appropriate symbols.

Since many net names have to be addressed with their hierarchical name, labelling all instances has been proved to be wise for simulation purposes. There is another simulation problem if an input signal is used both directly and registered. Stimuli applied using the name of the signal will only drive the direct, non-registered input since the register will be moved into the IOB. The registered input has to be fed directly from the pad signal which cannot be labelled in XBLOX since this signal is not available. You need to browse the *.lca file for the net name or you have to use conventional IBUF and IPAD symbols.

Due to the feed-throughs automatically inserted by PPR on heavily loaded signals, net names are changed so that they are not visible for post-layout simulation. So you again have to analyse the *.lca- or *.xnf-files after running PPR to get the appropriate information. This fact leads to the problem that the pre-layout simulation stimuli cannot be applied to the post-layout simulation without changes introducing possible design inconsistencies.

The design fits pretty nicely into a XC4006-PG156-5. The final routing lasted about half an hour on a SUN SPARC10. It took four weeks for designing the FPGA including tool installation, design corrections and bug fixing.

4 Conclusio

XBLOX has been proved to speed up the design definition but it does not relieve the designer of thinking about utilising special device structures. Bus manipulation is a tedious job. The usage of clock distribution circuits is not well supported until now. Since XBLOX parts use the fast carry path automatically, the resulting circuits are very fast. Some problems were detected concerning applying pre-layout stimuli in post-layout simulation due to changed net names.

Besides testing a new tool by means of a real life design instead of a benchmark circuitry, FPGAs have been proved to meet both the needs of advanced high speed signal processing and fast design cycles. Our further work will concentrate on the acquisition time control unit around the customised signal processor, which is now realised using analogue circuitry.

References

1. Westphal G.P.: A Preloaded Filter for high-rate, high-resolution gamma spectrometry. Nuclear Instruments and Methods in Physics Research, A299 (1990), pp. 261–267, 1990.

CONTINUOUS INTERCONNECT PROVIDES SOLUTION TO DENSITY / PERFORMANCE TRADE-OFF IN PROGRAMMABLE LOGIC

Nigel Toon

European Marketing Manager, Altera Corporation

This article will discuss the trade-off between segmented and continous interconnect in programmable logic devices and the effect on performance and ease of use in logic design. The article will describe the effects on interconnect delay paths of both styles of programmable interconnect and will show how these affect overal performance.

1. INTRODUCTION

The major challenge in high density programmable logic devices is providing high density logic together with high performance. One of the most significant differences between Mask Gate-Array and a programmable logic device is that in a gate-array the logic elements are connected by metal connections created as part of the customer specific manufacturing process, whereas in programmable logic devices the interconnect between the logic elements must be implemented with a user programmable connection. This article will discuss the two main forms of interconnect used in programmable logic devices Continuous Interconnect and Segmented Interconnect .

2.SEGMENTED INTERCONNECT.

 One concept used to implement programmable interconnect in high density programmable logic devices is to use short segments of metal lines that are interconnected by a programmable switch matrix which enables these short segments to be combined to create longer routing paths. This type of interconnect structure derives from the channel routed gate-array structures and has benefits in that these short segments can be combined in a wide variety of combinations making effective use of the metal lines available. As interconnect becomes used up in a particular path alternative routes can be found.

2.1 CUMULATIVE DELAYS.
Each switch matrix that a signal passes through in a segmented interconnect structure adds impedance to the path. As segments are joined together this loading builds-up and as a result interconnect paths will have different delays dependant on how many switch matrix elements the signal has passed through . In addition signal fanout will

also have an impact on the delay. Not only is an additional load impedance introduced by each signal destination but also additional load is introduced by the additional switch matrix elements that must be passed through to reach each signal end point.

3.CONTINUOUS INTERCONNECT.

In a simple PLD device the output from every logic element or macrocell is connected directly to the input of every other logic element or macrocell. In addition each signal is provided as both a true and a compliment - doubling the number of macrocells has the effect of increasing by a factor of four the amount of interconnect required. As a result it would be impractical in high density programmable logic devices to have logic element connected by a continuous line to every other element.

However it is possible to extend this concept of continuos interconnect by using a hierarchical structure where groups of logic elements are combined together into a block and then a continuous interconnect resource can connect signals from one block of logic elements to another. The Altera FLEX 8000 family of high density programmable logic devices use this type of a routing technique for the 3-dimensional 'Fast-Track' interconnect structure.

3.1 ALTERA FLEX 8000.

The FLEX 8000 devices are made up from a fine granularity Logic Element which consists of a four input look up table and a configurable flip-flop. Eight of these logic element are combined together into a Logic Array Block (LAB) which provides interconnect from any Logic Element to any other Logic Element contained with-in the LAB.

The Logic array blocks are connected by a Row and a Column Interconnect resource. These interconnect resources provide for each signal a continuous metal line that runs the complete length of the device in either the horizontal or the vertical. A multiplexing structure on the input to every Logic Array Block selects from the available signals in the Row or Column the signals required in that Logic Array Block. In this way any Logic Element can be connected to any other logic element with-in the device.

3.2 FAST-TRACK INTERCONNECT DELAYS

The continuous metal lines in both the Row and the Column interconnect resource run the complete length of the device. Each metal line is connected to every Logic Array Block through a multiplexer. When the device is configured the signal is either connected or not by the selection that is set on the multiplexer. Irrespective of whether the signal is selected or not into one or all of the Logic Array Blocks the loading on the line remains the same and as a result the delay remains the same irrespective of where the signal routes too or the signal fan-out. In the case of the 3-dimensional 'Fast-Track' interconnect - three predictable delays exist. A 1nSec delays exists for Logic Elements that are connected with-in a Logic Array Block. A 6nSec delay exists for Logic Elements that are connected from one Logic Array Block to another through the Row Interconnect Resource and a 9nSec delay for signals that must pass through both the Row and the Column Interconnect Resource.

In addition to these general purpose interconnect paths , specific connection paths are provided in the FLEX 8000 devices between adjacent Logic Elements. These paths are utilised to implement fast Carry-Chains and Cascade Chains. For Adders and Counters the Carry-Chain path is utilised to provide a dedicated path for the carry signal from one stage of the adder or counter to the next. For complex logic functions that require more then four variables the Cascade path is used to enable multiple Logic Elements to be cascaded together to provide for wider signal fan-in.

The combination of these routing resources enable an 8bit Registered Accumulator to be implemented with a delay of 8nSec and a frequency of 125Mhz, or a 16bit loadable up/down counter with a worst case delay of 13nSecs and a frequency of 75Mhz. A 24bit magnitude comparitor, a common function in video processing for example can be implemented with a worst case delay from data valid to the comparison result of 23nSec. Also a 24bit adder can be implemented with a worst case delay of 22nSec so it would be possible to implement a real-time 24bit image processing application that would be able to operate at approx. 40Mhz.

4. SUMMARY

Continuous interconnect enables high performance applications to be implemented with-in high density programmable logic devices. The predictable delays allow high performance to be achieved with the minimum of design effort. When designing with programmable logic devices with continuous interconnect it is not necessary to worry so much about the placement of the individual Logic Elements with-in the device because the overall performance will not be affected. As a result achieving high performance designs in these types of devices, commonly called Complex Programmable Logic Devices (CPLD's) is a much simpler process then with devices that utilise segmented interconnect such as FPGA's.

A HIGH DENSITY COMPLEX PLD FAMILY OPTIMIZED FOR FLEXIBILITY, PREDICTABILITY and 100% ROUTABILITY

Om P. Agrawal
Director of Strategic Product Planning
Advanced Micro Devices, Inc.
Sunnyvale, CA 94088

Abstract

This paper describes the silicon architecture of AMD's second generation Macro Array CMOS High Speed/High Density (MACH®) family of PLDs. With an advanced 0.65um technology and an innovative architecture, the next generation MACH family offers gate density up to 10,000+ gates with 100% routability, flexibility, and predictable worst-case pin-to-pin delays of 15ns.

Introduction

The MACH 3 & 4 family is AMD's second generation Macro Array CMOS High Performance High density (MACH®) family. AMD's first generation MACH 1 & 2 family, introduced in 1990, set the industry standard for 15ns worst case pin-to-pin delays for devices ranging from 900 to 3,600 gates. The 1st generation MACH family also pioneered the concept of fixed, predictable, deterministic, logic and routing independent signal delays. AMD's next generation MACH family raises the bar of "predictable speed standards" to 10,000 gates with significantly increased flexibility and 100% routability.

Second Generation MACH Family

MACH 3 & 4 family consists of 4 devices. This family begins at 3,500 gates and extends up to 10,000+ gates offering predictable, path-independent worst-case delays of 15ns. These devices offer between 96 and 256 logic cells with 96 to 384 registers, and are available in 84 to 208 pins PLCC and PQFP packages. Designed with proven, advanced 0.65- micron double metal CMOS electrically erasable technology the MACH 3 & 4 family devices are 100% testable and have 100% guaranteed programming and functional yields. The MACH 3 & 4 family also pioneers 5V incircuit programmability with full conformance to IEEE 1149.1 JTAG standard for PQFP packages beyond 84-pins. A single set of dedicated pins are used for both in-circuit programmability and JTAG compatability. Table 1 shows the members of the MACH 3 & 4 family and their capabilities.

Device	MACH 355	MACH435	MACH445	MACH465
Gate Count	3,500	5,000	5,000	10,000
Macro Cells	96	128	128	256
IO Cells	96	64	64	128
Registers	96	192	192	324
Speed	15ns	15ns	15ns	15ns
Package	144-PQFP	84-PLCC	100-PQFP	208-PQFP

Table 1 MACH 3xx/4xx Family

Like the first generation MACH 1xx/2xx family members, the MACH 3xx/4xx family consists of "multiple, optimized programmable logic blocks interconnected by high speed switch matrices." Retaining the fixed, predictable characteristics of the first generation MACH architecture, the 2nd generation MACH 3 & 4 family focuses on significantly increasing density, flexibility, routability and programmable connectivity without compromising speed. Major innovations in next generation MACH family include: multi-tiered, high speed switch matrices to provide 100% routability; and significant architectural enhancements to PAL blocks and macrocells to provide density, flexibility; and predictable speed.

MACH 3 & 4 Family Flexible Programmable Logic Block

Each programmable logic block of the MACH 3 & 4 family consists of: flexible Clock Generator, an enhanced AND-OR-XOR array, a more flexible logic allocator, and an array of logic and IO macrocells. With 16 macrocells each logic block is designed to handle wide gating functions up to 33-34 inputs. This makes it ideal for emerging 32-bit microprocessors bus interfaces and address decoding applications.

Each Logic Block has its own clock generator that can provide up to 4 different global, synchronous pin clocks for each block with programmable polarity. Accessibility to four different clock sources with programmable polarity helps to implement complex state machines inside a block. Each logic block contains a 33-34 inputs (true and complements) x 90 AND-OR-XOR product term array that form the basis of all logic implementation in a logic block. The product term array for each logic block consists of logic product terms and control product terms. Logic product terms are grouped in clusters. For control functions, product terms are not clustered. For MACH 3 & 4 family, each logic macrocell receives an average of 5 PT clusters from PT array. In addition, the logic allocator distributes up to 20 PTs of logic per macrocell in a more flexible fashion, with no speed penalty. The PT array also provides a separate output enable PT term for each IO macrocell. In addition, flexible asynchronous Reset product term and a separate asynchronous Preset product term are provided for all logic macrocells initialization within a logic block.

Flexible Synchronous/Asynchronous Logic Macrocells with XOR Functions

Macrocell enhancements for the MACH 3 & 4 family include more intelligent PT clustering, more PTs/macrocell, synchronous/asynchronous mode of operation, flexible clocking with pin or PT clocks with programmable polarity, flexible Reset/Preset swapping and built-in XOR capability.

Each logic macrocell provides a AND-OR array based sum-of-products with flexible XOR capability. Each logic cell can provide base logic capability up to 5 PTs. The 5 PT logic for each logic cell consists of two PTs clusters - one consisting of a 4 PT cluster and the other consisting of a single PT. The single PT is used as either logic PT for the OR gate or a logic input for the XOR. When used as a logic PT, the 5th PT can be steered to the 5-input OR-gate. In that situation it is not available as PT for the XOR gate. It can also be used as single PT controlling the XOR gate. The ability to use a single PT comes in handy for address decoder application - where single PT can be used and the 4 PT cluster can still be made available to adjacent macrocells.

Each logic cell has access to the logic clusters of its three adjacent neighbors: one from above and two from below, via the logic allocator. With accessibility to 3 adjacent macrocells, each cell can provide logic capability up to 20 PTs. *An unique strength of the MACH 3 & 4 family architecture is that no additional speed penalty is imposed for this additional logic flexibility.*

The storage element inside the logic cell can be individually programmed to operate in either a transparent flow-through latch or as an edge-triggered D or T-type register. Flow-through latching provides minimum input to output delays for speed critical functions such as chip-select decoding, while edge-triggering guarantees glitch-free outputs for applications needing synchronous counters and state machines. The built-in XOR gate in front of the macrocell, besides providing the polarity control of the signals going into the macrocell, can also be used for complex XOR arithmetic logic functions (comparators, adders etc) and for De Morgan's inversion for reducing the number of product terms (for logic optimization and synthesis). An important innovation of the MACH 3 & 4 family macrocell is its ability to support both synchronous and asynchronous logic capability on an individual macrocell basis, in the same macrocell with no speed penalty.

In synchronous mode, all logic macrocells are initialized together with the common asynchronous Reset or Preset product terms. However, each macrocell has the capability to swap the Set/Reset function on an individual macrocell basis. In the asynchronous mode, each macrocell can receive its own independent product term clock, plus independent set/reset PT. Further, each macrocell has the ability to swap the Set/Reset function, on an individual macrocell basis in both the synchronous and asynchronous modes. In synchronous mode, each macrocell selects its clock from 4 logic block clocks. In asynchronous mode, the clocking is more flexible. Each cell receives two logic block pin clocks (generated from its block clock generator), one individual PT clock (generated from its array inputs) and one the inverted PT clock. This provides an individual and separate PT clock with programmable polarity for each macrocell.

Flexible macrocell structures and intelligent PT allocation capability of the logic allocator with no additional speed penalty, is a key feature differentiating the next generation MACH family from other complex EPLDs. The architectural flexibility of the logic allocator and the flexible macrocell structure, combined with each blocks ability to handle up to 34 inputs and better than 2:1 input to output ratio allows each programmable logic block of the MACH 3 & 4 family architecture to pack quite a bit of synchronous and asynchronous logic in a single block.

The I/O macrocell for the MACH 3 & 4 family architecture consists of a three-state output buffer, control for the three-state buffer and an input macrocell. The three-state buffer can be configured in one of three ways: always enabled, disabled, or controlled individually by a separate PT. This gives designers the flexibility of configuring the I/O macrocell as an output, an input or bidirectional pin, or a three-state output for driving a bus. The input to the three-state buffer comes from the output switch matrix.

100% Routability

Another significant innovation of the next generation MACH family is its multi-tiered switch matrix structure. The multi-tiered switch matrices structures are designed to provide 100% routability with fixed, predictable delays. The second generation MACH architecture consists of three types of switch matrices: Input Switch Matrix (ISM), Central Switch matrix (CSM) and Output Switch Matrix (OSM).

With multitiered switch matrices, the second generation MACH offers a "true programmable connectivity" between logic block and IO pins. The multi-tiered switch matrix structure completely decouples the internal logic block from external physical IO pins and all internal feedbacks; and provides a uniform way for treating all signals with significantly increased routability. It significantly addresses the major concern of design changes effect on old-pinouts.

Each switch matrix structure has been optimized for speed, cost and flexibility. The ISM decouples the IO-pins feedbacks and logic block feedback signals from the internal logic block and provides multiple chances of signal entry to the global switch matrix. The CSM acts as the main signal routing structure to provide optimized global connectivity and is key for achieving fixed, predictable, deterministic, and path independent delays for the device. The OSM decouples the logic block from its IO pins and its own inputs and is key for addressing the concerns of design changes while retaining prior pinouts.

Software Support

AMD's MACHXL™ design development system fully exploits the density and flexibility of MACH 3xx/4xx family architecture and provides a low-cost software environment. The MACHXL development system includes PALASM® compatability, Boolean equations, State machines and High Level languages design entry. It also includes an automated logic compiler, logic synthesis package, automatic device partitioner, placer and router, functional unit delay simulator, JEDEC generator and a report generator. This low-cost design system provides designers with high level design capability for fast, hands-off, automatic routability plus fine control for fine-tuning the device. AMD's Besides AMD's MACHSXL design environment, the MACH 3xx/4xx family is supported by broad third-party PLD tools vendors such as DATA I/O, MINC, OrCAD, Logical Devices and ISDATA. Support in the CAE environments such as CADENCE, MENTOR, VIEWLOGIC and SYNOPSYS is made available via existing agreements between third-party PLD tools and CAE vendors. Programming support is made available by third-party vendors such as DATA I/O, Logical Devices, BP Microsystems, SMS etc.

Summary

The major strength of the next generation MACH 3 & 4 family are its architectural simplicity - multiple PAL-like AND-OR-XOR blocks interconnected by multi-tiered switch matrices with fixed, predictable, deterministic delays. A significant breakthrough for the MACH 3 & 4 family architecture is its combination of flexibility, programmable connectivity, and deterministic speed. The internal architecture of the MACH 3xx/4xx devices provides flexible global connectivity,with TRUE predictable speed. Delays are neither path dependent, placement dependent, fan-out dependent, nor logic and routing dependent. Logic block structures, logic macrocell structures, and multi-tiered switch matrices have been all optimized to give users a better optimal balance of speed, density, flexibility and programmable connectivity. Every macrocell communicates with every other macro cell along the same path and with the same, fixed predictable deterministic delays.With its simple, symmetrical, and optimal structure the next generation MACH 3xx/4xx family is structured to raise the bar of "predictable speed standards" to 10,000 gates.

Design Experience with Fine-grained FPGAs

P. Lysaght, D. McConnell and H. Dick

Department of Electronic and Electrical Engineering,
University of Strathclyde,
Glasgow G1 1XW

Abstract. The performance of fine-grained, cellular FPGAs is improving rapidly. In this paper, the experience of working with two relatively fine-grained FPGA architectures, the Atmel 6005 FPGA and the Dynamically Programmable Logic Device (DPLD) from Pilkington Micro-electronics Ltd, is described.

1. Introduction

This paper reports on the authors' experiences in designing with the Atmel AT6005 [1] and the Pilkington Micro-electronics Ltd. (PMeL) 3k6 Dynamically Programmable Logic Device DPLD [2]. The work began with a number of different logic designs implemented on the Atmel FPGA by undergraduate and postgraduate students at the University of Strathclyde. The subsequent porting of a representative sample of three of these designs to the DPLD architecture forms the basis of the paper. Aspects of the device architectures, CAD tools and their impact on the relative performance of the designs are considered.

2. CAD Tools and Design Porting

The authors have used a pre-release version of the CAD suite that PMeL is intending to release commercially. The software is PC based and consists essentially of physical design tools, i.e. placement and routing programs. Designs are imported via an EDIF interface and programs for device configuration and programming are also provided. At present, the only schematic and simulation libraries that are provided are for use with the Viewlogic CAD software. Post-layout delay information for back annotation is provided for the Viewlogic Workview simulator. The software also features a comprehensive on-line help facility.

The Atmel CAD software is also PC based and the components of its tool set are practically identical to those offered by the PMeL software. The main differences are that the Atmel software can exchange files directly with the Workview software without using EDIF and that a fully interactive, manual design editor is supplied.

The mechanics of design porting were quite straightforward. The Workview schematics of the Atmel designs were converted to DPLD designs by manually replacing all references to Atmel library primitives with the equivalent PMeL library primitives. The designs were then re-exported and the PMeL design tools were used to automatically place and route them. The success of this approach relied heavily on the availability of equivalent or replacement design primitives in the target library. No special, architecturally specific design primitives were used in either design. This point is developed further in the next section.

3. Architectural Issues

The cell structure of the AT6005 offers an unusually large number of permutations of logic and routing functions: for 13 of the 44 logic functions available in a single cell there are multiple configurations for the same function. In the manufacturer's literature, the logic functions are referred to as logical primitives while the associated cell configurations are referred to as physical primitives. The logical primitive representing the simple inverter, for example, has six different physical primitives associated with it. Figure 1 shows two of the more complex Atmel logical primitives. During synthesis the user may wish to use such cell structures to realise smaller and hence faster designs. At present, designers are obliged to recognise the opportunity for deploying the more complex logic primitives within their circuitry and then to enter them manually into their schematics.

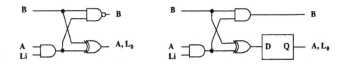

Fig. 1. Complex Logic Primitives

An analysis of the macro libraries provided an unexpected result. Neither of the complex logical primitives shown in Figure 1 are used in the construction of more complex macros in the system libraries. Furthermore, the same observation is true of approximately a quarter of the total of 44 logical primitives. One reason for the lack of use of the more complex logical primitives is that while they combine several logic functions into one cell, they can be extremely difficult to route efficiently. This restriction extends to some of the simpler logical primitives also.

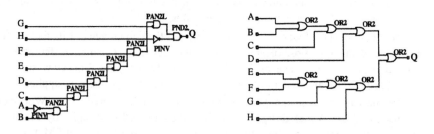

Fig. 2. Implementations of AT6005 and DPLD Wide–input OR gates

Consider the example of the logical primitive for the 2-input OR gate. It is constructed by inversion of the inputs and outputs of a two-input AND gate (as is the case in the DPLD) but is so difficult to route that it is rarely used. Figure 2 shows the construction of the 8-input OR gates in the Atmel and DPLD libraries. The Atmel macros do not use the 2-input OR primitive and are highly asymmetric with respect to one another. The 8-input OR gate requires nine cells to implement and contains eight levels of logic. In contrast, the symmetrical DPLD implementation of the same function uses seven gates with four levels of logic.

4. Designs

Three in-house designs have been ported from the Atmel AT6005 to the PMeL DPLD. They are a one-hot encoded finite state machine, a content addressable memory (CAM) and a hardware implementation of a queue model. The finite state machine (DESIGN_1) had ten states and represents the control sequence for a "walkman" portable audio player (without recording facilities). It was the smallest of the three designs and also the least regular circuit. It consisted of 101 nets, 10 flip-flops, 112 gates and had an estimated equivalent gate count of 218.5 gates.

The second design was a 4 by 4 bit content addressable memory (DESIGN_2). It was a highly regular design since each CAM cell is identical and was predominantly combinatorial in nature. It consisted of 206 nets, 8 flip-flops, 234 gates and had an estimated equivalent gate count of 328.5 gates. The final design to be described was a hardware implementation of a M/M/1 queue model (DESIGN_3). This design was comprised of two 32-bit, linear feedback shift registers (LFSRs) and two 16-bit, probabilistic bit-stream modulators which were very register-intensive. It also included an 8-bit up/down counter, a 24-bit ripple carry counter, three small finite state machines (encoded using one-hot techniques), an 8-bit data bus and some address decoding logic. It consisted of 312 nets, 147 flip-flops, 260 gates and had an estimated equivalent gate count of 1558.5 gates. The equivalent gate counts for the three designs are derived from the Atmel CAD tool reports.

5. Design Performance

The three designs were automatically placed and routed using both the Atmel and PMeL physical design tools as shown in Figure 3. The designs laid out on the Atmel FPGA occupy significantly more area that their counterparts on the DPLD array, even when the slightly larger zones of the DPLD are taken into account. This is partially the result of the autoplacement algorithm. It would appear that the Atmel algorithm adopts a strategy of dispersed component placement to make the task of the autorouter easier. In general, it was better to avoid using the autoplace software and to rely on manual placement. The designs were manually laid out on the Atmel array as shown in Fig. 4. An immediate improvement in the area utilisation is apparent for all three designs. In the case of the queue (DESIGN_3), the manually placed circuit did not successfully complete autorouting so it was manually routed. In the interests of comparison, DESIGN_1 and DESIGN_2 were also manually routed.

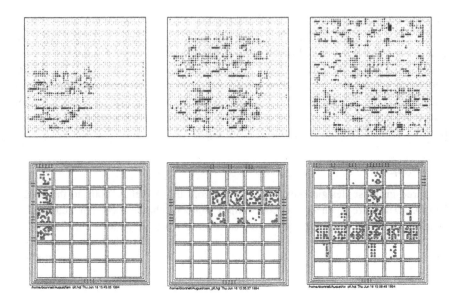

Fig. 3. APR of the FSM, CAM and Queue for the AT6005 and the DPLD 3k6

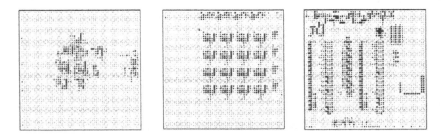

Fig. 4. Manually Placed FSM, CAM and Queue for the AT6005

For the CAM and the Queue, the autoroute software performed better than the human designer as is indicated by the design summary statistics in Table 1. In the case of each of the three designs the table indicates the resource utilisation first for the automatically placed and routed design (APR), then for the manually placed and automatically routed design (MPAR) and finally for the fully manual placed and routed design (MPMR). Even after manual placement and optimum routing, the DPLD designs are considerably more compact than the Atmel ones. Since greater area equates to increased delay in FPGAs, the designs are also substantially faster, by as much as a factor of four for the FSM and Queue designs. These figures were obtained from the Workview simulator for the Atmel designs and from the physical design tools' summaries in the case of the DPLD. The DPLD design tools on the PC do not support manual layout or interconnection, so no comparison is possible between the manual and automatic placement and routing.

AT6005	DESIGN_1: FSM			DESIGN_2: CAM			DESIGN_3: Queue		
	APR	MPAR	MPMR	APR	MPAR	MPMR	APR	MPAR	MPMR
No. of Turns	102	100	67	150	58	56	172	151	162
No. of Buses	337	267	265	765	327	355	613	701	628
Local Buses	305	242	230	628	314	286	549	555	576
Express Buses	32	25	35	137	13	86	64	146	52
No. of Cells	319	252	318	854	587	605	865	832	971

Table 1. Summary of Layout Statistics for AT6005

Neither set of summary statistics produced by the design tools report directly the number of cells used in through-cell routing. DESIGN _1 required 196 AT6005 cells and 115 DPLD cells for through routing, DESIGN_2 required 345 AT6005 and 278 DPLD cells and DESIGN_3 required 425 and 441 cells respectively. For both architectures the cell usage figures quoted by the software statistics is misleading, if taken in isolation. Though nothing useful can be achieved with a large proportion of the cells in otherwise fully committed sectors or zones the CAD tool statistics do not report them as being in use or *consumed.*

6. Conclusion

It should be pointed out that the device architecture and CAD tools for the PMeL FPGA are probably as much as a design generation ahead of the present Atmel FPGA. How much of the superiority of the PMeL device and CAD tools can be attributed to this factor is at present impossible to quantify. The use of default design parameters was assumed throughout the work reported here, though different optimisations have been tried. Future work will address these considerations.

Acknowledgements

The assistance of Juan-Carlos Azorin, Gordon McGregor and Jonathan Stockwood, and the support of the EPSRC, Pilkington Micro-electronics Ltd, and Atmel Inc. are gratefully acknowledged.

References

1. Atmel Corporation: Configurable Logic Design and Application Book, Atmel Corporation, San Jose, California, USA, 1994.
2. M.S. Jhitta: Introduction of a New FPGA Architecture, In: More FPGAs, W. Moore and W. Luk (edit.), Abingdon, England 1994.

FPGA Routing Structures from Real Circuits

Andrew Leaver

Oxford University Computing Laboratory, Wolfson Building, Parks Road, OXFORD,
OX1 3QD, England

1 Introduction

A typical field-programmable gate array (FPGA) consists of a mixture of routing resources and logic resources, ideally with a balance between the two that
matches that required by the circuits to be implemented. It is not obvious what
this balance is, or indeed whether there is a single ratio that is suitable for a
class or a set of classes of circuits. The work presented here attempts to identify
the common features of the routing in a set of example circuits. The approach
taken is to generate layouts from a set of circuits picked to represent those we
wish to implement on an FPGA and to analyse the routed layouts for common
features.

2 Experimental Procedure

Example Circuits The routing patterns seen in the layouts will be strongly
affected by the choice of example circuits. For the purposes of this work
the example circuits were made up of twenty-one test circuits from the the
Microelectronics Centre of North Carolina (MCNC), seven circuits generated
using Rebecca, a functional language hardware compiler, and five generated
using Handel, an imperative language hardware compiler. Both compilers
were developed in Oxford.

FPGA Logic Block In order to perform an experimental analysis of routing it
is necessary to fix on a logic block. A large variety of logic blocks have been
suggested and implemented in various FPGAs. It was desired that the logic
block chosen be simple and regular in order to facilitate automatic design
generation. This led to a logic block consisting of a four-input look-up table
and a D-latch, based on that proposed in [1]. The output of the D-latch can
be fed straight back into the look-up table. It is hoped that the simplicity
of this logic block will make it possible to map the results onto other logic
block architectures.

Technology Mapping The example circuits were technology-mapped for the
given logic block using SIS, a synthesis program from Berkeley. It would
have been preferable to use a technology mapper that takes account of the
routability of the mapped designs and not merely the number of nodes but
the lack of such a program prevented this.

Placement Hardware compilers require a fast placement algorithm in order to
reduce the compile-generate-edit time. This led to the use of a min-cut based

placement algorithm. When implemented this produced designs with 0% – 25% greater routing requirements than the same designs placed using APR, the Xilinx XC3000 series placement tool. It is important to note that this step commits the architecture design to a fast, simple placement stage. For FPGAs intended to be placed using a better but slower algorithm such as simulated annealing these results will give too much routing by a factor of around 20%.

Global Routing A typical automatic routing procedure splits the process into two stages: a global routing stage during which wires are assigned to channels followed by a detail routing stage in which the wires in each channel are allocated to the particular routing segments available. It is not clear that this is as appropriate for FPGAs as it is for conventional gate arrays but it has the advantage that the global routing stage can be made almost architecture-independent. For the same reason the routing channels are chosen to be over the rows and columns of logic blocks rather than along the empty space between the logic blocks. This moves some of the routing task from the global to the detail routing stage but frees the algorithm from dependence on a segmented channeled routing architecture.

3   Results from Global Routing

The procedure described above resulted in the generation of a globally routed layout for each of the thirty-three example circuits. For each circuit the distribution of segment lengths was plotted, a segment being a section of net between two pins. For example, a simple L-shaped net has two segments, a T-shape has three. A segment of length one spans the distance between two neighbouring logic blocks. The distribution seen in all cases was fitted well by the equation:

$$Number\ of\ wires = K \times Number\ of\ Logic\ Blocks\ \times Wire\ length^{-G}$$

The measured values of G and K for the example circuits are given below.

Circuits	G Mean	G Std. Dev.	K Mean	K Std. Dev.
All	2.27	0.42	3.40	1.82
MCNC	2.16	0.43	3.92	2.12
Rebecca	2.53	0.34	2.51	0.42
Handel	2.49	0.17	2.48	0.13

This formula supports the design of hierarchical FPGAs as it says that a circuit with $2n$ logic blocks has the same distribution of short wires as a circuit with n logic blocks plus some additional longer segments.

The parameter G measures the ratio of long segments to short segments. A small value of G indicates more longer segments. The results indicate that there is a strong similarity between the segment length distributions in a wide variety of circuits.

Values of K show a wider variation, particularly for the MCNC circuits. The compiled circuits are much more consistent.

The layouts can also be used to determine channel densities and switching patterns. Lack of space precludes discussion of this data here. It is presented along with more information on this work in [2].

4   Routing Architecture Generation

A simple segmented routing architecture can be designed as follows. K is chosen to be 3.4 and G to be 2.0. The design will be a square array whose width is a power of two, and with segment lengths of powers of two only. We also choose to make the routing symmetrical and regular, so we need only design routing for one row which is then replicated over the array. The formula can be rearranged to give $NS_L = 1.7 * \sqrt{NC}/L^2$ where NS_L is the number of segments of length L per row and NC is the number of logic blocks. K halves to allow for the segments used for column routing.

To determine the number of segments of each power-of-two length we break down the unused lengths, so that the number of segments of length 5 is added to the number of length 4 and the number of length 1. This gives for an 8x8 logic block array nineteen length 1, eight length 2, and four length 4 segments. These must be distributed among the switchboxes. Experiment suggests the segment arrangement is not critical. A randomly generated routing is shown in Fig. 1. This routing requires on average 40% more segments to connect two arbitrary pins than a fully-connected routing using only power-of-two segment lengths. In comparison the best hand-designed routing required 35% above the minimum. In practice the designer would wish to adjust this design before implementing to even out the number of connections per switchbox.

Fig. 1. Randomised Routing for an 8x8 Logic Block Array

A detail router for this architecture is currently being implemented. This will allow the propsed architecture to be tested against existing FPGA routing designs.

References

1. J. Rose, R.J. Francis, D. Lewis and P. Chow. *Architecture of Field-Programmable Gate Arrays: The Effect of Logic Block Functionality on Area Efficiency.* IEEE J. Solid-State Circuits, Vol. 25 No. 5, October 1990, pp. 1217 – 1225.
2. A. Leaver. *FPGA Design for Systems Compilation.* D.Phil. Thesis, Computing Laboratory, University of Oxford (to appear)

A Tool-Set for Simulating
Altera-PLDs Using VHDL

André Klindworth

University of Hamburg, Dept. of Computer Science
Vogt-Koelln-Str. 30, D-22527 Hamburg, Germany

Abstract. This paper presents a tool-set for simulating Altera-PLDs [1] using VHDL [2]. It has been successfully used in a graduate course on digital design with PLDs. The tool-set supports timing simulation as well as functional simulation of designs that have been designed with the Altera MAX+plusII development tool.

1 Background

Altera's PLD/FPGA design tool MAX+plusII comes with an own hardware description language (HDL) named AHDL [3]. Compared to standard HDLs like VHDL, AHDL is a low-level HDL with support for signals of type either bit or vector-of-bits only. The implicit signal types along with short-hand constructs for instantiating and interfacing of subdesigns allows very compact hierarchical design descriptions. In addition, AHDL gives the user a direct control over technology mapping and design partitioning. Both aspects let us choose AHDL as the means of design entry in a first year graduate course on digital design for PLDs in which most participants have little or no experience in using HDLs as well as PLDs.

2 Simulation Requirements

Validation by simulation is still the usual way verification of digital designs is done. The designer should have easy access to a powerful simulator and the development tools of PLD-vendors usually come with programs which generate a design description in a standard format like EDIF, Verilog, and/or VHDL. Such a description then can serve as an input to a third-party simulator.

When simulation is considered, it is highly desirable that the designers view of the design is structured in the same way as in the original design description, that is, design hierarchy, groups of signals (busses), and all symbolic names of instances, machine states, and internal as well as external signals should be preserved. MAX+plusII for workstations fails to meet these demands.[1] Netlist writers that produce an EDIF, VHDL or Verilog description of a design are available, but they merely write flattened netlists. These netlists contain full

---

[1] The PC-version of MAX+plusII comes with an integrated simulator, but the expressive power of the stimulus description language is limited.

timing information, but neither the design hierarchy nor names (except for the primary inputs and outputs) are preserved. Functional debugging of a design has to be done by analysis of the primary inputs and outputs only.

But before correct timing of a circuit is to be verified, the designer must check the functionality of the design. For this task, the exact timing of the circuit is quite irrelevant and a *functional* simulation suffices. This is especially true when the design is fully synchronous, as it is enforced by the structure of functional blocks in most PLDs. We believe that at this stage of the design process it is much more important to simulate a design that "looks" exactly the same as in the original description rather than using a modified model of the design with exact timing information.

3 Simulation Procedure

As a compromise between the needs of the designer and what can be fulfilled with a reasonable effort, we developped a tool-set which eases timing simulation and provides functional simulation of Altera PLDs that have been designed using AHDL and the MAX+plusII compiler. For both types of simulation, VHDL is used to interface MAX+plusII with a third-party simulator which in our case is the powerful Synopsys VHDL-simulator *vhdlsim* [4]. VHDL is also used by the user to define stimuli for the simulation. For this purpose, our tool-set embeds the design-to-test in an automatically generated VHDL testbench. This testbench is a VHDL entity with an architecture that contains a component declaration for the design, a signal declaration for each of its external connections, and an instantiation of the design, the ports of which are connected to the signals. Using such a testbench has been proposed in [5] and gives the user the full expressive power of VHDL to assign waveforms to the external inputs and to observe and evaluate the resulting responses of the design.

Our tool-set supports both types of simulation: *timing simulation* and *functional simulation*.

For *timing simulation*, the VHDL netlist written by the MAX+plusII compiler is slightly modified and embedded in the testbench. Two script files named *timegen* and *tsim* make the necessary calls to programs which

- check that the design project is up-to-date and has been successfully compiled in MAX+plusII.
- modify the VHDL-netlist that has been written by MAX+plusII. Modifications include regrouping of busses that have been resolved by the compiler.
- generate a VHDL testbench for the design.
- make the calls to the VHDL compiler to generate simulation models for the VHDL description of the design itself and its testbench.
- call the simulator.

For *functional simulation*, a VHDL model of the design is generated directly from the original AHDL description. This task is carried out by an AHDL-to-VHDL compiler named *a2vhdl*. The VHDL code produced by *a2vhdl* fully preserves the hierarchy of the design project as well as all signal groups (busses) and all symbolic names of instances, machine states and signals. This VHDL description of the design is embedded in the same testbench as the timing model, so that the same stimuli are used for both types of simulation. Changing between functional and timing simulation is done by simply configuring the testbench for using the timing architecture or the functional architecture of the design, respectively. Again, two script files named *funcgen* and *fsim* realize all the necessary steps to simulate an AHDL design that has been succesfully compiled with MAX+plusII.

To simulate an AHDL design using our tool-set, the designer has to follow the following procedure:

1. Design entry with AHDL.
2. Compile design with MAX+plusII.
3. Call *funcgen* or *timegen*, respectively.
4. Insert stimuli processes in the testbench.
5. Call *fsim* or *tsim*, respectively, to start the simulation.

The VHDL code generated for the design is transparent to the user. When doing a timing simulation, all he sees from the VHDL model is the testbench, containing the external connections of the design. In the case functional simulation is chosen, the same testbench is used. In addition, the interior of the functional VHDL model is structured in the same way as the original AHDL description. All design objects the user defined in the original design description do also exist in the functional VHDL model and have the same name. Internal signals may be traced during simulation using the original (hierarchical) node identifiers.

4 Conclusions

The tool-set as described in this paper proved to greatly simplify debugging of PLD designs. It combines the compactness of design description in AHDL with the expressive power of VHDL for the description of simulation stimuli and the monitoring capabilities of a powerful VHDL simulator.

References

1. *Altera Databook*, Altera Corporation, San Jose CA, 1993.
2. *IEEE Standard VHDL Language Reference Manual*, IEEE Standard 1076-1987, New York, 1988.
3. *Altera MAX+plusII reference manual*, Altera Corporation, San Jose CA, 1992.
4. *Synopsys VHDL Simulator Reference Manual*, Synopsys Inc., 1993.
5. Z. Navabi: *VHDL - Analysis and Modelling of Digital Systems*, McGraw-Hill, New York, 1993.

A CAD Tool for the Development of an Extra-Fast Fuzzy Logic Controller Based on FPGAs and Memory Modules

John Ant. Hallas [*], Evaggelinos. P. Mariatos [*], Michael K. Birbas [**],
Alexios N. Birbas [*] and Constantinos. E. Goutis [*]

[*] Electrical Engineering Dept., Patras University, Patras 26110, GREECE
[**] Synergy Systems Ltd., 68 Amerikis Str., Patras 26 441, GREECE

Abstract. A method for the development of an Extra-Fast Fuzzy Logic Controller is presented in this paper, using a *CAD tool to* utilize the potentials of *programmable hardware* [1]. This is accomplished by generating custom *VHDL synthesizable code* [2] that is targeted to an FPGA chip. The CAD tool produces, also, bit patterns that represent a compiled version of the fuzzy-logic controller [3]. These are stored in memory modules. The basic idea is simple but very efficient with respect to the achieved processing speed, the required hardware and the ease of programmability.

1   Introduction

Fuzzy logic control methods have been in wide use in a variety of application fields during the last years. The major advantage of these methods is the ability to develop a working system even when a strict mathematical model for the regulated process is not available. Observed patterns in the time-series of the input-output pairs or the knowledge of an expert can be exploited to formulate the membership functions and the rules of the fuzzy controller. Then, during the evaluation tests, the membership functions and the fuzzy rules can be trimmed to achieve the best results.

The presented method consists of a CAD tool that accepts information regarding the fuzzy controller and produces a bit pattern that is stored in ROM or loaded in a RAM module. Apart from the memory-based part, the CAD tool queries the user about several hardware aspects and develops a VHDL synthesizable code that will be targeted to an FPGA chip. The proposed procedure is depicted in figure 1. The memory-based fuzzy-logic control method, the required programmable hardware and the CAD tool are presented in the subsequent sections.

2   Memory-Aspects of Fuzzy-Logic Control

The overall architecture of a fuzzy-logic control system is presented in figure 2. The input signals are fuzzified according to their resemblance to certain membership functions. The fuzzified values are fed to the inference engine which examines the contents of the Fuzzy Rules Matrix (FRM). The outcome of the inference process is a set of values that represent the grades of the results of each individual rule. The last step is the defuzzification process, where the set of output grades is transformed, back, to a crisp value.

Many different methods exist both for the inference process and the defuzzification process and all of them are dependent on the number and the shape of the membership functions. It is obvious that to achieve an efficient hardware implementation for the aforementioned blocks, taking into account all the potential variations that might be requested, is a difficult task. Another approach for the development of a fuzzy-logic controller is to implement the algorithm in software running in a conventional microprocessor. Although this is a flexible solution, there is an obvious drawback: the developed system is many times slower that a pure hardware approach. The proposed technique overcomes both of the aforementioned problems by following a middle-road approach as is explained below.

The basic idea stems from the fact that most fuzzy-logic control systems need only two variables as input: a specific signal and its time-derivative. Since a conventional fuzzy controller's output is independent of its previous state (i.e. the controller presents combinatorial behaviour) the outputs for every input combination can be pre-computed (compiled) and stored in a contiguous part of a memory module [3]. A simple calculation shows that the required memory size for a two-input, one-output, eight-bit resolution (in both the input and output signals) is $8*2^{(8+8)}=$ 64 Kbytes.

Although the idea of using the memory module as a universal function approximator is simple, the advantages are manifold:

• There is no need to stick to a particular inference or defuzzification method.
• The processing speed is inversely-proportional to the memory-module access time.
• The inference time is independent on the size of the rule matrix or the methods for inference and defuzzification as well as independent on the particular input combination. This is essential for real-time systems.
• There are no compromises to the number, shape and overlap of the membership functions, the size of the rule matrix or the methods of inference.
• Each input or output can receive a different number of resolution bits.
• The number of resolution bits for the membership grade axis does not affect the size of the memory. The CAD tool computes the result using maximum accuracy.
• The fuzzy-logic controller designer can try various implementations without wasting resources either for writing software or for building hardware.

The memory module is assisted by a number of logic blocks in order to build a complete controller. All aspects of those hardware elements are described next.

3   FPGA-Aspects of Fuzzy Logic Control

Since the memory module that holds the input-output data of the controller must be embedded in a host digital system, some interface issues must be resolved in external hardware using FPGAs as vehicles for their implementation. The logic blocks that can optionally be included in the FPGA are:

• A divider of the host system's clock to meet the memory-module's timing.
• A subtractor that may be used for the calculation of the Δx variable i.e. the time-derivative of the x input variable if this is required. A similar method may be necessary if an integral of an input variable is requested.
• A multiplexing-demultiplexing scheme if the controller is used to regulate simultaneously more that one similar systems.
• An interface circuitry for the generation of *Start_of_Convertion (SOC)* and acknowledge of the *End_of_Convertion (EOC)* signals to assist the linking of the fuzzy logic controller with ADCs. Also, a *Convert* signal should be generated to drive a DAC if the output variable is in the analog domain.
• An adder that may be needed to trim the controller's output by a certain offset.
• Additional adder/subtractor modules may be necessary if the controller is required to produce an incremental output instead of a direct one.
• An additional clock divider may be useful if one of the inputs is used for timing.
• Glue logic to assist the interconnection and/or addressing when more that one memory modules are used to hierarchically increase the number of inputs [3].

It is obvious that the aforementioned logic circuitry can be implemented in a small

FPGA. This way, the whole fuzzy-logic controller can be constructed using only two components: a memory chip and an FPGA chip.

4 The *FuFMeV* (Fuzzy FPGA-Memory-VHDL) CAD Tool

The *FuFMeV* CAD tool is developed using the C programming language. Its output consists of two ASCII files:

a. A bit-pattern (*BP*) organized in words that are stored in ROM or loaded in RAM. This file contains the compiled version of the requested fuzzy-logic controller.

b. Synthesizable VHDL-code (*VC*) that is targeted to a family of FPGAs using a VHDL-based synthesis tool. This portion of VHDL code serves for both simulation and synthesis purposes.

The two-previous sections have covered most of the *memory-aspects* (associated with ASCII file *BP*) and the *fpga-aspects* (associated with ASCII file *VC*) that are handled by the CAD tool. The interactions between the *FuFMeV* tool and the designer are described below:

In the memory-part of the tool, the user is queried about the number and the resolution of the input variables, the resolution of the output variable, the number of the membership functions and their shape, the number and type of fuzzy-rules, the methods of inference and defuzzification and the base-address of the memory module. If the number of input variables is greater than two then the tool asks if a hierarchical approach is preferred. On affirmative answer, the number of required memory modules, their inputs and their base-addresses are also requested. This information is sufficient to produce the *BP* ASCII file.

In the FPGA-part of the tool the user is queried about the existence or not of the logic modules that are described in the previous section. Parametrical and synthesizable VHDL code generators for these simple logic blocks are hard-coded in the CAD tool. When the user has entered all required information, VHDL entities for the required blocks are produced and linked together in a structural format (netlist). The result is written in the *VC* ASCII file. This code is then processed by a commercial synthesis tool to generate gate-level netlist optimized according to the basic configurable logic block of certain FPGA families. An FPGA-family specific tool transforms this netlist to the format required for the hardware programming.

5 References

[1]   *The Programmable Gate Array Data Book*, XILINX Inc., 1992
[2]   *Autologic-VHDL User's Manual v8.2*, Mentor Graphics Corp.
[3]   Witold Pedrycz, *Fuzzy Control and Fuzzy Systems*, Research Studies Press Ltd., 1993

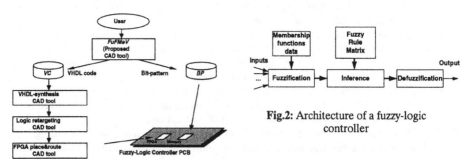

Fig.2: Architecture of a fuzzy-logic controller

Fig.1: Proposed procedure

Performance Characteristics of the Monte-Carlo Clustering Processor (MCCP) - a Field Programmable Logic based Custom Computing Machine

C.P. Cowen and S. Monaghan

Neural and VLSI Systems Group
Department of Electronic Systems Engineering
University of Essex
Colchester, CO4 3SQ England

Abstract. A special purpose processor originally designed for Monte-Carlo simulation using Metropolis type algorithms has been reconfigured to allow the use of a new improved class of Monte-Carlo algorithm without compromising the processor's performance.

1 Introduction

In Monte-Carlo simulations and digital signal processing applications it has often proved advantageous to use specially constructed processors in place of general purpose computers. SRAM-based Field Programmable Gate Arrays (FPGAs) appear to be suited to applications in this area [1, 2] as they can be used to reduce both the practical difficulties and the cost of building special purpose processors. Although FPGAs may be slower than non-programmable devices, they provide much greater flexibility in allowing the hardware to be readily altered. This paper describes how a piece of hardware based on Xilinx 3000 series FPGAs[3], originally built to perform Monte-Carlo (MC) computations and simulations based on the Metropolis[4] and related algorithms, has been reconfigured to implement a recently discovered and very different class of Monte-Carlo algorithms. These algorithms were not anticipated by the designers of the original hardware but the flexibility provided by the FPGAs allows the same board to adapt to these new algorithms. The computational hardware can be modified and the processor is able to keep up with developments in algorithm design that often occur in the kinds of areas to which FPGAs can be currently applied.

2 Hardware Platform

The architecture of a simple PC hosted processor board, built to implement variants of the Metropolis algorithm is shown in Fig.1.

In the original application this hardware platform was used as follows: an image represented by a string of pixels stored in SRAM1 is passed through sections of a FIFO implemented in FPGA2/SRAM2. The FIFO section in FPGA2 is tapped at the stages corresponding to a local processing window to provide the inputs to a processing element implemented in FPGA2/SRAM2. In the case of the Metropolis algorithm the processing

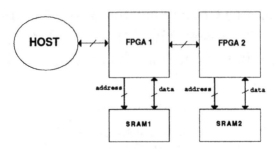

Fig. 1. FPGA/memory architecture

element typically consists of a random number generator (RNG) and a comparator in FPGA2 and a look-up table (LUT) in SRAM2. A data address generator (DAG) in FPGA1 controls the flow of data between SRAM1 and the FIFO and the address sequence of this DAG is completely known at run-time. The Metropolis algorithm, implemented in this way, executes an order of magnitude faster than a similarly costing Digital signal processor (DSP) for a number of different simulations of physical systems at non-zero temperatures [5](Simulated Annealing).

3 Cluster Algorithm

Cluster algorithms[6] consists of two stages. The first stage forms tree-like data structures (clusters), by linking neighbouring pixels according to a stochastic percolation process. The probability that adjacent cells will be linked is a function of the temperature of the simulated system. This clustering stage can be achieved via a pipelined arrangement similar to the above processor. The second, declustering stage consists of locating and updating each individual cluster of pixels separately by following the links to each pixel in the cluster. Since each pixel can be linked to more than one neighbour, a recursive method is used. The size, shape and number of these clusters will depend upon certain simulation parameters such as the system temperature and will vary throughout the simulation. This means that the sequence of the data addresses cannot be known prior to execution, and that a variable number of branches will take place during the declustering stage of the algorithm. It is not clear at first sight that these algorithms can be efficiently implemented in Fig.1.

3.1 MCCP architecture

The MCCP architecture can be placed into this arrangement for small problem sizes, given the small memories used in the prototype. Refer to [7] for details of the MCCP architecture. Again, SRAM1 is used to store the image, and the pipeline that performs the clustering stage is placed into FPGA2 along with the RNGs and comparators. SRAM2 is used as a stack which means that FPGA2 must also contain the stack controller. FPGA1 contains the DAG, which for cluster algorithms is larger, more complex and able to handle branch operations. FPGA2 also contains a finite state machine which generates the address sequence from cluster information.

4 Performance of MCCP architecture

To obtain an appropriate performance measure of the MCCP architecture, a version of the cluster algorithm, was written in hand-optimised assembly language on a popular Digital Signal Processor(DSP) [8]. This exercise confirms that the same order of magnitude performance gain is obtained[7] as with previous processors. This is because, although the DSP provides some of the data addressing capability required for efficient execution of cluster algorithms, there are a number of important functions, such as masking, stack control and random number generators (RNGs), that require large software overheads and must be carried out sequentially. Such operations can be readily implemented in hardware and carried out concurrently.

5 Conclusions

A processor implemented in 3000 series FPGAs, originally intended for Monte Carlo simulations based on the Metropolis algorithm has been reconfigured to run a new and very different type of Monte-Carlo algorithm while the performance gain of an order of magnitude over a similarly priced DSP is maintained. No changes have been made to the non-reconfigurable connections in the processor indicating that at least within a given area of application, FPGA-based computing machines are flexible. The reconfigurability of the FPGA-based platform allows the MCCP architecture to be used in conjunction with other Monte-Carlo processor architectures, providing a wider range of options for investigating Monte-Carlo problems. The use of field programmable interconnect devices is likely to improve further the applicability of this type of processor.

References

1. S. Monaghan, T. O'Brien, and P. Noakes. *FPGAs*, page 363. Abingdon CS Books, 1991. Edited by W. Moore and W. Luk.
2. S. Monaghan. Gate level reconfigurable Monte Carlo processor. *JVSP*, 6:139 – 153, 1993.
3. Xilinx. *The Programmable Gate Array Data Book*, 1991.
4. K. Binder, editor. *Applications of the Monte-Carlo Methods in Statistical Physics*. Springer-Verlag, 1984.
5. S. Monaghan and C.P. Cowen. Multi-bit reconfigurable processor for DSP applications in Statistical Physics. *Proc. FPGAs for Custom Computing Machines FCCM'93*, 1993. sponsored by IEEE Computer Society.
6. Robert H. Swendsen and Jian-Sheng Wang. Nonuniversal Critical Dynamics in Monte Carlo Simulations. *Physical Review Letters*, 58(2), January 1987.
7. C.P. Cowen and S. Monaghan. A Reconfigurable Monte-Carlo Clustering Processor (MCCP). *Proc. FPGAs for Custom Computing Machines FCCM'94*, 1994. sponsored by IEEE Computer Society.
8. Analog Devices. *ADSP-2101 Data Sheet*, 1990.

A Design Environment with Emulation of Prototypes for Hardware/Software Systems Using XILINX FPGA

Gerd vom Bögel[1], Petra Nauber[2], Jörg Winkler[2]

[1] Fraunhofer–Institute of Microelectronic Circuits and Systems, IMS Duisburg
Finkenstrasse 61, D–47057 Duisburg, Germany
[2] Fraunhofer–Institute of Microelectronic Circuits and Systems, IMS Dresden
Grenzstrasse 28, D–01109 Dresden, Germany

Abstract. The paper will present a *Design Environment with Emulation of Prototypes* (DEEP) for designing hardware/software systems. The CAE tool DEEP consists of an integrated workframe for designing hardware/software systems and the *Rapid Prototype Co–Emulator* (RPCE) for hardware/software co–emulation and verification. For a flexible realization of the designed systems FPGAs by XILINX are widely used within the RPCE. Experiences in prototype realizations will be presented.

1 Rapid–Prototyping and Design Verification by Emulation

For bringing the realized system functionality and the requirements of the embedding environment into line the design of microelectronic systems still requires the realization of the design as a prototype. The method of Rapid Prototyping allows a fast realization of prototypes meeting the demands of 'time to market' and product quality. DEEP supports this method by a 'soft–configurable' hardware.

For verification by emulation a physical replication of the designed system or sub–system is realized and embedded into its target application environment. The aim of such a physical replication is to get a functional and timing identical prototype of the system to be implemented later. The functional behaviour of the system can then be verified in the target environment under real time conditions. In DEEP the emulation of digital hardware runs on XILINX FPGAs and the emulation of software on standard microprocessors.

2 Components of DEEP

DEEP consists of PC based software tools for design entry, design translation and emulation management as well as the DEEP hardware called Rapid Prototype Co–Emulator.

2.1 DEEP Software

The DEEP software is fitted out with a MS–Windows workframe acting as user interface for invoking and controlling the different design steps. DEEP starts with a system specification by the notation of cooperating function components (FC), where hardware FC are specified using VHDL or schematics and software FC using the programming language C.

For the implementation on the RPCE the hardware FC may be divided into blocks which fit into the available FPGAs. For this a special software is used. It partitions the FPGA technology files and allocates the resulting blocks to the available devices. The partitioning process is user controlled with different selectable strategies.

The RPCE is controlled by the MS–Windows workframe. All commands like configure, emulation start/stop and trace are invoked from the emulation control menu.

The DEEP emulation system provides an integrated debugging environment to analyse the current system state of the emulated hardware and software. The hardware state is visualized by reporting the register values, got by reading back the LCA internal data, within the schematics. The software state is characterized in an extra window, showing the used registers and variables.

2.2 DEEP Hardware

The Rapid Prototype Co–Emulator (RPCE) is a VMEbus based system extended by different emulation modules (Fig. 1). Its modular structure facilitates a variety of extensions and a flexible adaption to widely differing application environments.

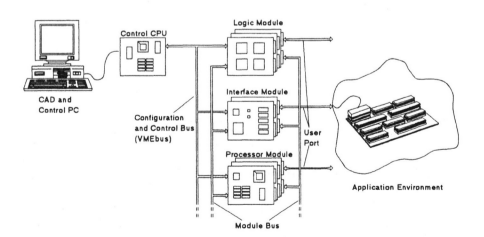

Fig. 1. DEEP Hardware Structure

The VMEbus itself acts as configuration and control bus mastered by a control CPU board. To link the emulation modules within the RPCE as well as to the target environment a flexible module bus and I/O bus structure is provided.

Until now, three types of modules are available:

- Logic modules for emulation of digital FC,
- Processor modules for emulation of software FC and
- Converter modules to tie up analogous components as well as sensors and actors.

In general the emulation modules consist of an *interface board* carrying the very *emulation board*.

The main functionality of the interface board is realized on the basis of two XILINX LCAs 3090PG175, one controlling the bus drivers of the module bus and I/O bus and the other interfacing the VMEbus to the emulation boards. The necessary control logic for the bus drivers is generated automatically during the design translation. For the function of the VMEbus interface LCA standard configurations are provided, which can be modified or extended by the user for special applications.

The *logic emulation board* consists of four XILINX LCAs 3090PG175 and 3195PG175, respectively, used for the circuit to be emulated. They are connected via a hard–wired array according to a special developed scheme. Thus digital circuits with a complexity of up to 15,000 gates (or 20,000 gates using 3195 LCAs) can be realized per module.

The *converter emulation board* allows the connection to analogous components via A/D and D/A converter (12–Bit, 100kHz samples, 16/8 A/D channels, 4 D/A channels). The specification of the converter interface is carried out within the DEEP workframe, where the configuration data for the module are generated afterwards. The configuration data are downloaded into the driver control LCA also facilitating various operating modes of the module.

Processor modules are used for the emulation of software FC. The module is set up on a VMEbus compatible CPU board, that has been extended for the RPCE by a special interface board. The CPU board contains a MC68070 CPU with RAM, EPROM and I/Os. The special interface board is mastered by a XILINX LCA 3090PG175 containing the control logic for the bus drivers which can be extended by logic for any necessary interface circuits. Additionally a plug–in interface for further CPUs (IMS2205, MC6805, I8051) is controlled by the LCA. In the same manner the integration of DSPs is under preparation.

3 Experiences in Prototype Realizations

Up to now three RPCE devices are used for prototype emulation at the IMS institutes. Experiences are gained by designing and emulating a highly parallel data decompression circuit and a knowledge based analytical controller ASIC both involving digital hardware as well as software components.

DSP Development with Full-Speed Prototyping Based on HW/SW Codesign Techniques

Jouni Isoaho[1], Axel Jantsch[2] and Hannu Tenhunen[2]

[1] Tampere University of Technology, Signal Processing Laboratory
P.O.Box 553, FIN-33101 Tampere, Finland, Email: jouni@cs.tut.fi
[2] Royal Institute of Technology, Electronic System Design
P.O. Box 204, S-16440 Stockholm, Sweden

1 Prototyping Method

Field Programmable Gate Arrays (FPGAs) have been established as efficient tools for prototyping of various DSP algorithms in [3] and [6]. Quite often prototypes are implemented, because the algorithms have to be tested with huge amount of different kind of real test data. Developing new algorithms and approaches using VHDL-synthesis based prototyping typically means that algorithm design, filtering analysis, VHDL coding, simulation, synthesis, prototyping implementation and measurements have to be done. This normally takes several weeks of time requiring a lot of different knowledge from the algorithm designer and this approach is also very sensitive to human errors.

To achieve more efficient system level design and analyses we propose a fast HW/SW codesign based method for Digital Signal Processing (DSP) algorithm development with "full-speed prototyping". The real world data is collected into the memory of workstation via an FPGA based board, filtered using bit-true C-models and sampled back to the environment via the same board with a realistic data rate. The communication mechanism between hardware interfaces and C-models is generated automatically during compilation. This approach allows an accurate analyses and an efficient system level optimization of DSP functions for the target application without an actual implementation. Our hardware, which is connected to Sun Sparcstation IPX via SBus, is based on two Xilinx 4000 series devices. The algorithm can be modelled first with the full word lengths (32 bits) to achieve the worst case scaling requirements. This model is later used to generate the reference file for implementation optimization. During the implementation minimization optimal scaling factors, internal word lengths and the arithmetic solutions can be selected. The output of the optimized model can be compared with the reference file, if no differences are found the model is acceptable and the optimization can continue. If differences are found the model has to be analysed in simulator or with prototyping to check the effects of distortion.

2 Development Environment

Filter modelling is done using C-language and thus allows the use of both floating-point and bit-true models. Basic signal analyses can be done in the Matlab graphical environment. The software models are connected to the target and prototyping environment by means of HW/SW codesign techniques which allows to prototype the DSP algorithm with a full operating speed. During the compilation an assembler code is generated for the parts that will be executed on the workstation, and VHDL for the parts that will be implemented on the FPGA board [5]. In the early phase prototyping this normally means that only the interface functions are linked to the corresponding components on our interface configuration library for Xilinx. Afterwards, the codesign environment together with High Level Synthesis tool, like SYNT [2] [4], can be used for implementing the system after validation and sharpening of the specification. The new modules of the system can easily be coprototyped with the parts that are already implemented in the earlier systems. To implement large systems the same communication protocol can be used with the RPM logic emulator [8]. The compiler generates the additional code for handling the communication between the software and hardware parts. On the software side the device drivers are called to open and close the logical communication channel, to configure the Slave FPGA, to initiate the application design and to serve interrupts from the board. For the hardware side a prespecified, parameterizable VHDL entity is generated, which takes care of the handshaking with the controller FPGA and the address generation for the local memory. The prototyping board is based on two Xilinx 4000 series devices, one of the devices contains the SBus interface (Master) [1] and the other one application configuration (Slave). The bottleneck which limits the overall data transfer rate to 4 Mbit/s for 8-bit data is located between the DVMA component and the XC4005 controller. If higher data rates are needed, the local memory, which is currently only 1 Mbit, can be used up to 25 MHz applications (32 bits).

We demonstrate the possibilities and advantages of the development system with an audio sigma-delta D/A converter example [7]. The interpolated data is forwarded to a sigma-delta noise shaper. The output bit-stream of the noise shaper can be forwarded to a 1-bit D/A and an analog postfilter to allow the listening experiments of the algorithm performance. As at maximum 64 times oversampling ratio for the audio data is possible with our prototyping environment due to the data transfer limitations, the second order noise shaper topology is needed to replace with the fourth order one [3]. Of course, if the second order topology is preferred, the other possibility is to implement the high sample rate parts of the running sum interpolator and the noise shaper on FPGAs. It is possible even within our current hardware configuration as the interfaces require only some logic modules in Xilinx 4000 series architecture.

The combined bit-modelling and prototyping approach is very advantageous with optimizing e.g. the nonlinear noise shaper, because the scaling always removes information and due to feedback loops this might cause also the system start to variate in the similar manner as a badly designed IIR filter. The optimized internal word lengths for the fourth order noise shaper topology are 20, 16, 13 and 8 bits. It means about 43% of savings in the size of the noise shaper compared to the full 25 bits of word length, which is the smallest usable common bit width if no input scalings before intergrators are used. Also, the effects of two's complement and saturative arithmetic can be researched. In this case, the test data is generated using a CD player and read into the Xilinx board at the speed of 44.1 kHz (2 x 16 bits of data). This data is sampled into the memory of the Sparcstation. After filtering and noise shaping the data is sampled out with the speed of 2.8 MHz (M = 64). The output bit-stream can be forwarded to the loudspeaker using a 1-bit D/A and an analog reconstruction filter. The interface system configurations for collecting data and writing out the samples are presented in Figure 1.

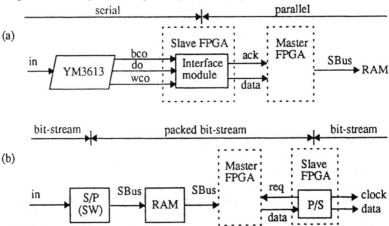

Fig. 1. (a) Data collection configuration. (b) Data output configuration with sample clock.

The prototyping board is connected to the CD player via a standard CD interface chip (YM3613B). The bit-serial data is further converted to the bit-parallel format in the Slave FPGA. The interface module acknowledges to the Master FPGA from which channel the data comes. As the system has different data lengths the 1-bit output of sigma-delta noise shaper is packaged in software for transmission into 16-bit dataframes before writing the data into the memory. From the memory the data is sent to the SBus according to the requests of the Slave FPGA. The dataframes are unpacked in the Slave FPGA before outsampling.

Using the fourth order noise shaper, we need to sample out filtered data at the rate of 176.4 kHz (16-bit frame). In the FPGA board, this 16-bit dataframe is decoded and 1-bit output data is sampled out from the system at the rate of 2.8224 MHz (maximum operating speed for one channel solution due to the current communication hardware). At this speed we need 1.4 Mbytes of 16-bit memory for an 8 seconds one channel sample. As in our configuration there is about 20 Mbytes usable memory for applications for one channel at maximum 112 seconds (M = 64) of sample can be filtered with the system presented. In this case, the bit-true drive of the C-program which performs the whole filtering of 8 seconds of sample can easily be completed on Sparcstation 10 within one hour (about two hours in a Sparcstation IPX) including the compilation and interface downloading, which is about the same time as needed for synthesizing and implementing a single filter which fits into a single XC4010. As comparison the FPGA implementation of this converter structure requires 3-4 XC4010 devices for each channel. As C-models handle only samples, the corresponding VHDL models also need emphasis on timing and control. Therefore code generation, design verification and debugging is much faster with C-models. The simulation time with bit-true models in C is e.g. about 10-20 times shorter than using VHDL models in Synopsys. Also, the hardware requirements are much slower. Synthesizing the filters required using our Sparcstation IPX with 24 Mbyte central memory is not possible in this case. 100 - 160 Mbytes of central memory is a proper amount for hierarchical synthesis approach in the Synopsys synthesis environment for our example converter system.

3 Conclusions

The automatic HW/SW tool provides an efficient tool for DSP system optimization. In addition to prototyping, the filtered data can also be closer analysed on Matlab with the real data collected from the system environment. The time to change the system takes only a few minutes to complete while the changing of hardware prototype on FPGAs takes usually even with synthesis tools some days in complex changes. The proposed approach allows a fast search of the best filtering solution with minimal effort for implementation. As no DSP functions are needed to implement it allows typically much faster operating speed than which is possible with normal FPGA prototypes or ASIC emulators and provides a very cost efficient environment for the system level development. Also, unlike in a full prototyping with FPGAs, the maximum operating speed and the system capacity is quite easy to approximate.

References

1. S. He and M. Torkelson, "FPGA Application in an SBus based Rapid Prototyping System for ASDP", *ICCD 91*, pp. 1 - 4, 1991.
2. A. Hemani, "High-Level Synthesis of Synchronous Digital Systems using Self-Organization Algorithms for Scheduling and Binding", *doctoral thesis*, Royal Institute of Technology, Sweden, 1992.
3. J. Isoaho, J. Pasanen, O. Vainio and H. Tenhunen, "DSP System Integration and Prototyping with FPGAs", *Journal of VLSI Signal Processing*, Vol. 6, pp. 155 - 172, 1993.
4. J. Isoaho, J. Öberg, A. Hemani and H. Tenhunen, "High Level Synthesis in DSP ASIC Optimization", In *Proc. 7th IEEE ASIC Conference and Exhibit*, Rochester, New York, Sept. 1994.
5. A. Jantsch, P. Ellervee, J. Öberg, A. Hemani, and H. Tenhunen, "A Software Oriented Approach to Hardware/ Software Codesign", in *Proc. of CC'94*, Edinburgh, April 1994.
6. J. Nousiainen, J. Isoaho and O. Vainio, "Fast Implementation of Stack Filters with VHDL-based Synthesis and FPGAs", In *Proc. IEEE Winter WS on Nonlinear DSP*, pp 5.2-4.1---5.2-4.6, Tampere, Finland, Jan. 1993.
7. T. Saramäki, T. Ritoniemi, T. Karema, J. Isoaho and H. Tenhunen. "VLSI-Realizable Multiplier-Free Interpolators for Sigma-Delta D/A Converters" In *Proc. of ISCAS89*, pp 60-63. Nanjing, China, July 1989.
8. J. Varghese, M. Butts, and J. Batcheller, "An Efficient Logic Emulation System", *IEEE Trans. VLSI Systems*, vol. 1, no. 2, pp. 171 - 174, June 1993.

The Architecture of a General-Purpose Processor Cell

Jiří DANĚČEK, Alois PLUHÁČEK, Michal Z. SERVÍT

Czech Technical University, Dept. of Computers,
Karlovo nám. 13, CZ - 121 35 Praha 2, Czech Republic

Abstract. A general-purpose processor cell, called DOP, is presented. The DOP is a 16-bit stack oriented processor designed to support efficiently imperative programming languages like C or Pascal. The architecture of DOP is a result of HW/SW co-design. The DOP is supposed to be used as a building block in a FPGA library.

1 Introduction

The architecture of a simple universal processor cell, called DOP, is presented. It is currently under implementation with the XILINX XC4000 family. Our motivation was to design a very simple yet still efficient processor which effectively supports High Level programming Languages (HLL's) like C or Pascal. We considered the design of DOP as an example of HW/SW co-design where the SW part is represented by a compiler (or rather by a code generator) and the HW part is the resulting processor. Our primary goal was to propose such a processor which would provide a simple and efficient compilation scheme, and which would keep the complexity of HW within reasonable limits. This is why the HLL requirements were investigated first [1], [2]. This analysis led us to the programming model and architecture of the DOP processor.

The DOP processor is designed to support efficient evaluation of *address expressions* and *arithmetic operations* with both signed and unsigned numbers of various length. The *2's complement* representation of signed numbers and the *small endian* representation of multibyte data stored in the main memory are supposed.

2 The processor organization

The DOP is a *stack oriented* processor with 16-bit internal bus (BUS) divided into low and high 8-bit part, 16-bit external address bus, and 8-bit external data bus (see Fig 1). The memory is byte organized and I/O devices are memory mapped.

The DOP contains six 16-bit programmer-visible *registers*: PC — program counter, SP — stack pointer, S — source operand address, D — destination operand address, PSW — program status word consisting of two 8-bit subregisters denoted L (Loop counter) and F (Flags), and W — working register.

The PC and SP are well known registers and they are used in a common manner which needs no comment.

The S and D registers are designated to store source and destination addresses (which can be evaluated in the ALU) — but they can be used for other purposes as well. The S and D registers have auto-post-incrementation capability to support multibyte operations.

The L register (the lower part of the PSW register) supports programming of loops. It has auto-post-decrementation capability and its content can be tested by conditional jump and interrupt instructions.

The F register (the higher part of the PSW register) contains the following flags: CF — carry flag, OF — overflow flag, SF — sign flag, ZF — zero flag, AF — auxiliary sign flag.

The W register can be considered as a part of the ALU (it can be loaded only through the ALU). The first operand (or the only operand) of an arithmetic/logic operation is always the content of the W register. As the second operand the data from the BUS are used.

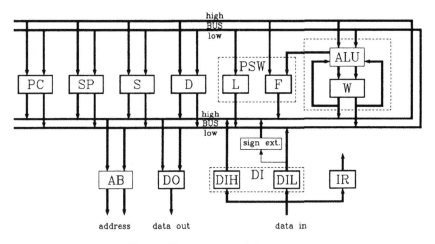

Fig. 1. The structure of the processor

The outputs of all the mentioned registers can be connected via BUS and Address Buffer (AB) to the external *address bus*. Using the BUS they can also be connected to the external *data bus* via DO (Data Out) register to write the data, or via DIL or DIH (Data In Low or High, respectively) registers to read data or instructions. One byte (8 bits) or one word (16 bits) data can be read or written. It is obvious that two bus cycles are needed to read or to write one word. When reading a word, the DIL register is used to store the lower byte temporally. When reading a byte, the sign- or zero-extension can be used.

The Instruction Register (IR) contains the just performed instruction.

3 The instruction set

The instruction set consists of 21 instruction types only covering all operations that can be expected in a processor of this class (for details see [1]). Further we will focus our attention on only "special" features of the DOP instruction set.

All *instruction codes* (consisting of instruction type identification, operand type identification and addressing mode identification) are packed into one byte. An instruction code can be followed by one- or two-byte immediate operand when appropriate.

The DOP supports the following *addressing modes*:
- *Register*, in which operand(s) and/or result reside in register(s) visible to the programmer. Registers are implicit or must be declared explicitly.
- *Immediate*, in which an operand follows instruction operation code.
- *Register indirect*, in which operand(s) and/or result reside in memory location pointed to by register content. In certain cases auto-incrementing or auto-decrementing of register content is feasible.
- *Stack*, in which operand or result resides on top of the stack or at the location determined by the offset to SP register, which is a part of the instruction.

Table 1. Possible data transfers

source	destination								
	PC	SP	S	D	PSW	W	[+SP]	[D+]	alu2
PC							2		
SP						w			w
S						w	1,2	1,2	w
D						w	1,2	1,2	w
PSW							1,2	1,2	
W	w	w	w	w		w	1,2	1,2	w
[PC+]	w		s,w,b	s,w,b	1,2	s,w,b			s,w,b
[SP−]	w		s,w,b	s,w,b	1,2	s,w,b			s,w,b
[S+]			s,w,b	s,w,b		s,w,b			s,w,b
[D]			s,w,b	s,w,b		s,w,b			s,w,b
[W]						s,w,b			
IntAddr	w								
SP−[PC+]			w	w		w			

The functionality of most instructions can be described as a single *data transfer* between registers, or between a register and memory, or between a register and ALU. Instructions *Load Local Address* (LLA) consisting of two or three transfers are exceptional from this point of view. Possible data transfers are summarized in the Table 1, where the following conventions are used:
- Square brackets denote the content of corresponding memory location(s).
- Signs plus and minus denote pre-/post- incrementation or decrementation, respectively; if the sign is in front of/behind a register name, pre-/post- is true, respectively.
- alu2 stands for the second ALU operand.

o s, b, and w stand for sign-extended byte (signed short integer), zero-extended byte (unsigned), and two-byte word, respectively.

o 1, and 2 stands for one byte, and two bytes, respectively.

o IntAddr stands for "Interrupt Address" that is from the set: {0xFFC0, 0xFFC4, ..., 0xFFFC}.

The last line in Table 1 corresponds to instructions LLA.

The following *arithmetic and logic operations* can be performed: addition, subtraction, AND, OR, XOR, rotate W left and right, NOT, and AAF (discussed below). The W register content is used as the first operand of a binary arithmetic/logic operation and as the only operand of an unary one. The possible sources of the second operand are given in the Table 1 (see column alu2). The result is normally stored in the W register. However, the storing can be suppressed by a special instruction *Suppress Write into* W (SWW) that must immediately precede the considered instruction. The only result is the new content of the F register. In this way such operations as "compare" or "mask" can be performed.

Any arithmetic/logic operation changes the *flags* (the content of the F register) [3]:

• The CF flag is set by any addition or subtraction instruction if the carry from the most significant position is equal to 1 (the subtraction is considered to be the addition of the opposite number — the CF flag is the inverse of the borrow). The rotate instructions are performed through CF which acts as the 17th bit of W. An instruction setting CF (called SCF) is included in the instruction set. By any logical operation the CF is cleared.

• The OF flag is set by any addition or subtraction instruction if the overflow occurs assuming the 2's complement number representation.

• The SF flag is changed to be equal to the most significant bit of the result.

• The ZF flag is set if the result is equal to 0.

• The AF flag is set by any addition and subtraction instruction if the second operand is negative or non-negative, respectively.

The CF, OF, SF and ZF flags can be used in branch and interrupt instructions. (Essentially, interrupt instructions are conditional fixed address calls.) The used flag is given as one parameter of the instruction. The second parameter is the value of the flag which implies a branch. In such a way, for example, the branch "if carry" can be executed as well as the branch "if not carry". Similarly, the branch "if $L = 0$" or the branch "if $L \neq 0$" can be executed. However, the L register is pre-decremented in this case.

The CF and AF flags serve for other purposes as well. A special instruction *Use Carry Flag* (UCF) supports addition and subtraction of multibyte operands. If this instruction is used immediately before any addition or subtraction instruction, the CF flag is added to the first operand when the corresponding operation is performed. In such a way, for example, the sequence UCF and ADD has the same functionality as the ADDC instruction known from other processors of this class.

The AF flag allows the addition and subtraction of signed operands of different lengths (e.g. addition of a 2-byte signed number to a 4-byte one). A special instruction called *Add extended* AF (AAF) serves this purpose. This instruction adds the CF value and "all zeroes" or "all ones" for $AF = 0$ or $AF = 1$, respectively, to the W register. When it is needed to add or to subtract a shorter operand to or from a longer operand, the appropriate addition or subtraction instruction (and UCF, if needed) is to be used for the lower parts (words or bytes) of both operands and then the higher parts of the first (longer) operand are to be loaded into the W register and to be modified using the AAF instruction. For unsigned operands, the UCF and the "add 0 to W" or "subtract 0 from W" instruction is to be used instead of AAF.

The important feature of the suggested use of the CF and AF flags and the UCF and AAF instructions is that the correct values of CF, OF and SF flags are received after performing the last instruction in the above mentioned sequence (e.g. after the last AAF). It holds for both signed (2's complement representation) and unsigned operands.

4 Conclusions

The XILINX XC4005 chip is used for the implementation of the DOP processor. It is supposed that the processor will be one of the XILINX XC4000 library elements. Certain parts of DOP, namely registers and ALU, were designed and simulated. Two VHDL models of the DOP have been written and debugged: the behavioral model and the register transfer model.

The software support of DOP has been developed. It includes the C compiler and the processor simulator. This software support enabled us to compare the DOP with other processors in terms of the code size [2]. Surprisingly enough, the DOP code is shorter than the code of much more complex processors.

The DOP processor is a result of HW/SW co-design. It seems to be a good compromise between the complexity of HW and the simplicity and efficiency of the compilation scheme.

References

1. Daněček, J., Drápal, F., Pluháček, A., Salčič, Z., Servít, M.: DOP — A Simple Processor for Custom Computing Machines. Journal of Microcomputer Applications (to appear)
2. Daněček, J., Drápal, F., Pluháček, A., Salčič, Z., Servít, M.: Methodologies for Computer Aided Hardware/Software Co-Design Using Field Programmable Gate Arrays. Research Report. Dept. of Computers, CTU Prague (in print)
3. Pluháček, A., Daněček, J.: The Manipulation with Flags on the DOP Processor. CTU SEMINAR 94, CTU Prague 1994, 55 – 56

This research was supported by the Czech Technical University under grant no. 8095 and by the Czech Grant Agency under grant no. 102/93/0916.

The Design of a Stack-Based Microprocessor

Michael Gschwind, Christian Mautner
{ *mike,chm* } *@vlsivie.tuwien.ac.at*

Institut für Technische Informatik
Technische Universität Wien
Treitlstraße 3-182-2
A-1040 Wien
AUSTRIA

Abstract. This paper describes the design of a stack-based CPU using field-programmable gate array technology. The architecture to be implemented was already defined by a compiler, which had been implemented previously. We describe what tools and strategies were used to implement different parts of the processor, as well as the final integration process.

1 Introduction

FPGAs offer a unique opportunity to prototype chip implementations. To more closely study this option, we have built a prototype board based on Xilinx FPGAs [Hub92] and conducted several implementation experiments. We implemented our first design, JAPROC, as part of the JAMIE project. JAPROC is a micro-controller upwardly compatible to the PIC16C57 [GJ92].

Field programmable gate arrays (FPGAs) can be used to allow fast implementation of chip designs [GAO92], [Gsc94]. This allows for a fast debug cycle, as designs can be altered and downloaded in a matter of hours. As FPGAs are pretested, only logic functionality has to be validated, reducing the time to get a workable implementation of a chip considerably.

Since this has proved to be a remarkable success, we have started to use FPGAs in student projects for logic design courses (building circuits such as multipliers and dividers) and to build more complex designs, such as the stack-based microprocessor presented here.

The advantage of this approach is that students do not have to deal with the electrical intricacies of silicon implementations or breadboarding. Also, the implementation cost is reduced dramatically.

2 Architecture

To maximize the understanding of the interaction of all levels of computer design (hardware, compilers, OS), we emphasize integration of system design consideration in student designs. Thus, the architecture presented here was used earlier in a compiler construction class [BF92].

The processor implements a stack machine, with all operands being addressed relative to the top-of-stack pointer or a frame pointer (local-pointer) which is used to access local variables [Mau94]. The memory model is that of a Harvard architecture, i.e. separate data and program memories. Memory addresses, as all other data, are 16 bit wide. Thus the processor can address $2^{16} = 64k$ words in each memory segment. The data memory is 16 bit wide, and instruction memory uses 24 bit. This allows each instruction word to encode a full 16 bit immediate constant or address.

Instruction	Description
NOP	no operation
PSHc const	push const
PSHl offset	push value at (FP + offset)
PSHli	pop offset, push value at (FP + offset)
STOl offset	pop value and store at (FP + offset)
STOli	pop offset, pop value and store at (FP + offset)
MVTc const	move top-pointer (SP) by const
MLT	local-pointer (FP) := top-pointer (SP)
PSL	push local-pointer (FP)
PPL	pop local-pointer (FP)
GET	read value from I/O port and push
PUT	write top stack element to I/O port
ADD	add
SUB	subtract
MUL	multiply
SWP	swap the top elements
JMP address	jump
JPE address	jump on equal
JPG address	jump on greater-than
JSR address	jump to subroutine – push PC+1, jump
RET	return – pop return address and jump
STP	stop execution

Table 1. Instruction set architecture

3 Implementation

3.1 Control Unit

The stack machine was implemented using a finite state machine (FSM) controlling the data path (see figure 1). The finite state machine was modeled using a microprogram-like mnemonic representation (see figure 2). We decided to automatically generate the controller part from a high-level description. This had several advantages:

– The high-level description can be used as specification of the controller behavior.

- No inconsistencies can arise between the specification and the actual implementation, as the implementation is automatically generated from the specification.
- If the specification changes, a new implementation can be generated with little effort, whereas a manual translation process such as used for JAPROC requires a complete re-design of the controller.

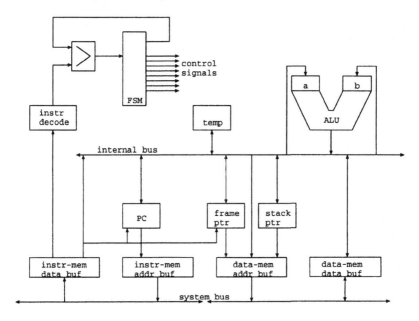

Fig. 1. Block level diagram of stack machine

```
LABEL(ADD)
  COM( top_to_dmemadd | dmemadd_le | dmem_rd | dbus_to_alu_a )
  COM( top_direct_down | top_clk_en )
  COM( top_to_dmemadd | dmemadd_le | dmem_rd | dbus_to_alu_b )
  COM( top_to_dmemadd | dmemadd_le | dmem_wr | alu_e_to_dbus | \
       alu_cntrl_add | pc_inc | fetch )
```

Fig. 2. FSM code for adding the top two stack elements

To describe the design, we used a simple language with two primitives, one to define the output signals to be generated in a particular state, a second primitive to symbolically name state numbers. The language primitives have C macros associated with them, so that the formal specification of the FSM can be executed. By executing the specification, a bit stream is generated which describes the control unit [Gsc93].

To implement the control unit, we use a ROM storing all state transitions and control signals. This ROM is implemented using the Xilinx **memgen** tool

[Xil92], which allows automatic generation of ROM- and RAM-like structures for FPGAs. The bit stream generated by the formal specification is used to initialize this ROM.

For JAPROC, we used espresso optimized random-logic to generate the FSM controlling the data path. As a result, the complete state machine had to be specified as a set of boolean equations and changes to the original control structure were much harder to achieve.

Due to the simplicity of the instruction set we implemented, each instruction is implemented by a linear sequence of two to eight states. Each state has exactly one successor state. The only time when control is not transferred to a well defined next state is during the decoding stage of a new macro-instruction: to decode a macro instruction, the opcode is fed to a decoder (also implemented as ROM and automatically generated from the same executable specification) which decodes an instruction by setting the controller state to the beginning of the state sequence which implements the macro instruction.

3.2 Data Path

The design of the data path was straight-forward, using Xilinx-supplied macros (soft- and hard-macros), the TTL emulation library and our own, generic bit-slice ALU.

Integration of the design was seamless, but the the usage of hard macros and of multiple XNF modules complicated things somewhat: to generate an FPGA description which can be simulated, the design has to be translated first to XNF level where all XNF modules were merged.

The merged design was then translated to the LCA level where hard macros could be integrated. Then the whole translation process was reversed to generate a VSM-type file for simulation. This lengthy translation process showed a number of interfacing bugs in the Xilinx software and between the Xilinx and ViewLogic environments which have to this date not been resolved.

The simulation was largely successful, but exhibited occasional unexpected behavior, like erroneous incrementing of the PC – this was tracked down to hazards in the automatically generated ROM. The control signals had been stabilized by latching the current state, allowing hazards to propagate to all functional units in the data path. By latching the control signals of the current state instead, these hazards were masked out. After this final verification, the original compiler was adapted to reflect the changes made to the architecture at the beginning phase of the project. Thus, a fully functional microprocessor environment was available, including a compiler and a hardware prototype, implemented on one Xilinx XC4006 FPGA.

4 Results and Experiments

We simulated a whole system by integrating this CPU design in a ViewLogic schematic which also contains instruction and data memories, and all the necessary glue logic. This board level design was then simulated using ViewSim.

Simulation shows that the CPU designed here will run at 12.5 MHz, and that the processor speed is limited by the memory subsystem. The circuit itself could operate at a much higher clock rate.

The XC4006 FPGA showed 100% utilization of CLBs, with a huge degree of flip-flops being unused. It is interesting to note that more than a third of the available CLBs were used for implementing the two ROMs used in the control unit. For larger designs, CLB-based FPGAs should probably not be used to implement large look-up tables. An alternative is to used dedicated parts for memory-type resources, as described in [KNZB93].

5   Related Work

Intel Corp. used 14 Xilinx-based Quickturn RPMs to fully simulate its current top-of-the-line Pentium™ microprocessor as part of the Pentium™ pre-silicon validation process [KNZB93]. The simulated Pentium™ microprocessor achieved an emulation speed of 300 kHz and booted all major operating systems for Intel's x86 processor family.

6   Conclusion and Future work

We have shown that FPGAs are a useful tool for CPU prototyping. We are currently embarking on a project to model the MIPS R3000 CPU using FPGAs as target technology and VHDL for design specification. This design will be targeted towards and enhanced board featuring multiple Xilinx FPGAs and local static RAM.

7   Acknowledgement

We wish to thank Alexander Jaud for his help with the ViewLogic and the Xilinx design environments.

References

[BF92]   Manfred Brockhaus and Andreas Falkner. Übersetzerbau. Vorlesungsskriptum, TU Wien, 1992.

[GAO92] T. Gal, K. Agusa, and Y. Ohno. Educational purpose microprocessors implemented with user-programmable gate arrays. In *Proc. of the 2nd International Workshop on Field-Programmable Logic and Applications*, Vienna, Austria, August 1992.

[GJ92]   Herbert Grünbacher and Alexander Jaud. JAPROC – an 8 bit microcontroller and its test environment. In *Proc. of the Second International Workshop on Field-Programmable Logic and Applications*, Vienna, Austria, August 1992.

[Gsc93]   Michael Gschwind. Automatic generation of finite state machines for data path control. Technical report, TU Wien, 1993.

[Gsc94]   Michael Gschwind. Reprogrammable hardware for educational purposes. In *Proc. of the 25th ACM SIGCSE Symposium*, SIGCSE Bulletin, pages 183–187, Phoenix, AZ, March 1994. ACM.

[Hub92]   Ernst Huber. *Eine Einsteckkarte für den IBM-PC/AT zur Programmierung von Xilinx FPGAs*. Diplomarbeit, Institut für Technische Informatik, Technische Universität Wien, Vienna, Austria, September 1992.

[KNZB93] Wern-Yan Koe, Harish Nayak, Nazar Zaidi, and Azam Barkatullah. Pre-silicon validation of Pentium CPU. In *Hot Chips V – Symposium Record*, Palo Alto, CA, August 1993. TC on Microprocessors and Microcomputers of the IEEE Computer Society.

[Mau94]   Christian Mautner. Entwurf eines Stackprozessors als Konfiguration eines Xilinx FPGA Serie 4000. Technical report, Institut für Technische Informatik, Technische Universität Wien, April 1994.

[Xil92]   Xilinx. *XACT Reference Guide*. Xilinx, October 1992.

Implementation and Performance Evaluation of an Image Pre-Processing Chain on FPGA

Mohamed AKIL and Marcelo ALVES DE BARROS
Groupe ESIEE - Laboratoire IAAI - BP 99, Cité Descartes, 2, Bd. Blaise Pascal
93162 - NOISY LE GRAND CEDEX - FRANCE - email: akilm@esiee.fr

1. Introduction

In image processing domain, many applications need their implementation respect both flexibility and real time constraints. Tasks in low level image processing are characterised by a great operation regularity and recursivity, as well as a large data density (image pictorial format). These low level tasks, according to their high computing requirements, have led to design specific architectures and Application Specific Integrated Circuits (ASICs) as hardware solutions.

The programmable technology today gives FPGA circuits with considerable performance and integration capacity [Xil 92]. The Xiinx XC4010 circuit used in this work is a two-dimensional array with 400 Configurable Logic Blocks (CLBs), 160 In-Out Blocks (IOBs) and a programmable interconnection network. The CLB is a basic element of the circuit effective area.

This work is based on MODARC (*MODular reconfigurable ARChitectures*) approach [Alves93]. MODARC methodology was conceived to allow synthesis, implementation and system test of specific operators for low level image processing real time applications, with a low cost of development. In this environment an application is represented by a data flow graph and a set of graph transforming procedures are used to adapt algorithm and architecture graphs to the MODARC hardware support graph. The hardware model consists of a cascade of basic operators placed on a linear array of physical modules composed of SRAM based FPGA circuits and interconnected memory resources.

2. Evaluation of performances and area costs

The evaluation approach is adapted to the technology characteristics and to the proposed architectural model, i.e. pipelining basic operators in a data-flow operation mode. Let *Ecc*, *Emm* and *Eint* be the quantities representing respectively the "energy" of *computing, memory* and *interconnections* of the specific architecture. The term "energy" is used to express the amount of the several different components present in the architecture, belonging to these three classes of architectural resources. Let *Acc*, *Amm* and *Aint* be the quantities representing respectively the area cost of the computing, memory and interconnection elementary components. In MODARC approach an architectural elementary component library is defined in order to represent low level image processing (8 or 16-bit adders, comparators, bus, registers, etc.). The cost of these components is given as a number of CLBs. The algorithm architecture whole area cost can be calculated by the expression:

$$A = \left(\sum_i Ecc_i . Acc_i \right) + Emm.Amm + Emr.Amr + E\,\text{int}.A\,\text{int} \qquad Eq.\ 1$$

where Ecc_i is the energy of the *i-th* computing component of the architecture, Acc_i its the area cost and *Emr* and *Amr* represent respectively the energy and the area of an elementary register component. The critical data-path of a basic operator or a pipeline stage is evaluated by:

$$T_C = N_{CCLB}(T_{CLB} + L.T_{INT}) + T_{CKS} + T_{CKSK} \qquad Eq.\ 2$$

where: N_{CCLB} : number of CLBs layers (logic levels); T_{CLB} propagation delay of a CLB; L : interconnection length in the circuit for each CLB layer; T_{INT} : propagation delay per unit of interconnection length; T_{CKS} : propagation delay from clock to registers outputs; T_{CKSK} : dispersion delay (skew) of the clock in the circuits.

For the proposed model, the architecture latency is determined by the number of pipeline layers used. The maximum system frequency, *Fmax*, is defined by the slower pipeline stage propagation delay. *Fmax* is given by: *Fmax=1/max[Tc]* (Eq. 3), where Tc is the *i-th* pipeline stage propagation delay.

3. Implementing an image pre-processing chain

The pre-processing chain is part of an image-recognition system showed in Figure 1. The aim of the pre-processing tasks (shaded block in Figure 1) is to generate accurate edges for the extraction of the object features such as area, perimeter and curvature, after contours closing and region labelling. These features are compared to objects models by a pattern matching algorithm to identify the current object. The desired implementation on FPGA include the algorithms from noise reduction until contour detection.

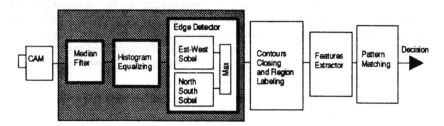

Figure 1. An image pre-processing chain (shaded area).

The median filter is used to reduce the image noise. The algorithm proposed is a variation of the original median algorithm, named separable median, for 3x3 windows. This non-linear algorithm reduces noise preserving edge information [Nar 85]. For an image *I(m,n)*, the processed pixel *O(m,n)* is given by the expression:

$$O(m,n)=median[median[I(k,n-1),I(k,n),I(k,n+1)]],$$

where: $k=m-1$, m, $m+1$, $n=1$, ..., L, and L is the size of the image. This operator requires $\sqrt{N}(\sqrt{N}-1)$ comparing/exchanging elementary operations (UCE), where N is the number of pixels in the window ($N=9$ for 3x3 windows). The histogram equalizer reduces sensitivity to illumination variations. It is performed by a look-up table. This function corresponds to a memory reading operation. The Sobel operator is used to enhance the edges of the images. The algorithm corresponds to convolving the image by two ortogonal differential masks and choice the maximum absolute value of the two convolutions. This operator can be performed by 12 additions/subtractions and a comparison/exchanging step.

The median is composed of two one-dimensional pipelined filters performed by two 3-way sorters (Median3 modules). The vertical filter sorter processes the 3 pixels of a column in the 3x3 window and gives the median. The second filter (horizontal median3) sorts the last 3 outputs of the vertical median3 module and computes the median.

The linear-logarithmic converter look-up table is performed by a 256x8-bit "ROM" (Read Only Memory) integrated in the FPGA. The input image pixels are used as the address of the "ROM". In order to avoid implementation of the convolvers, the Sobel architecture is adapted to a 3-stage pipeline made of adders and subtractors. This adaptation is based on filter separability and power of two coding of its coefficients. Separability, as in the case of the median filter, reduces the original algorithm to the application of two one-dimensional filters ($[1 \ 0 \ -1]$ and $[1 \ 2 \ 1]$), respectively to the columns and to the lines of the 3x3 windows. Power of two coding reduces the multiplications to a shifting of a bit in the input of the adders.

From equations 1, 2 and 3, and considering the XC4010-5 circuit, we have area costs (A) and time estimation (T) as shown in Table below, for each function. Third column shows results from implementation by use of Xilinx place and routing tools.

Algorithms	Estimated	Implemented
Separable Median (MED)	A=102 CLBs T=38 ns	A=102 CLBs T=39 ns
Histogram Equalizer (EQU)	A=69 CLBs T =35 ns	A=69 CLBs T =35 ns
Sobel Operator (SOB)	A=67 CLBs T=MAX[Tc1;Tc2;Tc3]= MAX[26;49;38]=49ns	A=68 CLBs T=49ns

The overall pre-processing chain architecture uses 239 CLBs corresponding to 60% of the available area in the FPGA. 72 IOBs are needed for communications with external FIFO memories (46% of available I/O resources). The maximum internal frequency allowed is $Fmax=1/49$ ns ≈ 20 MHz.

4. Conclusions

An efficient implementation of a low level image processing algorithm on FPGA requires to consider the particularities of the device architecture in the design of the application architecture and different optimization levels. The proposed evaluation method allows to study the feasibility on Xilinx circuits to implement low level image processing tasks at video rate.

5. Bibliography

[Alves 93] M. ALVES DE BARROS, M. AKIL. "Circuits Reconfigurables et Traitement Bas Niveau d'Images en Temps Réel". Ann. 14ème GRETSI, Juan les Pins, France, Sept. 1993.
[Xil 92] *The XC4000 Programmable Gate Array Data Book*, Xilinx, San Jose, USA, 1992.
[Nar 85] P. M. NARENDA " A separable median filter for image noise smoothing", In: Digital Image Processing and Analysis. Vol 1, IEEE Computer Society, 1985, 450-459.

Signature Testability of PLA

E.P.Kalosha[12], V.N.Yarmolik[1] and M.G.Karpovsky[3]

[1] Computer Science Department, Minsk Radioengineering Institute
6, Brovki Str. Minsk 220600, Belarus. Tel. (7 0172) 39 86 66
[2] To moment: Fachbereich Datenverarbeitung, Fachgebiet Elektrotechnik,
Universität-GH-Duisburg
[3] Department of Electrical, Computer and Systems Engineering
Boston University, Boston, MA 02215. FAX (617) 353 6440

Abstract. This paper deals with the design of a Built-In Self Test (BIST) environment for the Programmable Logic Arrays that minimizes the aliasing probability. The signature testability condition is developed that prove criteria to compare the BIST environment aliasing. An important feature of the developed approach is that the criteria proved by signature testability allows to design both pseudo-random test pattern generator (PRPG) and signature analyzer (SA).

1 Introduction

The aliasing is an important problem in the compact technique. The aliasing occurs when the fault signature is identical to the fault-free one. The aliasing probability of the compact technique has been studied by coding theory framework [1]. In [2], the detection properties of the BIST environment for the errors that can be expressed as a single product term has been discussed. It has been shown that the BIST schemes based on the PRTG and the SA with the same feedback polynomial detect almost all cases of the above errors. Furthermore, the BIST environment with reciprocal polynomials have the poor error detection capability.

Here, we propose a signature testability condition that allows to determine the error detection capability for all combinations of PRTG- and SA-polynomials. We also propose a new error model generalized the one in [2]. The signature testability condition is derived to analyze the aliasing for proposed error model.

2 Signature Testability

Consider an algebraic model of a testing configuration that is shown in Fig. 1. Let $g(x)$ denote the primitive feedback polynomial of the PRTG and let $h(x)$ denote the irreducible feedback polynomial of SA. It is assumed that $deg\ g(x) = deg\ h(x) = m$. In this paper, α denotes the primitive element over $GF(2^m)$ defined by the primitive polynomial $g(x)$. The irreducible polynomial $h(x)$ defines the element β over $GF(2^m)$. Let $\beta = \alpha^k$. We assume here that the initial state of SA is zero. Initial state of PRTG is $0\ldots010$ and can be described by the element α. Let PRTG generate $2^m - 1$ test patterns, that are applied to the CUT.

Now consider the signature value for an Boolean function f that describes CUT. Let $f(\gamma)$ denote the value $f(g_{m-1}, \ldots, g_0)$ of the Boolean function f, where

$\gamma \in GF(2^m)$ and $\mathbf{G} = (g_{m-1}, \ldots, g_0)$ is the vector form of the element γ in the bases $1, \alpha^1, \ldots, \alpha^{m-1}$. Then [2],[3]

$$Sg(f) = \sum_{i=0}^{2^m-2} f(\alpha^i)\,\alpha^{-ik}. \tag{1}$$

In this paper, the criteria to estimate the aliasing of the BIST environment is discussed that based on the following error model.

Let $f(x)$ and $f_e(x)$ denote the Boolean function of the fault free and faulty CUT, respectively. The error function is given by $e(x) = f(x) + f_e(x)$, where $+$ denote the modulo two sum. We consider the Reed-Muller canonical form of the error function that is the modulo two sum of the terms. Let r be the literals number in the term with the maximum multiplicity.

Let the fault be called a signature testable fault if

$$Sg(e) = \sum_{i=0}^{2^m-2} e(\alpha^i)\,\alpha^{-ik} \neq 0. \tag{2}$$

Let us consider a set F of Boolean functions that have no more then r literals in the term with maximum multiplicity. For each function $f \in F$ the following vector can be formed: $v[f] = [f(\alpha^{2^m-2}), f(\alpha^{2^m-3}), \cdots, f(\alpha^0)]$. The set of vectors forms a vector subspace over $GF(2)$. This vector subspace is the r-th order Reed-Muller code $R(r,m)$ [4]. For each vector $v[f]$ can be formed polynomial $P(f) = \sum_{i=0}^{2^m-2} f(\alpha^i)x^i$. Common to all polynomials roots form a set of code nulls. The nulls of the rth-order Reed-Muller code have the following property.

The α^k is a root of the generator polynomial of the rth-order Reed-Muller code $R(r,m)$ if and only if $0 < w_2(k) < m - r$, where $w_2(k)$ is the binary weight of integer k [4].

Let error multiplicity be r. Then, the error sequence belongs to the rth-order Reed-Muller code $R(r,m)$. The set of the nulls of the code corresponds to a set of the SA for that the code word is an error causes the fault masking. Thus, the error with the multiplicity r is undetectable for all SA that have the feedback polynomial with root $\beta = \alpha^k$ and $m - w_2(-k) > r$.

Let $\sigma[g,h] = m - w_2(-\log_\alpha \beta)$, where α, β are roots of $g(x)$ and $h(x)$, respectively.

Signature testability condition. *The error with multiplicity r is signature testable if $\sigma[g,h] \leq r$.*

Hence, the BIST environment has the good error detectability when $\sigma[g,h]$ is small. The PRTG-SA pair has the minimum aliasing if $\sigma[g,h] = 1$. That is when $g(x) = h(x)$. The worst error detectability is when $\sigma[g,h] = m - 1$. The feedback polynomials of PRTG and SA are reciprocal in this case.

Now, we apply the signature testability condition to design the BIST environment. Let us analyzed the circuit error distribution over the error multiplicity r. We assume that in the digital circuit the errors with multiplicity $r > \rho$ occure only. Thus, feedback polynomials with $\sigma[g,h] \leq \rho$ must be used.

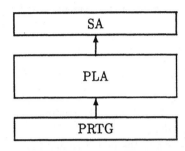

Fig. 1. Testing Configuration

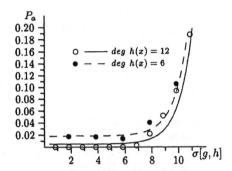

Fig. 2. Aliasing probability of BIST

3 Experimental Results

Some simulation experiments were performed to test the signature testability condition. Fig. 2 presents the results of one such experiment. The simulation experiments were done on the (12,103,12)-PLA. The aliasing probability was calculated for all cross-point-faults. The plot shows that primitive polynomials of the same degree can have the significantly different aliasing. The obtained σ-dependence of the aliasing probability supports the signature testability condition proposed. The experimental results show that the BIST environment with the $\sigma[g, h]$value close to m have the low error detection capability. The feedback polynomial $h'(x)$ of the 6 degree can have the lower aliasing than the 12-polynomial $h''(x)$ when $\sigma[g, h'] > \sigma[g, h'']$.

4 Conclusion

In this paper, the signature testability technique is developed that can be used to analyze aliasing of the BIST environment for Programmable Logic Arrays. The proposed approach is a first technique to estimate aliasing of both PRTG and SA, what have not been available yet.

The group of the BIST environment with the good error detection capability can be obtained by the signature testability condition. By using the property of the error distribution and the testability condition, the design procedure for BIST environment is proposed.

References

1. Pradhan D.K. and Gupta S.K. "A New Framework for Designing and Analyzing BIST Techniques and Zero Aliasing Compression", IEEE Trans. on Computers, vol. 40, N 6, June 1991, pp. 743-763.
2. Nagvajara P., Karpovsky M.G., "Coset Error Detection in BIST Design", IEEE VLSI Test Symposium 1992, pp. 79-83.
3. Yarmolik V.N., Kalosha E.P., "Signature-Testable LSSD-Circuits", Avtomatika i Vychislitelnaya Technika, 1990, N 1, pp. 94-95.
4. MacWilliams F.J. and Sloane N.J.A., "The Theory of Error-Correcting Codes", NewYork: North-Holland, 1977.

A FPL Prototyping Package with a C++ Interface for the PC Bus

Ibrahim bin Mat

Malaysian Institute of
Microelectronic Systems
Exchange Square, Damansara Heights
50490 KUALA LUMPUR
Malaysia
telephone: 03 2552700
ibm@rangkom.my

James M. Noras

Department of Electronic and
Electrical Engineering
The University
BRADFORD BD7 1DP
United Kingdom
telephone: 0274 384036
J.M.Noras@bradford.ac.uk

Abstract. Even though student projects generally have severe constraints of time
and other resources, they should aspire to reach ambitious targets. One way of
giving suitable projects a flying start is to build on the PC bus. In this paper
we outline a package which aids the rapid development of FPL-based systems
within a reliable and powerful interface of hardware and software. This
removes the need for an initial phase studying the details of bus protocols.

1 Introduction

One important practical use for field programable logic devices is the rapid
prototyping of systems. This is particularly useful in education, where low cost and
re-usability are gratefully accepted. It is very helpful to have simple demonstration
download boards, for example with push-buttons and LEDs, but greater scope is
readily available: [1,2]. Quite sophisticated, "serious" designs can be investigated with
FPL-based systems, and for research and advanced undergraduate projects it is
desirable to have a prototyping framework that supports the full potential of emerging
technologies, with real-time testing. Many complete systems can be targeted to cards
sitting on the PC bus, so that their rapid and easy development would be a
contribution to a wide range of research and learning. Also student projects should
be ambitious, and not curtailed by routine detail, re-inventing and testing wheels.

With these aims in mind, we have produced a set of hardware templates and C++
interface routines to aid the reliable development of hardware attached to the PC bus.
We chose this environment because it is affordable and is in common use, with many
application areas. This work started as an M.Sc. project at Bradford University.

2 System overview

In this section we set out the attributes that we perceive as most useful in a general-
purpose prototyping and evaluation system [3], describe the design and test process,

and the features available at present. The programmable hardware used is manufactured by Xilinx [4]. This is not the only possibility, but was readily available and provided satisfactory flexibility, speed and power.

2.1 System specification
We aimed to supply the following:
>1) A PCB template with integral bus-lines, interfacing and block addressing, to which users could add their particular designs;
>2) A library of macros, programmable in FPL hardware on the PC card to give interface functions for host-application communication;
>3) Software routines to link high-level programs with application cards;
>4) A user guide to the above features.

This permits users to implement designs without first having to study the details of PC interfacing, but to progress directly to interesting problems. If subsequent work requires deeper knowledge of the interface, by then the user will have a grasp of the main task in hand. Designs are testable from the PC host at realistic clock and data rates - a great advance on the use of static testing for hardware development.

2.2 Implementation
A major hazard of development of original PC peripheral cards is damage to the host computer during the early learning phase, so our first design used basic discrete components in order to prove the hardware ideas, only then going on to a second system using programmable logic. This provided an useful tutorial aid, as the card using FPL had simpler routing and much smaller area, as expected. PCB design using double-sided construction was done with the Boardmaker package and Orcad schematic capture software was used for the programmable logic. Software development used Turbo C++ routines in modular form for ease both of testing and of subsequent adoption by users. Much time went into testing to avoid damage due to errors and to ensure that this work would be a reliable platform for later designs.

2.3 Present features
Two demonstration cards have been produced and tested: the first with discrete logic only was a prototype for the second card. This contains some discrete logic to enable card initialisation, but all other interfacing functions are provided using a FPL device. These permit bi-directional data transfer with 8-bit or 16-bit word lengths. Together with interface routines written in C++ the card has the following features:
>1) Setting the block I/O address with a 4-bit DIP switch;
>2) Initialising for 8-bit or 16-bit operation with a SPST switch;
>3) Configuring the FPL chip on power-up;
>4) Resetting the FPL chip by:
>>a) push button,
>>b) a PC system reset, or
>>c) a software reset;
>5) Activating a reprogramme signal to clear the FPL configuration, prior to reprogramming.

For tutorial purposes the card has 16 LEDs and 16 DIP switches. Under menu-driven commands, 8-bit or 16-bit data can be read to the host PC screen from the DIP bank, data can be written to the LEDs from the host PC.

For prototyping use we provide a PCB template in Boardmaker form. This has the tracking for the hardwired initialisation and block-addressing logic, connections to all the required bus-lines, and for the correct location of one Xilinx chip. We supply the Xilinx code to allow users to programme the chip with the interfacing functions specified above. The user is also given information about available free resources within the programmable chip, and about which pins on this chip are uncommitted.

3 Summary and continuing work

We have built and tested a system of hardware and software which permits the rapid and reliable development of PC-based peripheral hardware for I/O processing and custom coprocesser cards. With this system, we have carried out student projects which concentrate on applications, taking our environment of hardware and software as a reliable and pre-existing starting point. Present development concerns FPL library components to permit DMA and interrupt communication between card and host. The card will retain a hardwired 8-bit part for system initialisation, but users will have Xilinx macros for the full range of bus features. A new PCB design template has connections for the additional PC/AT bus signals. Supporting software is being tested. All communication software will be written in assembly code device-drivers for efficient control and data transfer, as well as convenience for users.

This prototyping framework permits users to concentrate on the novel features of their applications, and removes early delays in hardware construction and testing. This environment will encourage novice designers to tackle projects that would otherwise be too daunting.

References

1. Brown, G.M. and Vrana, N., "A Computer Architecture Laboratory Course Using Programmable Logic", Proc. 3nd Intl. Workshop on Field-Programmable Logic and Applications, Oxford, 1993.

2. Kebschull, U., Schubert, E., Thole, P. and Rosentiel, W., "The Design and Implementation of an Educational Computer System Based on FPGAs", Proc. 4th Eurochip Workshop on VLSI Design Training, Toledo, 1993.

3. Mat, I. and Noras, J.M., "A system for rapid prototyping with field programmable logic", Proc. Intl. Conference on Robotics, Vision and Parallel Processing For Industrial Automation, Ipoh, 1994.

4. Xilinx, "The Programmable Logic Data Book", 1993.

Design of Safety Systems Using Field Programmable Gate Arrays

Juan J. Rodríguez-Andina, J. Alvarez and E. Mandado

Departamento de Tecnología Electrónica, Universidad de Vigo
Apartado Oficial, 36200 Vigo, Spain

Abstract. This paper presents a design methodology to implement fail-safe circuits (i.e. circuits whose output is at any moment correct, or fails in a safe manner) in Field Programmable Gate Arrays.

In order to be fail-safe, a circuit must include redundant elements. We consider the time redundancy technique called alternating logic, that is based on the use of a given function and its dual function in two consecutive time intervals.

1 Introduction

Fail-safe systems can be defined as those that, in the presence of faults, avoid the propagation of errors to other systems and, in addition, their outputs are in a correct state or can be forced to a safe one.

A circuit is called totally self-checking (TSC) if, for every fault from a prescribed set, there exists at least one input which produces a noncode-space output, and there exists no input which produces an erroneous code-space output. In other words, a TSC circuit is self-testing and fault-secure. TSC circuits can be used as basic blocks in the design of fail-safe systems. Fault-security provides the fault containment capability, and the self-testing condition allows the system to force its outputs to a safe state when a noncode-space output is to be produced.

This paper presents a method to implement fail-safe logic controllers by using time redundancy. The method is based on the interconnection of several predefined *safety parts* to form a fail-safe circuit.

2 Alternating Circuits

Alternating logic [1] is a design technique that utilizes time redundancy to achieve its fault detection capability. It is based on the implementation of self-dual combinatorial functions. A function F of a set of binary variables X is self-dual if $F(X') = F'(X)$, where X' and F' are the complemented values of X and F respectively.

An alternating variable is one that takes its true value during one time interval, followed by its complemented value during the next time interval. An alternating circuit produces a logic function F at time t, and the function dual of F at time $t+1$ [i.e, a pair (y, y') of binary complemented values] if the inputs are alternating. Therefore, the output of the circuit is an alternating variable if the inputs are alternating. Time intervals are determined by the true and false intervals of a clock *ck*. An alternating variable is equivalent to a logic value '1' if it equals *ck* and to a logic '0' if it equals *ck'*.

Alternating circuits are TSC for single permanent and intermittent stuck-at faults if they are irredundant (self-testing condition) and do not include any line that takes a non-alternating, but correct, value for an input vector *(X, X')* (fault-secure condition). To demonstrate the latter condition (that is alternative to the ones proposed in [1], and more suitable for our purposes) let us assume, by way of contradiction, that the circuit produces an erroneous alternating output [*(y',y)* instead of *(y, y')*] when line *n* within it is stuck-at-*d*. If an erroneous alternating output is to be produced, both *X* and *X'* must sensitize the fault, but then the fault-free value of line *n* is *d'* for both input combinations. Therefore, line *n* takes a non-alternating, but correct, value for *(X, X')*, a contradiction by hypothesis, q.e.d.

This condition implies that all functions that depend on both *X* and *X'* must be alternating. For instance, the combinatorial functions implemented in a look-up table in an SRAM-based FPGA [2] must be alternating, while the values stored in individual SRAM cells can have a correct non-alternating value without compromising fault security, because each one of them depends only on *X* or *X'*, not on both.

3 Design Methodology

Figure 1 shows the general structure of a fail-safe alternating logic controller, whose components are available in an user library, to facilitate the design.

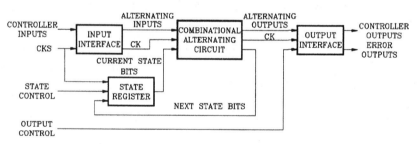

Figure 1: Fail-safe alternating logic controller.

The input interface transforms input variables in alternating ones. It is a modular block, consisting of as many *input adapter cells* as the number of inputs.

The state register consists of one *two-stage shift register* (clocked at twice the frequency of clock *ck*, by clock *ck_s*, to accomplish the proper alternation of the current-state variables) and a *load* block (that allows to force the system to an arbitrary state) per state variable.

The output interface transforms alternating variables in non-alternating ones by means of *output adapter cells*. It includes additional logic to force the outputs to a safe state, if necessary, and a TSC code checker that indicates, by means of one or more error signals, if all the outputs of the combinatorial circuit alternate. Both the adapter and checker blocks are modular.

Finally, the combinatorial alternating circuit generates the next-state and output functions of the controller.

The transformation of a behavioural representation of a logic controller in an alternating fail-safe circuit is straightforward. The idea is to automatically add some elements to the *unsafe* design, to transform it in a fail-safe one. The necessary information to be obtained from the behaviour of the system is detailed on the following paragraphs.

Number of inputs. This information is used to generate the input interface using the library part *input adapter cell*.

Number and safe state(s) of the controller. This information allows us to generate the state register from the library parts *two-stage shift register* and *load state*. In order to force the state of the system (for instance, at power-on) it is necessary to generate the *state control inputs*. This function is realized by an external auxiliary circuit, whose consideration exceeds the limits of the present work.

Number and safe state(s) of the outputs. With this information it is possible to build the output interface block utilizing the library parts *output adapter cell*, *safe output logic* and *alternating TSC checker*. The *output control inputs* are generated by the above mentioned external auxiliary circuit.

Next-state and output functions. These functions are generated by a combinatorial alternating circuit. The first step in the design of this circuit is to obtain the truth tables of the functions. If a function F is self-dual, it remains unchanged. In other case, it is modified to build an equivalent self-dual function F^* by adding the clock ck as an input variable [1]. If the function is to be implemented in a set of n-input LUTs, the function is partitioned, if necessary, in a set of self-dual m-input ($m \leq n$) sub-functions, to meet the fault-secure condition presented in the previous section.

It is important to notice that no changes are needed in the behavioural description of the system in order to implement it as a fail-safe circuit, except that a list of safe states has to be provided. The *fail-safe* design procedure can be totally automated and, therefore, transparent to the designer.

4 Conclusions

A method has been presented that allows to automatically implement fail-safe logic controllers in FPGAs from standard behavioural descriptions. The increment in design time is negligible, because the only additional information needed consists of a list of safe states. The method is based on the use of time redundancy, but it can be extended to the use of other forms of redundancy, such as information redundancy.

References

1. D.A.Reynolds and G.Metze: Fault detection capabilities of alternating logic. IEEE Transactions on Computers 12, 1093-1098 (1978).

2. Xilinx Inc. The Programmable Gate Array Databook (1991).

A Job Dispatcher-Collector
Made of FPGA's
for a Centralized Voice Server

J.C Debize and R.J Glaise

Compagnie IBM France CER La Gaude (France)
Transport Network Node Development

Abstract. It is the purpose of this paper to describe a specific application made of XILINX XC4000 series Field Programmable Gate Arrays (FPGA's).
The application takes advantage of a feature of this RAM based device where logic is implemented under the form of an array of small look-up tables which may be as well used as an array of small RAMS. The paper shows how this array of RAMs is well suited to do the function that dispatches and collects, over a bunch of Digital Signal Processors or DSP's, Digitized Voice Packets that need to be compressed (decompressed) before (after) transmission on a Tele Processing line to realize a Voice Server function that saves a significant amount of the transport medium bandwidth.
The paper tends to demonstrate that the availability of x1 organized independent internal RAM devices permits to simply carry out the function, at the nominal speed, in a couple of FPGA's while the equivalent function would only be achievable in a standard gate array at the expense of the use of many Flip Flops.

1. What is to be done:

The function to realize, depicted on the Figure 2, assumes it is possible at a node of a digital network to dispatch over a bunch of DSP's, at regular intervals, "packets" of digitized samples of phone conversation in progress. The packets are compressed within the DSP's then, transported through the digital network thus saving a significant portion of the medium bandwidth. Decompression of the received voice packets is done by the same bunch of DSP's so to sustain a transparent full duplex phone conversation between two parties at both ends of a network as shown in Figure 1
The packets are typically 160 byte large before compression representing 20 milliseconds of a phone conservation (digitized voice channel at 64 kbits/sec). After processing the source sampling of the phone conversation is compressed into only 32 bytes. The compression and decompression algorithm are running on a bunch of Motorola DSP's 56166 capable of handling several voice channels each. Thus, the role of the dispatcher is to be able to continuously feed the DSP's from an Inbound Buffer and collects the compressed or decompressed packets to the Outbound Buffer while more than 100 phone conversations are in progress.
The source and sink of the packets brought to the Voice Server is a port of a switch representing a particular node in the digital network through which the phone communications are established.

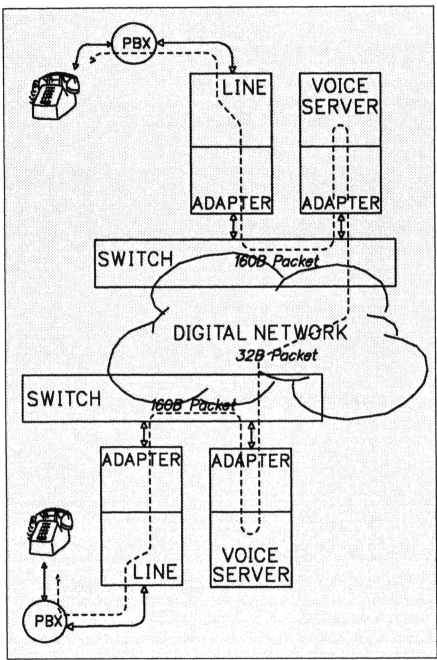

Figure 1. **A phone conversation through a Digital Network**

2. How it is done:

The Motorola DSP 56166 is equipped with two serial ports to load or fetch packets. To permit a complete decoupling between loading and fetching one port is used to load the data to process while fetching may go on independently on the second port when a processed packet is ready to go. Therefore, there is no contention possible between the two operations. The collector is informed that a job is ready to go because an interrupt is raised by a DSP while loading rate of the jobs, from the dispatcher, must stay compatible with the total processing capability of each DSP (35 MIPS).

The other end of the dispatcher is a 4 byte interface (DMA-like) with a buffer memory, managed by a specialized controller. It receives through the inbound switch port the remote compressed packet (to be de-compressed) or the local packets that need to be compressed before they are sent at the remote node through the network.

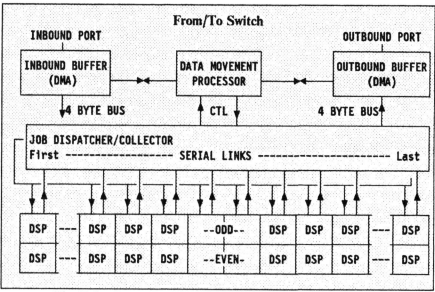

Figure 2. The Voice Server function

2.1 A Buffer Serializer for Dispatching the Jobs

The heart of the dispatcher function is a buffer serializer whose principle is shown in Figure 3. A 4 byte word fetched from the buffer memory is first latched into an interface register then immediately after temporarily stored, at two consecutive addresses, into a set of sixteen 16x1 internal RAMs. The 32 bits are then, soon after, unloaded serially to a DSP through a serial port in two consecutive 16 bit frames as DSP 56166 serial port mode of operation calls for.

The parallel loading and the serial shifting are done alternatively. Every second cycle (the system clock is running at 60 Ns or 18 Mhz.) there is an opportunity to load in parallel two bytes into one buffer while the other half of the cycles are used to serialize the data on all the serial links at a rate of 1 bit every 120Ns which is the maximum frequency at which the DSP serial ports may be operated.

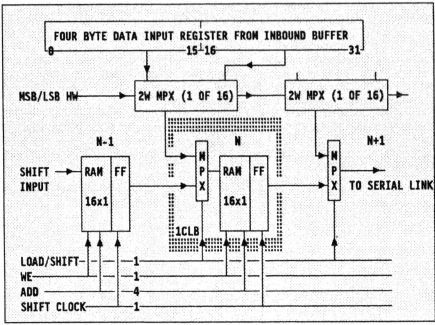

Figure 3. **Dispatcher Buffer Serializer**

2.2 Buffer Serializer Cell Mode of Operation

The way a buffer-serializer cell (in the dotted area of Figure 3) is operated is summarized in Table 1. The 16x1 RAM may be loaded either from the upper or lower half of the data input register.

To actually write into the RAM 'WE' strobe must be active. Otherwise the buffer serializer is in standby although the shift clock is free running. Shifting of the data occurs by writing the RAM contents from one device to the next one through the latches with following timing:

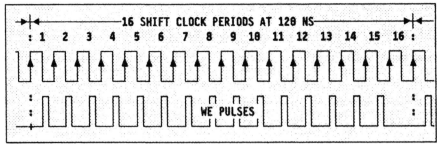

Figure 4. **Shift Timing**

The address where the shift occurs must obviously be stable. 'WE' strobe must be ON and the RAM input multiplexer selects the previous RAM output into the chain so the data shift from left to right from one 16x1 RAM to the other at clock rate. The first bit to go out on the serial link toward a DSP is the one from the most right device. The most left device is filled with 0's (tie down). At the end of the shift the corresponding row (16 bit wide) is cleared since 0's have been pushed in while a 16 bit word has been transferred to the DSP's.

Table 1. Serializer mode of operation			
+ LOAD	+ MSB/-LSB	+ WE	STATUS:
X	X	0	*STANDBY*
1	1	1	*LOAD BITS 0-15*
1	0	1	*LOAD BITS 16-31*
0	X	1	*SHIFT*

2.3 RAM Buffering Organization

The temporary buffering of data is organized within the RAMs as shown in Table 2. For a given user 8 bytes are pre-fetched before serialization may occur. This leaves enough time to pre-fetch data while serialization goes on on all the links. Serialization and prefetching are alternatively done on a per (60 Ns) cycle basis in such a way that both interfaces are active together thus implementing a pseudo dual port (one parallel, one multi serial) scheme.
The Figure 2 and Table 2 show that there are two DSP rows one EVEN and one ODD that are alternatively fed through a common set of serial ports. The chief reason for this is that DSP's are unable to sustain a continuous stream of bits. A pause must be observed between two strings of 16 bits. Thus the application rather toggles between ODD and EVEN DSP's to keep the serial links continuously busy.
Therefore, the temporary buffering is organized in a such a way that while data are fetched on the upper interface and stored in the "FLIP" port serialization goes on on the bottom serial links from the "FLOP" portion. And vice versa.
Whenever there are no data to be transferred to a DSP (just because there are no Voice Packets to process) the corresponding temporary buffering portion is not fed and the validation bit contained in a 17th bit is not set. Thus, when serialization will later occur the Serial Port for this data string will not be actually started (Frame Synchro line is OFF. See references [1] and [2]).

Table 2. RAM organization				
1	16 bits			
VALIDATION BIT MUST BE ON	DATA FOR ODD DSP's	FLOP	15	(Bits 16-31)
			14	(Bits 0-15)
			13	(Bits 16-31)
			12	(Bits 0-15)
		FLIP	11	(Bits 16-31)
			10	(Bits 0-15)
			9	(Bits 16-31)
			8	(Bits 0-15)
	DATA FOR EVEN DSP's	FLOP	7	(Bits 16-31)
			6	(Bits 0-15)
			5	(Bits 16-31)
			4	(Bits 0-15)
		FLIP	3	(Bits 16-31)
			2	(Bits 0-15)
			1	(Bits 16-31)
			0	(Bits 0-15)

2.4 Configurable Logic Blocks (CLB) count

Logic implemented in XILINX FPGA must fit into a certain number of Configurable Logic Blocks or CLB's. A basic dispatcher cell is shown in Figure 3 fit in one XC4000 series CLB. This includes the input multiplexer, the 16x1 RAM itself and the output flip flop. Thus, an array of fourteen 17 bit buffer serializer (the 17th bit is the validation bit) which is the core of the function, requires only:

One serializer .... $17 \times 1 = 17$ CLB's
Fourteen Serializer .... $14 \times 17 = 238$ CLB's
Plus the input reg with MPX .... $238 + 16 = 254$ CLB's

which easily fits into a XC4013 (576 CLB's) with the rest of the control logic not discussed in this paper. Then, the buffer serializer array represents a total 14x17x16 or almost 4 k bits of static RAM that would need to be done entirely with F/F in a standard gate array.

Furthermore the wiring of the cells is straightforwards from the CLB structure. Each CLB needs only to be connected to its neighbor one (down in a column) with only one wire as shown in Figure 5. All the other controls signals and the data inputs are distributed using the metal "Long Lines" another feature of Xilinx FPGA's. *Keeping the wiring simple is the key factor to get the product running at 60 Ns in worst case conditions* .

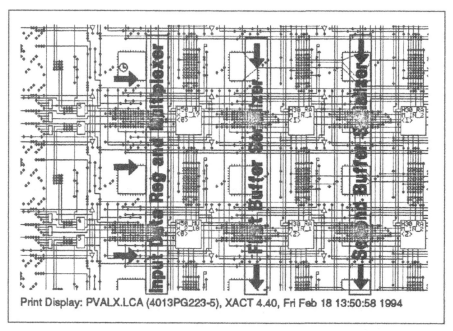

Print Display: PVALX.LCA (4013PG223-5), XACT 4.40, Fri Feb 18 13:50:58 1994

Figure 5. A few Buffer serializer cells

3. Job Collector

The collection of the packets after processing by the DSP's is just the opposite of what has been described up to now. Reading is done one bit at a time on the serial links and enter a similar array of RAMs in the Collector FPGA. Whenever enough data has been assembled they are transferred, 4 byte at a time, to the outbound buffer. When a complete packet is ready it is routed to the network through the Voice Server outbound port.
This part is very similar to the dispatcher and is not be further described.

4. Summary and Conclusion

It is a purpose of this paper to demonstrate that, taking advantage of the internal RAMs of a RAM based FPGA, like the 4000 Series of XILINX, it becomes possible to carry out a function that otherwise would require a too large amount of flip flops to be feasible in a FPGA or would require a standard gate array as large as 50 kcells to do the equivalent.
Although FPGA are expensive devices the approach retained permits to go through the engineering and pre-production phases with a re-programmable device while the production phase will use a less expensive hard wired solution. *The important point is that both real estate and performance of the FPGA phase need to keep up with the hard wired solution.*
The use of the internal RAMs was the answer to this challenge in this particular application.

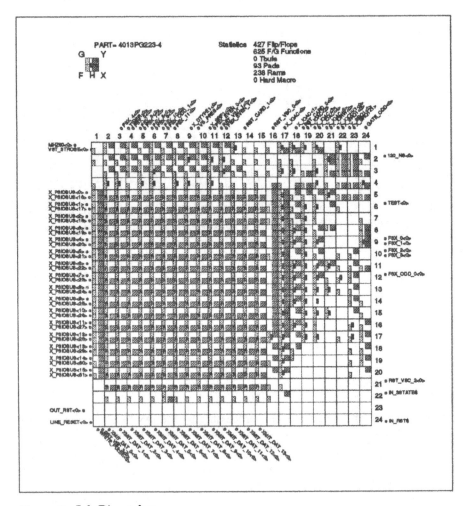

Figure 6. **Job Dispatcher**

References

[1] "16-bit General Purpose Digital Signal Processor" DSP 56166 *MOTOROLA INC.* Technical Data Sheet, 6/15/93.
[2] "DSP56166" Digital Signal Processor User's Manual *MOTOROLA INC.* 1993.

An Optoelectronic 3-D Field Programmable Gate Array

J. Depreitere[1], H. Neefs[1], H. Van Marck[1], J. Van Campenhout[1], R. Baets[2], B. Dhoedt[2], H. Thienpont[3] and I. Veretennicoff[3]

[1] University of Ghent, Electronics and Information Systems Department
St.-Pietersnieuwstraat 41, B-9000 Ghent, Belgium
[2] University of Ghent Electronics, Information Technology Department
St.-Pietersnieuwstraat 41, B-9000 Ghent, Belgium
[3] Free University of Brussels, Applied Physics Department
Pleinlaan 2, B-1050 Brussels, Belgium

Abstract. Traditional Field-Programmable Gate Arrays suffer from a lack of routing resources when implementing complex logic designs. This paper proposes two possible improvements to the FPGA structure that could alleviate these problems. We suggest extending the FPGA class to 3-D architectures. The 3-D architectures could be constructed of a stack of optically interconnected 2-D planes. Furthermore, we suggest a hierarchical distribution of routing resources that closely matches the wire length distributions of the intended class of applications.

1 Introduction

Field-Programmable Gate Arrays (FPGAs) are a rapidly growing class of electronic components. They offer a low-cost, off-the-shelf solution for implementing or prototyping a broad range of digital designs. This is achieved by using programmable logic blocks interconnected via programmable routing resources.

Although routing resources consume the major part of the chip area, complex designs remain difficult to implement. Due to routing problems, logic block utilization seldom achieves more than 50% and critical paths are forced into more indirect routes [1]. This has a negative impact on the performance of the implemented circuit. We are looking at ways to overcome these problems.

One way to model circuit complexity is by using Rent's rule [2]. Rent considers hierarchical models of circuit interconnection graphs. At every level, nodes are grouped into modules, to form the nodes of the next higher level. The grouping is such that the total number of pins emerging from the modules is minimal. This results in the following relation between the average number of pins emerging from the modules and the average number of (basic) nodes inside the modules:

$$P = C\,B^r, \;\; 0 \le r < 1 \;, \tag{1}$$

where P is the average number of pins emerging from the modules, B is the average number of (basic) nodes inside the modules, C is a constant related to the average fanout of the nodes and r is the Rent exponent. The value of $r = 0.5$

represents 'easy' to route, planar-like circuits. Complex, highly interconnected designs are characterized by larger values of r. Studies about placement of digital logic [3, 4] show that circuit complexity – characterized by the Rent exponent – and wire lengths of the implemented circuits are strongly correlated. Furthermore, it is clear that the wire length of the implemented circuits and the routing area are also correlated.

An analysis of the interconnection lengths of designs implemented in three dimensional structures [5] shows that interconnection length is significantly lower compared to two dimensional implementations. The gains are most pronounced with high-Rent designs. The availability of integrated optoelectronic components, allowing massive parallel free-space interconnections, holds the promise of constructing three dimensional systems. Hence, we consider building a 3-D optoelectronic FPGA structure consisting of a stack of 2-D electronic planes that are optically interconnected.

Furthermore, the design of routing resources in traditional FPGAs seems rather ad hoc. We state that routing resources could be more efficient when they are more closely matched to the needs of the logic designs that would be implemented in FPGAs. Based on a study [3] we make some suggestions that could offer an alternative for the seemingly ad hoc nature of existing interconnection structures.

2 Improvements to the FPGA Architecture

2.1 Using 3-D FPGA interconnections

The construction of isotropic three dimensional FPGAs, having the same interconnection density in all three dimensions, would lead to a massive increase of the number of interconnections. It seems obvious that routing problems would thus be decreased. However, the technical requirements may preclude the realization of such architectures in the foreseeable future.

Therefore, we propose an architecture with an anisotropic interconnection structure (see Fig. 1). We create a 3-D system as a stack of interconnected 2-D electronic planes. Each horizontal plane consists of a large FPGA structure. The FPGA's logic blocks are grouped into modules. In between these modules, we place interconnections to the next plane. These interconnections between planes could be either electrical (through vias [6]) or optical. It is clear that such architectures will exhibit a smaller gain in routing resources than truly isotropic 3-D systems. Hence, one should carefully investigate the cost/benefit ratio of varying degrees of sparseness of the interconnections in the third dimension.

Van Marck [5] has studied the average interconnections lengths of designs implemented into 2-D and 3-D architectures. He examined isotropic Manhattan grids – i.e., grids where the interconnection density is the same in all three dimensions – as well as anisotropic grids (see Fig. 2). Although the model is still quite simple and does not take routing problems into account, the study shows that the reduction in average wire length is sufficiently large to arouse one's interest. This indicates that anisotropic 3-D systems will offer advantages over 2-D systems.

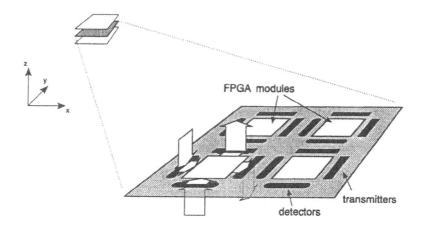

Fig. 1. Proposed architecture for the 3-D FPGA

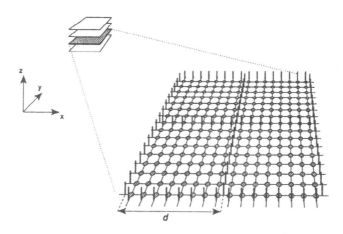

Fig. 2. Anisotropic grid: one layer of a cubic grid with sparse interconnections to the next layer

2.2 Improved 2-D interconnection topology

General-purpose FPGAs (e.g. Xilinx FPGAs [7]) have a rather small number of local interconnections, i.e., interconnections between near-by logic blocks. Long-line interconnections provide large fanout interconnections that cross the entire width of the chip. All other routing resources can be programmed to interconnect any two blocks. However, this flexible way of routing is not very area-efficient and causes large delays. Furthermore, when implementing complex designs in such structures, one very often finds that there is a lack of routing resources.

To avoid this problem, we should approach the routing resource architecture form a different point of view. The routing architecture of an FPGA should match the "needs" of a large class of logic designs. Studies about wire lengths [3, 4] have examined placements of digital logic on square grids. From these studies it follows that optimally placed logic designs have a wire length distribution function that is given by:

$$f_k = g/k^\gamma, \quad 1 \le k \le L$$
$$\approx 0, \qquad k > L \ , \tag{2}$$

where f_k is the fraction of wires with length k; g is a normalization constant; L is a constant related to the size of the array and the adequacy of the placement; and γ is related to the complexity of the circuit architecture (based on Rent's rule [2]).

Consequently, an FPGA should have a hierarchy of routing resources leading to an interconnection distribution given by (2). First, FPGAs should have more local routing resources, which could be hard-wired and will thus be faster. Secondly, there should be a smaller number of longer routing resources. These longer interconnections need not be general; i.e., they need not to be able to interconnect any two blocks, but they should have a distribution according to (2).

The structure of the routing resources of the Triptych FPGA [1] has emerged from reflections about the inherent fanin/fanout trees of logic designs. It therefore comes as no surprise that the resulting structure satisfies (2).

These suggestions should lead to architectures in which implemented designs are faster and easier-to-route. A first step towards such systems has been made and has resulted in the design of a 3-D optoelectronic FPGA demonstrator.

3 The Demonstrator

3.1 Goal

The current complexity of the optoelectronic FPGA demonstrator and its physical form are not geared towards demonstrating the attainable routing gains. It is aimed at establishing the feasibility of free-space optical interconnections – at the logical circuit level – between traditional planar subsystems. We have chosen to use optical interconnections because of the increasing availability of optoelectronic interconnection devices such as LEDs (Light Emitting Diodes) and VCSELs (Vertical-Cavity Surface Emitting Lasers). Among others, optical interconnections hold the promise of high-bandwidth data transfers in galvanically isolated subsystems. Furthermore, the use of **free-space** interconnections allows a much easier cooling of the subsystems.

3.2 General Structure

In view of the technological capabilities at hand in our research teams, we have chosen to construct the demonstrator as shown in Fig. 3. It consists of a stack of three robust, metal frames. Each metal frame holds a PCB and a small glass plate. The PCB carries the FPGAs, while the glass carries the optics-related components, i.e., the LEDs, driver chips, detector diodes and receiving amplifiers. The metal frame has two fittings through which a reproducible positioning of the planes relative to each other is realized. The lateral positioning accuracy between two adjacent planes should be better than 10 μm. Proper functioning of the optical links necessitates an accuracy of 50 μm. The layer separation is 5 mm, while the lateral measures are 150 × 140 mm. On each PCB we put 4 CMOS FPGA chips. These chips are fabricated in a 1.5 μm ES2 process. They measure 8518 μm × 6300 μm. On the glass there are a LED array, two driver chips, one detector array and two receiver chips. The LED array has integrated diffractive lenses [8] and contains 16 LEDs. The infrared LEDs are used at a bitrate of 50 Mbit/s. The LED arrays are bonded using solder bumps or gold bumps. The detector arrays are fastened with anisotropic conducting adhesive film (ACAF). Receiving lenses, enhancing light collection and reducing cross-talk, are provided on the bottom of the glass.

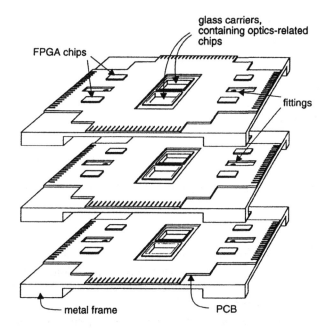

Fig. 3. Physical form of the demonstrator

3.3   The 2-D FPGA structure

Each layer of the stack contains a 2-D FPGA structure, consisting of an array of two by two smaller FPGAs. One small FPGA contains an array of four by four logic blocks. Hence, one layer of the three dimensional FPGA contains an array of eight by eight logic blocks. So, the full 3-D FPGA contains 3 × 8 × 8, or 192 logic blocks.

Routing Resources. The routing resources of the small FPGAs were inspired by the structure proposed in the Triptych architecture [1]. The purpose of the routing scheme is to match, as closely as possible, the routing requirements of a hierarchically designed circuit, as explained in section 2.2. The routing resources consist of interconnections through the logic blocks, direct interconnections, the so called long line interconnections and interconnections to the optical devices.

Routing through logic blocks. Beside computing a logic function, the logic blocks of both the Triptych and our architecture can be used to route signals (see Fig. 4). This allows a more versatile use of the routing resources. Furthermore, since a trade-off can be made between routing and logic, less area remains unused due to routing problems.

Direct interconnections. The structure of the FPGA consists of logic blocks having unidirectional diagonal direct interconnections between logic blocks (see Fig. 5(a)). The resulting scheme is then mirrored bottom to top and overlaid on the original scheme. In this way we get an upward and a downward data flow. Finally, the data flow direction can be changed by using the so called feedback connections. The direct interconnections are hard-wired and therefore should be used for local high-speed communication between logic blocks.

Long line interconnections. The long line interconnections are longer range connections and have a larger fanout (see Fig. 5(b)). As with the direct interconnections, the long line downward data flow is obtained by mirroring the scheme of Fig. 5(b) bottom to top and overlaying it on the original scheme. There are two types of long line interconnections:

- intra-chip long line interconnections, which connect one logic block with the next row of logic blocks inside the chip. This type of long line interconnections is hard-wired. Due to the limited number of logic blocks in the chip, these long line interconnections have a fanout of two.
- inter-chip long line interconnections, which can be **programmed** to connect one logic block with the next row of logic blocks inside the chip as well as with the next row of the neighbouring chips. These long lines can have a larger fanout.

Optical interconnections. The output of the logic functions of the logic blocks are hard-wired to the optical transmitters. The data of the four logic blocks of one row are multiplexed. By multiplexing the data from the four logic blocks on the same row, we can take advantage of the large bandwidth of the optical connections. After transmission and demultiplexing, the data can be interchanged, in fact providing us with a simple space/time routing switch. By doing so, the data from a logic block can be connected with any block of the row of logic blocks above it. This leads to factor four decrease of the number of optical components (see Fig. 5(c)).

Logic Block Architecture. The structure of the logic blocks is shown in Fig. 4. This LUT-based logic block is capable of simultaneous function calculation and signal routing, as in the Triptych architecture. Two functions of four inputs can be calculated or one function of five variables. Studies on the routing resources of LUT-based FPGAs indicate that 4-input LUTs are the most area-efficient [9]. Nevertheless, it is not clear if this holds for our type of routing resources.

Every input of the logic block can be programmed to feed the two LUTs. The outputs of the LUTs can either be latched or unlatched. Furthermore, the outputs of the flip-flops can be fed back and used as inputs of the LUTs. By doing this, we can implement finite state machines.

At present, the electrical and mechanical parts of the demonstrator are being assembled and tested; the final integration with the optoelectronic components is expected to take place within the coming months.

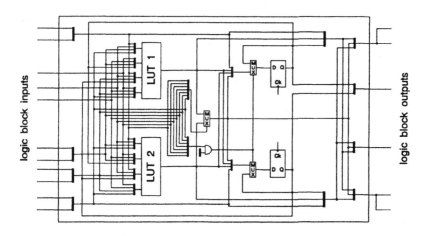

Fig. 4. Routing and Logic Block

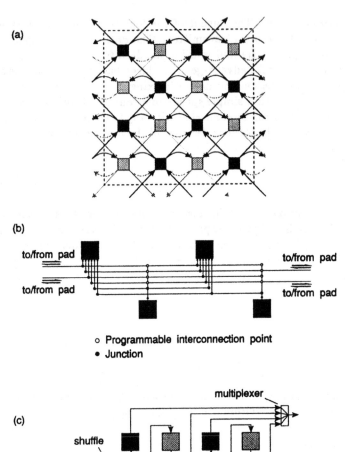

(a)

(b)

to/from pad   to/from pad

to/from pad   to/from pad

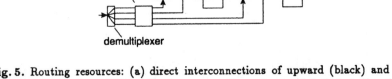

o Programmable interconnection point
• Junction

(c)

Fig. 5. Routing resources: (a) direct interconnections of upward (black) and downward (grey) data flow; (b) long line interconnections of upward data flow direction; (c) interconnections to optoelectronic devices for one row of logic blocks

4 Conclusions

In this paper we have presented a possible way to alleviate routing problems in FPGAs. We have proposed to introduce a third interconnection dimension, effectively reducing estimated interconnection length of implemented complex designs. We have suggested to realize this additional interconnection dimension using free-space optical interconnections. We have also touched upon more appropriate ways to provide suitable interconnection structures in future FPGAs.

Our theoretical research effort is being complemented by the design and realization of a prototype 3-D optoelectronic FPGA demonstrator, the purpose of which is to establish the feasibility of the proposed optical interconnection technique.

5 Acknowledgements

The above text presents research results of the Interuniversity Attraction Poles Program IUAP24, initiated by the Belgian State, Prime Minister's Service, Science Policy Office. Additional support was provided by an IMEC research program.

References

1. G. Borriello, C. Ebeling, and S. Hauck. Triptych: An FPGA architecture with integrated logic and routing. In J. Savage T. Knight, editor, *Advanced Research in VLSI and Parallel Systems*, pages 26–43. The MIT Press, 1992.
2. B. S. Landman and R. L. Russo. On a pin versus block relationship for partitions of logic graphs. *IEEE Trans on Computers*, C-20:1469–1479, 1971.
3. W. E. Donath. Wire length distribution for placements of computer logic. *IBM J. Res. D.*, 25:152–155, 1981.
4. M. Feuer. Connectivity of random logic. In *Proceedings of the workshop on large-scale networks and systems*, pages 7–11. IEEE 1980 Symposium on Circuits and Systems, 1981.
5. H. Van Marck and J. Van Campenhout. Modeling and evaluating optoelectronic designs. In R. T. Chen and J. A. Neff, editors, *Optoelectronics II*, pages 2153:307–314. Proceedings of the SPIE, 1994.
6. R. C. Eden. Capabilities of normal metal electrical interconnections for 3-D MCM electronic packaging. In R. T. Chen and J. A. Neff, editors, *Optoelectronics II*, pages 2153:132–145. Proceedings of the SPIE, 1994.
7. Xilinx Inc. *The Programmable Logic Data Book*, 1994.
8. B. Dhoedt, P. De Dobbelaere, J. Blondelle, P. Van Daele, P. Demeester, H. Neefs, J. Van Campenhout, and R. Baets. Arrays of light emitting diodes with integrated diffractive microlenses for board-to-board optical interconnect applications: design, modelling and experimentel assessment. *Accepted for publication in CLEO '94 technical digest*, 1994.
9. S. D. Brown, R. J. Francis, J. Rose, and Z. G. Vranesic. *Field-Programmable Gate Arrays*. Kluwer Academic Publishers, 1992.

On Channel Architecture and Routability for FPGA's under Faulty Conditions

Kaushik Roy[1] and Sudip Nag[2]

[1] Electrical Engineering, Purdue University, West Lafayette, IN, USA
[2] Electrical Engineering, Carnegie-Mellon University, Pittsburgh, PA, USA

Abstract. The Field Programmable Gate Array (FPGA) routing resources are fixed and their usage is constrained by the location of Programmable Connections *(PC's)* such as antifuses. The routing or the interconnect delays are determined by the length of segments assigned to the nets of various lengths and the number of *PC's* programmed for routing of each net. Due to the use of *PC's* certain unconventional faults may appear. This paper models the *PC* faults and analyzes the performance of FPGA channel architecture under faulty conditions to achieve 100% routability with graceful degradation in performance. A channel architecture has been synthesized to achieve routability and performance even under faulty conditions. Results on a set of industrial designs and MCNC benchmark examples show the feasibility of achieving routability and performance under a large number of faults in the channel.

1 Introduction

Field Programmable Gate Arrays (FPGA's) combine the flexibility of mask programmable gate arrays with the convenience of field programmability. Figure 1 shows the row-based FPGA architecture [1, 2]. Each row of logic modules is separated by channels. Each channel has a fixed number of horizontal routing tracks which are segmented. For example, Figure 1 has 3 tracks per channel. The topmost track is divided into two segments *a* and *b* separated by a *horizontal* antifuse (*hfuse*). In the unprogrammed state the antifuse offers a very high resistance, and hence, there is no electrical connection between the segments. A low resistance electrical connection between the segments can be established by programming the antifuse. Dedicated vertical lines through each input and output pin of a logic module connect the pins to the routing tracks. Vertical feedthroughs pass through the modules, serving as links between different channels. There is a *cross* antifuse (*cfuse*) located at the crossing of each horizontal and vertical segment. Programming these antifuses produces a bi-directional connection between the horizontal and vertical segments. Let us again consider Figure 1. Due to the different choices available during routing, it may be possible to achieve 100% routability even under the presence of a large number of faults. One can also notice that there are a large number of both cross and horizontal antifuses present in the channel to achieve flexibility in routing. However, most the antifuses remain in the unprogrammed state even after FPGA programmation.

An alternate scheme replaces antifuses by switches, making the architecture *reprogrammable*. This is achieved at the cost of larger area required for the switches. An antifuse is of the size of a *via*, and requires very small area.

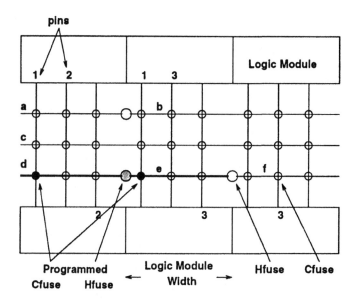

Fig. 1. FPGA Channel Architecture

For such an FPGA architecture, it is not adequate to consider only the stuck-at faults for the logic modules. The unconventional *PC* short and open faults in the channel will also have to be considered to determine the complete functionality of the chip. In this paper we model the *PC* faults and address the design and routability of the FPGA channel architecture to achieve 100% routing with minimum performance penalty in the presence of *PC* faults.

The paper is organized as follows. Section 2 considers the possible faults in FPGA's and models the faults. The routability of a channel and the timing performance associated with routing are described in Section 3 along with the routing algorithm. Section 4 considers channel routability under the presence of faults. Section 5 deals with automatic synthesis of channel architecture to achieve routability and performance under the presence of faults. Section 6 presents the results of injecting faults into the synthesized architecture and the conclusions are drawn in Section 7.

2 Fault Modeling and Fault Location

Let us first consider antifuse technology for *PC's*. Routability of the FPGA's requires the presence of a large number of *cfuses* and *hfuses* in the channel. An antifuse requires multiple programming pulses to successfully form the electrical connection between logic modules or segments. And depending on the circuit size and device type, 3,000 to 20,000 antifuses are typically programmed [3].

Two types of faults in the antifuses are considered. The first type of fault can be diagnosed *a priori* i.e. before programmation and are called *type1* (or short) faults. A shorted (or already programmed) *cfuse* or *hfuse* is an example of type1 fault. Such faults can be easily detected by the programming circuitry which is able to address each antifuse individually. For example, programming a *cfuse* X_1 in Figure 1 requires charging the vertical line crossing X_1 to a voltage VPP and the segment d to GND (0V). VPP is determined by the antifuse technology under consideration. A voltage stress of VPP across the antifuse for a certain period of time creates a low resistance (technology dependent) connection between the vertical line and the horizontal segment. All other vertical and horizontal lines are charged to a voltage of $VPP/2$. It can be noted that the other unprogrammed *cfuses* experience a voltage stress of 0 volts or $VPP/^o$ volts, and hence remain unprogrammed. The above programming concept can be used to stress any unprogrammed *cfuse* to a $VPP/2$ volts to detect if a low resistance short exits or not. *Type2* (or open) faults cannot be diagnosed *a priori*. Let us consider an antifuse which shows a normal behavior (open) in the unprogrammed state but after programmation does not produce a low resistance connection. Such a fault cannot be diagnosed before the FPGA is programmed. It can be noted that if *reprogrammable PC's* are used instead of antifuses then both open and short faults are detectable before FPGA programmation. After programming, a faulty antifuse may produce an electrical connection between the horizontal and vertical line or between two horizontal segments having a resistance higher than the nominal value. In such cases, the path which includes that programmed antifuse in it may experience larger than normal delay, producing delay or timing error.

The *hfuse* faults can also be classified into *type1* and *type2* categories. The *type1* faults can be detected by precharging the horizontal segment adjacent to any unprogrammed *hfuse*. A *type1* fault on the *hfuse* would also charge up the other adjacent segment. As in the case of *cfuses*, both *type1* and *type2* faults may produce a moderately high resistive connection between two adjacent segments producing a timing fault. It should be noted that each horizontal segment can be separately precharged by the programming circuitry. This helps in *vertical line* and *channel track testing*. The test charges up each track, and after a predetermined time needed to maintain the level, the charge must still be high to allow a pass.

All the inputs and ouput to each logic module can be individually addressed and accessed through serial shift registers. And the inputs can be toggled through all the test vectors required to test each combinational and sequential modules completely for any stuck-at faults.

3 Routing

The channel routing problem is formulated as an assignment problem where each net within a channel is assigned to one or more unassigned segments. A net in a channel can use at most one track due to a technology constraint which does not allow programming of antifuses connected in an L-shaped fashion [1].

Consider K-segment routing for a net x of length L_x which uses p segments ($1 \leq p \leq K$) and H_x horizontal PC's. We define the cost of routing net x as

$$C_x = w_1.\alpha + w_2.\beta, \quad where \quad \alpha = \frac{(\sum_{j=1}^{p} L_j) - L_x}{(\sum_{j=1}^{p} L_j)} \quad and \quad \beta = \frac{H_x}{K}$$

α and β are penalties for segment length wastage and horizontal PC usage, respectively, and are both positive and less than 1. The factor α is associated with both routability and performance because the unprogrammed antifuses add to capacitive loading, while β is associated with routing performance. The weights w_1, w_2 assigned to the wastage factor, and the horizontal PC usage factor respectively, are technology-dependent.

Green et. al. [5] have shown that K-segment ($K > 1$) channel routing problem is NP-complete. For K-segment routing, each net is allowed a maximum of K adjacent segments (on the same track) for routing. For our purposes, we use a fast, greedy routing algorithm. The nets within a channel are ordered in decreasing order of length. We assume that the longer nets are more critical, and hence, they are routed first. However, each net can be assigned a criticality value and depending on that the nets can be ordered. Let the leftmost and the right most coordinate of a net x (or a segment i) be given by $left_x$ and $right_x$ ($left_i$ and $right_i$) respectively. Net x is routable using segment i if the segment has not been previously assigned to any other net and the following conditions are met:

$$left_x > left_i \;, \; right_x < right_i$$

Such conditions can be easily extended when two or more adjacent segments are required for routing the net on a given track.

4 Routability under the Presence of Faults

For the FPGA architecture of Figure 1, 100% routability may be achieved even under the presence of faults in the channel. Let us first consider the cross PC and the horizontal PC faults in the channel. The cross PC's are located at the crossing of each vertical line, which connects to a pin of a logic module, and the horizontal routing tracks. Typically, the router connects each pin to one horizontal track. The rest of the cross PC's on that vertical line remain unprogrammed. Similarly, most of the horizontal PC's also remain in the unprogrammed state because only a few of the nets require more than one segment for routing. So it may be possible to route nets such that the PC faults do not cause an error to occur during the normal operation.

After detecting the type1 faults as shown in Section 2, routing can be performed around the faulty cross PC or the horizontal PC, if possible. *Type2* faults for antifuses cannot be diagnosed *a priori*, and hence, routing reconfiguration is not possible. However, if reprogrammable PC technology is used, then both short and open faults are detectable before programmation (in the test mode), and hence, routing around both the open and the short faults might be possible.

Let the probability that a cross PC is faulty (both *type1* or *type2* faults) be given by f. Each vertical line in a channel has T number of cross PC's, where T is the number of tracks. There are T number of segments going across any vertical line in a channel. The vertical line is connected to either a pin i in a logic module or is a feedthrough across channels. The number of available tracks that pin i, and therefore net i, can be assigned to is given by N_i, where $N_i \leq T$. N_i is a function of the number of nets routed in the channel before routing net i and the segments assigned to those nets. We assume that pin i can be assigned to any one of the available N_i tracks. Therefore, the probability that net i uses the jth available track is given by $1/N_i$. If we assume exactly one cross PC fault per vertical line within a channel, then the probability that net i gets assigned to a track segment having a faulty cross PC on that vertical line is given by

$$p_1 = \frac{1}{N_i}.f.(1-f)^{N_i-1}$$

It follows from the above discussions that the probability of 100% routability of the net in the presence of upto n faults on a vertical line in a channel is

$$R = 1 - \frac{1}{N_i} \sum_{j=1}^{n} j.f^j.(1-f)^{N_i-j} \tag{1}$$

$$= 1 - P \tag{2}$$

where P is the probability of an error occurring with at most n faults on a vertical line. It should be observed that a *type1* fault on a *cfuse* can be handled by assigning net i to the corresponding track segment. Any *type2* faults on unprogrammed cross PC's can also be tolerated. Therefore, in reality, the probability of 100% routability of a net in the presence of faults is greater than R.

The above fault types can also be considered for horizontal PC's to come up with analytical expressions for routability. Let the probability that a net i is routed using one segment be given by α_{i1}. α_{i1} is a function of net length and its spatial location within a channel for a given channel segmentation. Let α_{ip} be the probability of routing net i with p segments requiring the programmation of $(p-1)$ horizontal PC's. Let us also assume that a maximum of $(K+1)$ segments (K *hfuses*) be allowed for routing of any net within a channel. The number K is user defined, and is associated with the routing performance, because each programmed antifuse contributes positively toward critical path delay. Such a routing scheme is defined to be *(K+1)-segment routing*. If we only consider horizontal PC faults, then the probability of faulty routing for net i using exactly 2 adjacent segments (one horizontal PC) is given by

$$Q_{i2} = \alpha_{i2}.f_2$$

where f_2 is the probability of *type2* fault on a horizontal *PC*. Note that it is possible to tolerate a *type1* fault if two or more segments are used for routing. Similarly, for exactly 3-segment routing for net i

$$Q_{i3} = \alpha_{i3}[f_2^2 + C_1^2.f_2.(1 - f_2)]$$

where C_m^n represents n *choose* m. It can be shown that

$$Q_{i,p+1} = \alpha_{i,p+1}.\sum_{j=1}^{p} C_{p-j}^p.f_2^j.(1 - f_2)^{p-j} \tag{3}$$

From the above analysis it is clear that the probability that net i can be routed using one or more segments (upto K *hfuses*) is

$$V_i = 1 - \sum_{j=1}^{K} Q_{ij}.\alpha_{ij}$$

The unprogrammed horizontal *PC's* can potentially have *type2* faults without causing any routing error. As the majority of the nets use a single segment for routing, most of the *type2* faults can be tolerated. The *type1* horizontal *PC* faults associated with unassigned segments can also be tolerated. However, in order to achieve 100% routability, nets might get assigned to longer and/or large number of segments which in turn can increase critical path delays.

5 Fault-Tolerant Channel Architecture

A routing solution is dependent on the *existing placement* which defines the routing requirements, the *channel architecture* which defines the available routing resources and the *routing algorithm* which efficiently uses these routing resources so that the final routing solution meets some performance requirements. Apart from the usual performance requirements of 100% routability with critical path delay constraints, our router described in the previous section also addresses the issue of routing under faulty conditions. Evidently, such a router's ability to meet such performance criteria will be dependent on the channel architecture in addition to the routing algorithm used. Therefore, in order to improve routing solutions under faulty conditions, it is imperative to design the architecture with such a performance requirement in mind.

The primary difference between handling faults at the routing level and at the architecture design level is that while at the routing level, the information regarding the faults is used to determine the routing solution, at the architecture design level, no information is available regarding the location of faults. Therefore, we assume that the *PC* faults are randomly distributed across the channel. One of the other characteristics of the faults are that they are found to be clustered around some particular areas of the circuit [8]. Hence, we also consider clustered nature of faults in our analysis.

As pointed out in Section 1, the routability, performance, and fault handling capacity of a channel depends largely on the channel segmentation scheme. Intuitively, a strong correlation between the segment length and the net length distributions within a channel is very desirable so that the single channel architecture is able to handle different types net distributions. However, the mere existence of a unique segment of acceptable length for every net in a channel does not guarantee 100% routability, or required performance and fault tolerance. This is due to the fact that an additional factor, the *location* of a segment with respect to a net span in a channel is also important in determining whether that segment can be used for routing that net. It is imperative, therefore, to consider the *spatial* distribution of nets. The set of benchmark net distributions were obtained from Texas Instruments' gate array designs.

We extend our architecture design scheme originally targeted towards wirability and timing [6] so as to include the fault-tolerant capability for random and clustered fault distribution models. The basic approach used in our scheme was to generate an optimal architecture with respect to wirability and timing for a large set of sample net-lists. Our extension therefore also addresses the fault handling capability of such an architecture with respect to the large number of sample net-lists. The optimization technique used was simulated annealing [9] which explores a plethora of possible architectures and selects the optimal one based on a cost-function.

The routing cost for channel i, having the set of nets N_i, and a set S of already laid out segments is given by

$$C_i = \nu_r \cdot \mid N_{iu} \mid + \nu_w \cdot \sum_{k \in N_{ir}} \alpha_k + \nu_o \cdot \sum_{k \in N_{iu}} \frac{\theta_{kg}}{g_r - g_l} +$$

$$\nu_f \cdot \gamma + \nu_l \cdot \sum_Z \sum_G f(P_{zg} - M_g) \tag{4}$$

The original cost-function targeted wirability and timing, using the first three terms of Equation 4. We have added a fourth and a fifth term for addressing the fault-handling capability. We will describe these terms following a brief discussion of the first three terms which address the issues of wirability and timing. $\mid N_{iu} \mid$ in Equation 4 represents the cardinality of set N_{iu}, and is equal to the number of 1-segment unroutable nets in channel i. The set N_{iu} is the set of unroutable nets in channel i. For the set of 1-segment routable nets N_{ir} in channel i, the segment wastage factor α_k for each net k is calculated (refer to section 3). θ_{kg} corresponds to the maximum overlap of net k with an unassigned segment g, and is a measure of n-segment routability of a 1-segment unroutable net k. The left and the right coordinates of segment g are given by g_l and g_r respectively, and hence, $g_r - g_l$ represents the length of segment g. The weights ν_r, ν_w, and ν_o are associated with the corresponding factors. The routability weight ν_r is much higher than ν_w or ν_o as it relates to both *routability* and *performance*. The *1-segment routable* nets with very low segment wastage usually have lower interconnect delays than nets requiring two or more segments for routing due to the presence of programmed horizontal antifuse(s). In fact, the exact routing delay depends on the number of unprogrammed *cfuses* on the segment(s), the length of the segment(s), the resistance of any programmed *hfuse(s)*, and *cfuses*. The weight ν_o associated with the overlap factor is a small negative number. For unroutable nets we consider a larger overlap to be better – the net has a higher probability of getting routed using two or more segments. For p different channels (p sets of nets, N_i, N_2, .. N_p), the total cost of routing, C, using the same set of segments S is given by $C = \sum_{i=1}^{p} C_i$. The usage of *1-segment routability* and the *overlap factor* represents a novel way of efficiently predicting the K-segment routability of a segmented channel architecture.

The fourth term measures the fault-handling capability of the architecture. For a particular net-list, 1-segment routing is done. This results in a set of nets that are unroutable (using one segment) and a set of segments that are *free* or available for routing. The information on free tracks for every zone is stored. The unrouted nets would have to be routed using multiple segments. These nets would require tracks at least in the zones denoting the nets' span. We use this measure to estimate δ : a lower limit on the number of free tracks at each zone after complete routing. If δ is negative for a particular zone, this would guarantee unroutability. If δ is zero for a particular zone, there is a high probability of routability problems in that zone. These cases are therefore penalized (although indirectly) by the wirability related terms. However, an interesting observation here is that the larger δ is for a zone, the larger would be the number of tracks free for that zone. Therefore after complete routing, if some faults exist in that zone, larger δ implies larger fault-tolerance. In other words, a larger δ in a particular zone implies a larger probability that a channel router can achieve 100% routability in that zone despite the existence of faults in that zone since it would have extra free tracks to use in place of the faulty one.

Assuming random faults, the smaller the δ is in the zones, the less fault-tolerant the architecture would be. Specifically, the zonal fault-intolerance FT_z of a zone z is a function of δ for that zone. This function was derived empirically based on the observation that routability problems started appearing with δ values less than 3. γ, which is the fault-intolerance of the architecture is calculated as $\sum_{z \in Z} FT_z$.

Assuming clustered faults (the clusters themselves being randomly located anywhere in the chip), existence of clusters of adjacent zones with small δ would result in a less fault-tolerant architecture. Specifically, assuming clusters of length L, and a total of C such possible clusters in a channel, the cluster fault-intolerance FTC_c of a cluster c is a function of the FT_z of the zones forming the cluster c. If the FT_z values for multiple zones are large, then FTC_c is a large number depending on how many zones in cluster c have high FT_z values. γ, in this case, is calculated as $\sum_{c \in C} FTC_c$. ν_f is the weight factor used for the fourth term.

In any zone, if a fault causes the inability to use a particular segment, the usage of an identical segment of similar length would result in the minimum deviation of timing and segment wastage due to the fault. Therefore it is intuitively desirable to have in each zone, a few large segments, a few medium segments, a few small segments etc. To achieve this, the segments are divided into G groups based on their lengths. each segment group comprises segments of length between a certain group-specific range. Constraints are provided in each zone in the form of a minimum number of elements (M_g) required to be present for group g. Evidently there could be G such constraints.

For each zone, it is tested if all these constraints have been met. The fifth term measures the summation of all the violations of these constraints in different zones. For zone z, P_{zg} is the number of segments belonging to group g in that zone. If $P_{zg} \geq M_g$, it implies the constraint k has been met for zone z and function f returns 0. However if $P_{zg} = M_g - a$, it implies that constraint k has been violated by an amount a in zone z, and f returns a^2 in this case. The superlinear function is used to heavily penalize large deviations. ν_l is the weight factor for the fifth term.

Due to the complex nature of the cost function, simulated annealing was used. Given a large set of sample net distributions, annealing starts by assigning an arbitrary segmentation for a channel of given width and a given number of tracks. Two moves are allowed in this specific annealing algorithm — merging of two adjacent segments in a track and breaking of a segment within a track into two segments such that the broken segments add up to the original segment length. The segments are randomly selected for either merging or breaking. Merges or breaks are also determined randomly. It is not possible to break a segment of length equal to the width of a single logic module. After each move the cost C_i is calculated for each of the given sets of net distributions, $N_i,....N_p$. If C decreases from its previous value the move is accepted. However, a move with a higher C is accepted with a probability $e^{\frac{-|\delta(C)|}{Temp}}$, where $|\delta(C)|$ is the absolute value of the change in cost C and $Temp$ is the annealing temperature.

6   Results

The algorithms for fault-tolerant routing and architecture synthesis were implemented in C on a Sparc 10 workstation. We present below the results of our experiments with 7 MCNC and industrial examples from Texas Instruments with 900 to 2300 gates. The designs were logic synthesized and placed on TPC1010 [3] type template having 44 logic modules per row. Each channel had 25 segmented routing tracks. Table 1 shows the effect of injecting cross PC short faults in the channels. The number of logic modules (LM's) and the number of channels required to implement each design is also shown in the table. The columns show that 0, 15, 20, and 25 cross PC short faults were randomly injected into each channel of each design. The total number of cross PC's present in each channel is $NumberTracks \times NumberVerticalLines$, which is 25 x 572 = 14300 in our case. Results show that as the number of injected cross PC short faults decreases, the routability increases and so does performance. The total number of horizontal Programmable Connections $(PC's)$ used and the average percentage segment wastage over all the channels (refer to Section 3) are measures of routing performance and routability and are shown in the table. With 15 randomly injected faults per channel, routability was obtained for all the designs. For the unroutable designs there is a "–" entry in the column for percentage segment wastage.

Table 1. Routing results with cross PC short faults

Design Name	No. LM's	No. Channels	Number of Unrouted Nets				Horizontal Prog. Conn.				Avg. % Seg. Wastage			
			0	15	20	25	0	15	20	25	0	15	20	25
bw	144	8	0	0	0	0	0	0	0	1	38.4	38.7	38.9	39.0
duke2	318	9	0	0	1	2	6	7	6	12	42.7	42.6	–	–
f104667	262	9	0	0	0	0	0	0	1	1	46.8	46.9	46.9	47.2
f104243	512	15	0	0	0	0	0	0	0	0	41.8	42.3	42.1	42.7
f104780	671	19	0	0	0	1	12	14	16	23	40.5	41.2	41.5	–
f103918	782	19	0	0	0	0	4	5	5	4	38.8	39.0	39.1	39.4
cf92382a	668	19	0	0	0	0	0	0	0	2	46.6	46.7	46.7	46.6

Horizontal PC short faults were also randomly injected into each channel. Table 2 shows the routing results with 0, 7, 8, and 10 horizontal PC short faults injected randomly into each channel. All designs were routable with 7 randomly injected faults in each channel. However, routability and/or segment wastage deteriorated with the increase in the number of horizontal PC short faults in each channel.

Table 2. Routing results with horizontal PC short faults

Design Name	Number of Unrouted Nets				Horizontal Prog. Conn.				Avg. % Seg. Wastage			
	0	7	8	10	0	7	8	10	0	7	8	10
bw	0	0	0	0	0	0	0	0	38.4	38.7	40.1	40.1
duke2	0	0	1	1	6	8	10	11	42.7	42.7	–	–
f104667	0	0	0	0	0	0	1	2	46.8	46.9	46.9	48.0
f104243	0	0	0	0	0	0	0	0	41.8	41.8	41.9	41.9
f104780	0	0	0	1	12	14	18	18	40.5	40.7	41.2	–
f103918	0	0	0	0	4	6	8	6	38.8	38.8	38.8	39.0
cf92382a	0	0	0	0	0	0	0	2	46.6	46.9	46.9	47.2

Table 3 shows the routing results when cross PC open faults were randomly introduced into the channel. The routing results are compared with 0, 4%, 12% and 20% random cross PC faults. Even with 12% (1716) cross PC faults per channel, the router could route all the designs. It can be observed that as more cross *PC* faults were introduced, the routability and performance deteriorates which is reflected by the larger number of horizontal antifuse usage, and higher segment wastage. It should be noted that for antifuses such open faults are not *a priori* known and can only be detected while trying to program the antifuse.

Table 3. Routing results with cross *PC* open faults for reprogrammable PC's

Design Name	Number of Unrouted Nets				Horizontal Prog. Conn.				Avg. % Seg. Wastage			
	0	4%	12%	20%	0	4%	12%	20%	0	4%	12%	20%
bw	0	0	0	0	0	0	3	6	38.4	38.7	38.9	42.0
duke2	0	0	0	7	6	8	11	38	42.7	42.9	43.1	–
f104667	0	0	0	0	0	0	3	4	46.8	47.3	47.4	48.9
f104243	0	0	0	1	0	2	7	12	41.8	42.7	42.9	–
f104780	0	0	0	9	12	19	27	92	40.5	40.9	43.2	–
f103918	0	0	0	3	4	9	27	49	38.8	39.2	40.1	–
cf92382a	0	0	0	2	0	0	4	32	46.6	46.8	46.9	–

Table 4 shows the results of introducing open faults for the horizontal *PC's* when reprogrammable technology is used. There were 20 and 40 open faults respectively introduced in each channel for experimentation. Results show that a large number of such open faults can be tolerated. This is due to the fact that most of the nets in a channel are routed with single segments. The routing results for all the designs remain unchanged from the 0 fault case except for design *f104780* and *duke2*. However, all the nets were routable for that design even with the presence of a large number of open faults.

The architecture synthesized for routability, performance, and fault tolerance was compared to the architecture that we developed only for performance and routability [6]. The results show that the new architecture can handle about 20% more random faults with similar performance for the designs that we experimented with. Experiments were also conducted with random clustered faults. Similar results were also obtained for clustered faults.

7 Conclusions

This paper shows the feasibility of achieving routability and performance under the presence of *PC* faults in FPGA channel architecture. A channel architecture has been synthesized which not only considers routability and performance, but also enhances the routability of the architecture under the presence of *PC* faults without sacrificing performance. Results show that a large number of faults can be tolerated in the new architecture using a channel routing algorithm which can

Table 4. Routing results with horizontal *PC* open faults with reprogrammable technology

Design Name	Number of Unrouted Nets			Horizontal Prog. Conn			Avg. % Seg. Wastage		
	0	20	40	0	20	40	0	20	40
bw	0	0	0	0	0	0	38.4	38.5	38.5
duke2	0	0	0	4	6	8	42.7	42.7	42.6
f104667	0	0	0	0	0	0	46.8	46.8	46.8
f104243	0	0	0	0	0	0	41.8	41.8	41.8
f104780	0	0	0	12	12	12	40.5	40.5	40.7
f103918	0	0	0	4	4	5	38.8	38.8	39.0
cf92382a	0	0	0	0	0	0	46.6	46.6	46.6

route nets under the presence of faults. The antifuse faults have been characterized into two categories - those that can be detected before programmation, and those that can be only detected after programmation. It has also been shown that some of the *PC* open and short faults may also appear as delay faults due to open or short resistances being moderately large.

References

1. A.E. Gammal et. al., "An Architecture for Electrically Configurable Gate Array," *IEEE Journal of Solid State Circuits*, Vol. 24, No. 2, pp. 394-398, April 1989.
2. J. Birkner et. al., "A Very High Speed Field Programmable Gate Array Using Metal to Metal Antifuse Programming Elements," *IEEE Custom Integrated Circuits Conf.*, pp 1.7.1-1.7.6, May 1991.
3. *Field Programmable Gate Array – Application Handbook*, Texas Instruments, 1992.
4. K. Roy, "A Bounded Search Algorithm for Segmented Channel Routing of FPGAs and Associated Channel Architecture Issues," *IEEE Trans. on Computer-Aided Design*, pp. 1695-1705, November 1993.
5. J. Green, V. Roychowdhury, S. Kaptanaglu, and A. Gammal, "Segmented Channel Routing," *IEEE/ACM Design Automation Conf.*, pp. 567-572, 1990.
6. K. Roy, S. Nag, and S. Datta, "Channel Architecture Optimization for Performance and Routability for Row-Based FPGAs," *IEEE Intl. Conf. on Computer Design (ICCD)*, pp. 220-223, 1993.
7. K. Roy and M. Mehendale, "Optimization of Channel Segmentation for Channelled Architecture FPGAs," *IEEE Custom Integrated Circuits Conf.*, pp. 4.4.1-4.4.3, 1992.
8. C. Stapper, "The Effects of Wafer to Wafer Defect Density Variations on Integrated Circuit defect and Fault Distribution," *IBM Journal of Research and Development*, pp. 87-97, January 1985.
9. S. Kirkpatrick, C. Gellat, and M. Vecchi, "Optimization by Simulated Annealing," *Science*, Vol. 220, N. 4598, pp. 671-680, May 1983.

High-Performance Datapath Implementation on Field-Programmable Multi-Chip Module (FPMCM) *

Tsuyoshi Isshiki and Wayne Wei-Ming Dai

Computer Engineering, University of California, Santa Cruz CA 95064

Abstract. In this paper, a new design style for multi-FPGA system is proposed. It fills the large gap between high-level synthesis and the FPGA logic design by providing datapath circuit module library which can contain high-level simulation models as well as low-level circuit netlists. Some bit-serial circuit modules have been designed which are easy to partition and place within multiple FPGAs. Also, we have describe our novel work on Field-Programmable Multi-Chip Module which demonstrates its ability in reducing hardware size, reducing power consumption, reducing packaging cost and providing with high density chip-to-chip connections.

1 Introduction

The potential of Field-Programmable Gate Array technology (FPGA) has been demonstrated by many researchers in this field to provide an alternative approach to computation intensive applications. Custom chips which are optimized for some specific applications are possible only if they promise a high volume production. Parallel processing using general purpose microprocessors or digital signal processors can be effective if a wide range of applications is targeted. FPGAs take in the advantages from both sides: an efficient and high-performance datapath implementation of a custom chip and the programmability of a microprocessor. Researches on this FPGA-based *custom computing* have been active on both aspects [1][2][3], however, often failing to merge the two together. Engineers can implement an efficient, high-performance design on this FPGA-based system with a design methodology similar to custom chip design requiring extensive knowledge of digital system and experience [3]. On the other hand, efforts in making the FPGA-based systems easy to program has come to a point where designers can write VHDL-like programs and the tools will automatically generate the FPGA configuration data [4][5]. Here, the problem is that the automatically generated designs are often inefficient in terms of resource usage and performance. Also, these hardware description language still requires basic digital design skills.

This paper attempts to solve these problems by proposing a new design style which integrates the digital system design knowledge and experience into the

---

* This work is supported in part by ARPA under ONR Grant N00014-93-1-1334.

design environment to guarantee the quality of performance and efficiency, and still regain the user-friendliness for the programmers without the knowledge of digital system designs which the majority of the potential users would not have. Our design style consisting of bit-serial arithmetic module library eliminates the large gap of high-level synthesis and logic design on FPGAs, bringing the design decision process to the highest level of abstraction while providing accurate tradeoff measures.

One other aspect in FPGA-based system is that in order to effectively and efficiently tackle such computation intensive applications, a large amount of logic resources are needed, in the order of millions of gate counts. Decreasing the physical size of the hardware of the FPGA-based systems is not only important by itself, but also critical in increasing the performance, cutting power consumption, and cutting the overall cost. Our novel work on Field-Programmable Multi-Chip Module (FPMCM) clearly demonstrates a feasible and effective solution to this hardware compaction problem of the FPGA-based custom computer systems.

2 FPMCM - An Integration of FPGA and MCM Technology

FPGAs suffer from low logic density and slow circuit speed. Fortunately, the recent advancement in device technology is making these problems less critical. The gate capacity of a single chip is growing close towards 100K gate counts. The circuit speed has being improving where 4-input logic functions can be computed in less than 3ns. Whereas for the multiple-FPGA systems, there are some more problems:

- Existing FPGA chips may be IO limited when used in multiple-FPGA configuration.
- IO drivers are designed to drive large load, therefore often slow. Chip-to-chip communication penalty is large.
- Multiple-FPGA hardware results in large size and requires large power.

We have address the first problem, discussed in the later section, by using bit-serial datapath modules where IO limitation is not a problem. We have actually demonstrated in our multiplier design example that the chip-to-chip communication penalty is indeed critical.

In the following section, we will describe our current work on Field-Programmable Multi-Chip Module (FPMCM) which will help us deal with those problems using the new packaging technology.

2.1 Overview of MCM

A Multi-Chip Module (MCM) has several bare chips or dice mounted and interconnected on a multi-layer substrate which functions as a single IC. Usually, a silicon substrate consists of a ground plane, a power plane, and two signal routing layers. The power and ground plane form a good decoupling capacitance.

If necessary, additional processing steps can be added to produce intergrated floating capacitors, thin-film resistors, inductors, and bipolar transistors.

The most promising assembly technique for MCM is the flip-chip attachment [7]. In flip-chip, dice are attached with pads facing down via solder bumps which form the mechanical and electrical connections. The flip-chip technology provides area pads through solder bumps which are distributed over the entire chip surface, rather than being confined to the periphery as in conventional packaging.

The resulting assembly is attached, wirebonded, and encapsulated in a second level package. The second level package, used for insertion into the final system, is typically a pin or ball-grid array (BGA). This process is well suited to FPMCMs because the substrates can accommodate a high density of interconnect; There are two layers with a wire pitch on the order of 1.5 mils. The process has also been carefully optimized to minimize cost without sacrificing the performance of digital circuits.

2.2 FPMCM-I Architecture

We have designed and are in the process of manufacturing a first generation FPMCM (FPMCM-I) [9]. The purpose of the first generation device is to fully exercise the MCM fabrication and assembly technology, quickly familiarize ourselves with the technical problems of FPMCM design, uncover any pitfalls, and try out an initial architecture.

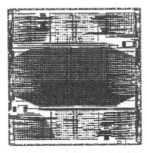 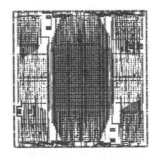 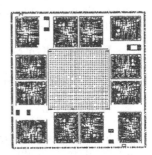

Fig. 1. Physical Design First Generation FPMCM. Outer dimensions of the module are 30.6 mm square. The device contains 12 Xilinx XC3042's and a single Aptix FPIC. Four ceramic chip capacitors are provided to stabilize the on module power and ground planes. The figure shows the three metal routing layers. The first two (X and Y) layers shown are on the silicon substrate, while the third consists of an identical I/O redistribution pattern for each XC3042 and the solder bump grid of the FPIC which is necessary because the bond pad pitch on the 3000 series is too small for inexpensive, reliable flip chip assembly.

The FPMCM-I consists of 12 Xilinx XC3042s encircling a Field-Programmable Interconnect Chip (FPIC), the Aptix PIC-R. This FPIC die has an array-IO of

32 × 32 resulting in 1019 IO pins including 976 user-IOs which can be treated almost like a crossbar network. Four surface mount chip capacitors provide power decoupling on the substrate. Each FPGA has 12 pins connecting to each nearest neighbor in the ring and 22 pins connecting to external bondpads. Four global signals are shared by all FPGAs. For testability purposes every signal net on the MCM substrate connects to a PIC pin. There are a small number of direct connections from the pins of the MCM to the PIC for diagnostic purposes; this is in addition to the connections needed for the configuration and control lines. To summarize, the FPMCM-I has 268 user IO in addition to numerous power, ground and control signals [9].

2.3 Advantages of MCM

MCMs have many advantages over traditional packaging: lower electrical parasitics between packages, smaller size, increased interconnect resources, and reduced packaging cost.

1. Power consumption reduction:
   All other things being equal, the lower capacitive loads of shorter MCM interconnect provide substantial power savings. The power savings for MCMs are typically fifty percent lower than conventional packaging. This power savings is increased dramatically when the drivers are specifically designed for the MCM interconnects that they drive. Low power is an important feature of MCM, especially when applied to portable systems and add-on cards which are power-limited.
2. High-speed high-density chip-to-chip communication:
   High wiring density of the silicon substrate allows communication between chips to be very dense. And since the parasitic capacitance is significantly lower compared to on-board communications, fast drivers can be built in the die. AT&T Bell Laboratory has designed a set of low voltage, high-speed IO buffers optimized specifically for MCM. The results show that these new buffers reduce power consumption 6× and increase performance 2.5× compared to conventional CMOS buffers and they can operate at up to 400MHz [8].

3 FPMCM Design Environment

3.1 Problems of Current FPGA Design Environment

The existing FPGA design tools are provided with the thought that the FPGA users are engineers with experience and patience. And they are specially tuned to implement random logics efficiently as possible. Tools such as logic minimization, logic partition, technology mapping, and automatic placement and routing are therefore designed under the assumption that the circuit has a random structure. There are several problems in applying this design environment to datapath implementation:

1. Datapath circuits such as ALUs, registers, ROMs, RAMs, adders, counters and multipliers are regularly structured and their statistical characteristics vary between each other. Synthesizing datapath circuits as random logics using the existing tools will often result in poor implementation in terms of logic size and circuit speed.
2. Design change, in order to make the design fit in the available FPGA hardware or to speed up the circuit to meet the given specification, is difficult and time consuming.
3. Design verification including functionality and timing is also time consuming.
4. Design entry requires digital system design skills which will greatly discourage the programmers without the skill to use the system. Simply attaching a high-level synthesis tool which translates high-level behavioral description like C and PASCAL into structural description would result in poor implementation since the decisions made in the high-level synthesis can only be based upon some unreliable information about the hardware size and performance of the datapath components [6].
5. Each design iteration takes too long. A time to compile a structural description into configuration data takes tens of minutes, possibly hours. This also discourages the users.

Based on these observations, we now propose a new FPGA design style very similar to the high-level synthesis approach.

3.2 A New Design Style Based on High-Level Synthesis

High-level synthesis, first called silicon compilation, have gained a great attention since the early 1980s, where VLSI technology has advanced to a point where the time it took to design chips became as long as chip lifetimes, leading to a bottleneck in the product development cycle. High-level synthesis are programs that generate layout data from some higher-level description.

The essential building block for the high-level synthesis is the *module* components. Modules is defined in high-level synthesis environment as microarchitectural entities that perform one or a few specific functions and consist of one or more arrays of cells or tiles of a specific type. Examples of modules are PLAs, ROMs, RAMs, register stacks, multipliers, ALUs, and counters. Modules are compiled from *cells* which are single-bit logics or storage functions of some microarchitectural components or circuits of SSI or MSI complexity. The important feature here is that modules are associated with high-level models such as functional, logic, timing, power, and testability models to be used by verification, analysis, and optimization tools such as functional and logic simulators, timing and power analyzers, and layout compactors. In this way, designers are able to identify at high and abstract level which part of the design is causing the error, or which parts need to be reworked in order to meet the requirements of performance and hardware size.

Fig.2(b) shows the high-level synthesis design style applied on FPGA designs. The description can either be behavioral or structural. From a behavioral

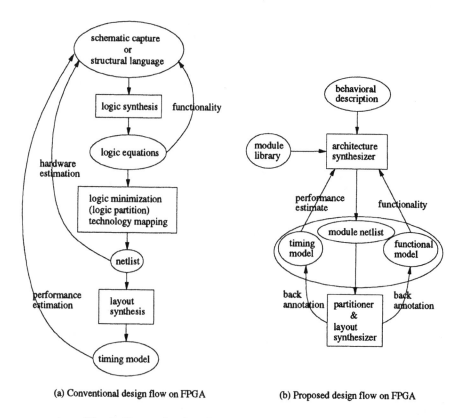

(a) Conventional design flow on FPGA   (b) Proposed design flow on FPGA

Fig. 2. Conventional and proposed design flow on FPGA

description, a *architecture synthesizer* generates structural description. This part of high-level synthesis consists of data dependency analysis, register allocation, scheduling, memory allocation, and so on. This process requires accurate informations about hardware complexity and performance measures in order to generate area-efficient high-performance datapath structure. In conventional FPGA design style, these information are not provided until the design has been already mapped, placed and routed, at the lowest level of abstraction. This made it almost impossible to incorporate high-level synthesis in the FPGA design environment.

A structural description consists of ALUs, registers, multipliers, counters and other arithmetic functions. These modules are defined in the module library with wide varieties of functions. Each module is described in logic block functions for a particular FPGA, therefore the amount of logic blocks needed for the design is already known at this stage. A timing model is also provided for each module, therefore the performance of the design is also predictable. By completely relying on the module library, the designers are able to eliminate the lengthy process of logic minimization, logic partition, and technology mapping which are tuned to work well on random circuits but not on datapath circuits. A variety of design verification is done at this structural level, which makes it very easy to detect design errors, timing errors, feasibility of actual implementation, and performance.

In order for this design style to be realistic, the module library has to meet the following requirement.

1. Module library should provide with wide choices of components, at least include decoders, register stacks, multipliers, ALUs, counters, etc.

2. Following models for each module have to be provided in high-level description.

   (a) Functional model
   (b) Timing model
   (c) Testability model

3. Each module has to be physically compatible with other modules in terms of placement and routing.

4. The routability and the propagation delays of inter-module connections have to be highly predictable.

We have to note that the prediction of routability and propagation delays is extremely difficult in FPGA architecture. In custom chip design which is one of the original targets for the high-level synthesis, routing propagation delays are only caused by wiring capacitance. Also, routing has maximum flexibility in custom chips, designer has the freedom of assigning enough routing resources to congested regions. Therefore, the nets are almost always routable. Whereas in FPGA architecture using pass-transistors for connecting routing channels, drain capacitance of the pass-transistors is significantly larger than wiring capacitance. Routing propagation delays totally depend on the number of pass-transistors the net has to go through. Furthermore, the limited routing resources per channel not only make the routing more difficult but also force routing nets in the congested region to scatter to other routing channels. This makes the prediction of the routing delays even harder.

We have realized that the last two critical requirements are the keys to high-level synthesis. Our decision of using bit-serial arithmetic which will be discussed in the next section is the result of this observation.

4 Bit-Serial Arithmetic Modules

Computer arithmetic schemes and datapath implementation techniques have a large impact on performance, circuit complexity and power consumption. We have to be particularly careful in designing datapath circuits on FPGA. Since the logic density is lower and circuit speed is slower, we cannot afford to waste the resources or misuse them which may severely affect the performance. And we also have to be more careful about the routing than we would normally do on custom chip designs since the routing penalty is very high due to the high capacitance of the pass transistors on the routing channels. This directly links with the concerns on module-based methodology of FPGA design.

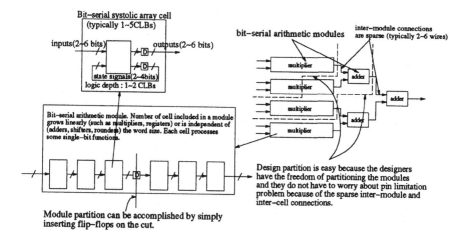

Fig. 3. Systolic array implementation of bit-serial modules

4.1 Examples of Bit-Serial Arithmetic Modules

Bit-serial arithmetic modules can be implemented in systolic array fashion (Fig.3). Each cell contains several combinational logics implementing single bit function plus several storage elements. We have designed some bit-serial modules on Xilinx XC3000 architecture. Each systolic array cell is typically implemented using 1 to 5 configurable logic blocks (CLBs) with logic depth of 1 or 2 CLBs. Each bit-serial module consists of a single cell as in adders and shifters, or multiple cells proportional to the word length as in multipliers and registers. Inter-cell connections and inter-module connections are sparse, typically 2 to 6 wires, which are independent of the word length (Table 1).

4.2 Comparison Between Bit-Serial and Bit-Parallel Modules

Let us compare this bit-serial arithmetic modules with a bit-parallel modules in terms of partition, placement and routing which are the key factor in creating a realistic module library for high-level synthesis.

1. Partition:
   (a) Bit-parallel modules are often hard to partition over multiple FPGAs since partitioning such circuits often leads to I/O pin limitation problem. Also, performance is critically affected by the partition, thus making the performance measures at the high-level unreliable. It is therefore not practical to partition bit-parallel modules. Also, design partitioning is difficult and may result in poor logic resource usage because of the coarse granularity of the module circuit composed of tens of CLBs and pin limitation problem.
   (b) Bit-serial modules are easy to partition since inter-cell connections are sparse. Performance degradation by partition can be totally eliminated

Table 1. Statistics of bit-serial datapath modules. Word size $= N$

Modules	Area	Logic depth
Multiplier		
(1-input 1-constant)	$4N$ CLBs	1 CLB
(2-inputs)	$5N$ CLBs	2 CLBs
Adder		
(single precision)	1 CLB	1CLB
(double precision)	3 CLBs	1 CLB
Rounder		
(truncate)	1 CLB	1 CLB
(round-to-nearest-even)	4 CLBs	1 CLB
Absolute operator	$2 + N/2$ CLBs	1 CLB
Max-min selector		
(least-significant-bit-first)	$1 + N/2$ CLBs	1 CLB
(most-significant-bit-first)	4 CLBs	1 CLB

by inserting additional flip-flops on the partitioned inter-cell connections under the assumption that the chip-to-chip delay is smaller than the internal critical path of the systolic array cells. Design partitioning is also easy because of the fine granularity of the cell circuit composed of only several CLBs and no pin limitation problem, and result in a very high logic resource usage, as high as 100%.

2. Routing:
   (a) Bit-parallel modules tend to be large in hardware, may have a large fan-in and fan-out, may have a vast area of dense connectivity. And as a result, wiring distance can be considerably long for some nets. Routability of such modules are hard to predict, and their routing delays are also unpredictable. Therefore, in order to construct a useful module library for the high-level synthesis, bit-parallel modules have to be placed and routed as seen in Hard-Macros. This physical restriction of the module will affect the routability of the other parts of the chip.
   (b) Bit-serial modules consisting of systolic array cells only has local connections. Since the distance of those wires are all short, the propagation delays of those wires can be highly predictable. Therefore, unrouted modules can still provide with reliable performance measures. Also because of the local connections, routing wires tend to be evenly distributed throughout the chip, naturally avoiding routing congestions.

3. Placement:
   (a) Since bit-parallel modules have to be placed and routed in order to provide the information needed in high-level synthesis, physical compatibility between modules tend to be poor. Locking the relative position of CLBs within the module, locking the routing wires, locking the position of I/O nodes, all these reduce the feasible placement search space significantly.

(b) Bit-serial modules can afford to be unrouted as long as the CLBs within a cell and adjacent cells within a module is placed close enough, to make the routing wires short and to make the routing delays predictable. This flexibility increases the physical compatibility between other modules, which minimizes the affect of the routability from the other circuits in the same chip.

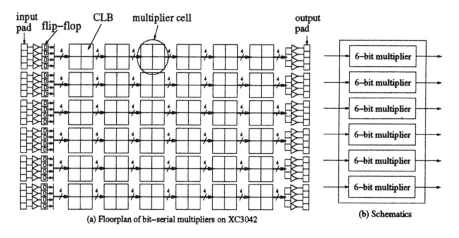

(a) Floorplan of bit-serial multipliers on XC3042

(b) Schematics

Fig. 4. Bit-serial multiplier layout on XC3042

To support the above argument, we have provided one actual design example to demonstrate the high logic resource usage of the bit-serial arithmetic modules. In Fig.4, actual layout example of a bit-serial multiplier on XC3042 is shown. We were successful in mapping 6 × 6-bit multiplier on one chip using up *all* of the 144 CLBs. The internal critical path (flip-flop to flip-flop) was 17.0ns. The placement was done by hand, and routing was done by Automatic Placement and Routing (APR). The maximum internal routing delay was 5ns. However, the external critical path (flip-flop to flip-flop via chip pads) was 31.5ns. Taking in account of clock skew between chips (~2ns) and chip-to-chip interconnect delay (<~10ns), the overall critical path is around 45ns. We can see that by using bit-serial datapath modules whose logic depth is very small, clock period can be reduced significantly to a point where the delays from the IO pad drivers become dominant in the critical path. In Table 2, estimated performance and area of several applications on our FPMCM chip are shown. There are two clock frequencies assumed. 20MHz is the upper bound imposed by the IO pad buffer delays (~ 45ns). 50MHz is the upper bound when the IO pad buffer delays are ignored, and the internal critical path is the overall critical path (17ns). The reason for ignoring the IO pad buffer delays is that by using MCM packaging technology for multiple-FPGA system which is discussed in the next section, IO pad buffers can be made significantly smaller and faster.

Table 2. Estimated performance measures on FPMCM chip. Multiplication, addition, rounding, max-selector, and min-selector are all counted as 1 operation. min-max-selector are counted as 2 operations. Others are not counted as operations. The assumed clock frequency is 20MHz and 50MHz.

	# of CLBs	word size	sampling rate		Ops /cycle	Ops per second	
			(20MHz)	(50MHz)		(20MHz)	(50MHz)
8-point IDCT	1064	16 bits	1.25MHz	3.125MHz	56	70 MOPS	175 MOPS
FIR filter (25 taps)	1675	16 bits	1.25MHz	3.125MHz	51	63 MOPS	159 MOPS
FIR filter (49 taps)	1715	8 bits	2.5MHz	6.25MHz	99	247 MOPS	618 MOPS
8-point bubble sort	126	8 bits	2.5MHz	6.25MHz	40	100 MOPS	250 MOPS

5   Partition, Placement and Routing for FPMCM

For our proposed design flow, we have to provide an automatic partition and placement algorithm to map the bit-serial arithmetic modules onto multiple FPGAs.

1. Partition algorithm would consist of assigning bit-serial systolic array cells to each FPGA chip while assuring a feasible placement inside the chip and a feasible routing inside the FPIC, retiming of data due to the insertion of flip-flops at the chip boundary and back-annotating to the functional model.
2. Placement algorithm would consist of assigning each CLB to physical location while assuring that CLBs of the same cell is placed adjacent to each other, placing adjacent cells as close as possible to make the routing delay below the predicted margin, and placing boundary cells, cells which is connected to cells in a different FPGA chip, closest to the I/O blocks.
3. Routing algorithm would consist of routing the FPIC to provide the required connection between FPGAs while keeping the chip-to-chip delay within the tolerant margin, that is, the critical path within the FPGA. Routing of the individual FPGA has to be done by the routing tools provided by the FPGA vendors since this requires extensive knowledge of the FPGA routing architecture.

These three tasks has to occur simultaneously in order to obtain the best results since these tasks are dependent of each other. However, since we are dealing with a very special class of circuits with only local connections and very limited fan-in and fan-out, simultaneous partition, placement and routing should be a lot easier than the ones for random circuits.

These algorithms are currently under development.

6 Summary

We have purposed a new design style for multi-FPGA system which fills the large gap between high-level synthesis and the FPGA logic designs. This design style is aimed to attract a large number of potential users, the programmers for computer intensive applications, who may have little knowledge about digital circuit design. We have demonstrated the efficiency of using bit-serial arithmetic modules which result in very high usage of logic resources, of up to 100%. We have given several design examples which the numbers of operations per second can exceed 600 MOPS.

Also, we have described our work of FPMCM which is current being fabricated. We have demonstrated the potential of the MCM technology in multi-FPGA system for its ability in reducing hardware size, reducing power consumption, reducing packaging cost and providing with high density chip-to-chip connections.

References

1. M. Gokhale, W. Holmes, A. Kopser, S. Lucas, R. Minnich and D. Sweely, "Building and Using a Highly Parallel Programmable Logic Array," *IEEE Computers*, pp. 81–89, Jan. 1991.
2. Dzung T. Hoang, "Searching Genetic Databases on Splash 2," *Proc. IEEE Workshop on FPGAs for Custom Computing Machines*, pp. 185–191, April 1993.
3. C. E. Cox and W. E. Blanz, "GANGLION–A Fast Field-Programmble Gate Array Implementation of a Connectionist Classifier," *IEEE Solid-State Circuits* Vol. 27, No. 3, pp. 288–299, March 1992.
4. M. Wazlowski, L. Agarwal, T. Lee, A. Smith, E. Lam, P. Athanas, H. Silverman and S. Ghosh, "PRISM II: Compiler and Architecture," *Proc. IEEE Workshop on FPGAs for Custom Computing Machines*, pp. 9–16, April 1993.
5. David E. Van den Bout, "The Anyboard: Programming and Enhancements," *Proc. IEEE Workshop on FPGAs for Custom Computing Machines*, pp. 68–77, April 1993.
6. Patrice Bertin and Herve Touati, "PAM Programming Environments: Practice and Experience," *Proc. IEEE Workshop on FPGAs for Custom Computing Machines*, April 1994.
7. R.C. Frye, K.L Tai, M.Y. Lau and A.W.C. Lin, "Low-cost silicon-on-silicon MCMs with integrated passive components," *Proc. 1992 International Electronics Packaging Conference*, 1992.
8. T. Gabara, W. Fischer, S. Knauer, R. Frye, K. Tai and M. Lau, "A I/O CMOS Buffer Set for Silicon Multi-Chip Modules," *Proc. IEEE Multichip Module Conference*, 1993.
9. J. Darnauer, P. Garay, T. Isshiki, J. Ramirez and W. M. Dai, "A Field Programmable Multichip Module," *Proc. IEEE Workshop on FPGAs for Custom Computing Machines*, April 1994.

A Laboratory for a Digital Design Course Using FPGAs

Stephan Gehring   Stefan Ludwig   Niklaus Wirth

Institute for Computer Systems, Federal Institute of Technology (ETH)
CH–8092 Zurich, Switzerland
{gehring ludwig wirth}@inf.ethz.ch

Abstract. *In our digital design laboratory we have replaced the traditional wired circuit modules by workstations equipped with an extension board containing a single FPGA. This hardware is supplemented with a set of software tools consisting of a compiler for the circuit specification language Lola, a graphical layout editor for design entry, and a checker to verify conformity of a layout with its specification in Lola. The new laboratory has been used with considerable success in digital design courses for computer science students. Not only is this solution much cheaper than collections of modules to be wired, but it also allows for more substantial and challenging exercises.*

1 Introduction

In order to demonstrate that what had been learnt in the classroom can actually be materialized into useful, correctly operating circuits, digital circuit design courses are accompanied by exercises in the laboratory. There, students select building elements from an available collection and assemble circuits by plugging them together, by wire-wrapping, or by soldering. We have replaced this setup by workstations used in programming courses [1] and equipping them with an FPGA on a simple extension board. Not only is this replacement substantially less expensive, but it allows for the implementation of considerably more realistic and challenging designs. This is due to the large number of available building elements in the form of FPGA cells. Instead of plugging units together, cells are configured and connected using a graphical circuit editor. Indeed we consider this laboratory as *the* application of SRAM-based FPGAs, where their inherent flexibility is not merely an advantage, but a simple necessity. After all, a design is not only changed for correction or improvement, but also discarded upon successful completion, whereafter the FPGA is reused for a next exercise. Our experience also shows that learning effect and motivation surpass our expectations, and that simulation by software can no longer be justified as a substitute for actual circuit implementation. Furthermore, the concurrent design of test programs on the host computer helps to bridge the perceived gap between hardware and software, and is a strongly motivating factor, in particular for Computer Science students.

Whereas the construction of the FPGA-board was a rather trivial matter, most of the project's efforts were spent on the design of adequate software tools. They comprise not only a graphical layout editor, but also a small circuit specification language called *Lola* and its compiler (Sect. 2). A typical exercise starts with the formulation of the informally described circuit in terms of this (textual) notation. The second step consists of mapping it onto the FPGA, i.e. of finding a layout and entering it with the aid of the *layout editor* (Sect. 3). Before testing the circuit with test programs, a second tool, the *Checker* is applied to verify the consistency of the layout with the circuit's specification in terms of Lola (Sect. 4).

We stress the fact that these tools have not only proved most useful in digital design courses, but also adequate and effective in practice.

2 The Circuit Specification Language Lola

In the design of Lola we have made a deliberate effort to let the basic notions of digital circuits be expressed as concisely and as regularly as possible, making use of constructs of programming languages, while omitting unnecessary and redundant features and facilities. The similarity of its appearance (syntax) with that of structured programming languages is intentional and facilitates the learning process. However, the reader is reminded that "programs" describe static circuits rather than algorithmic processes. Although the entire language is defined in a report of some six pages only, we here choose to convey its "flavor" by showing a few examples rather than by presenting a comprehensive tutorial.

2.1 Declarations, Expressions, and Assignments

Every variable (signal) is explicitly declared. Its declaration specifies a type (binary, tri-state, open-collector) and possibly a structure (array dimension). Variables occur in expressions defining new signal values. The available operators are those of Boolean algebra: not ($\sim$), and ($*$), or ($+$), and xor ($-$). Expressions are assigned to variables, thereby defining their value depending on other variables. The frequently encountered multiplexer operation is defined as

$$MUX(s: x, y) = \sim s * x + s * y$$

The following basic operators allow the specification of storage elements and registers, and thereby of (synchronous) sequential circuits.

$SR(s', r')$	set-reset flipflop
$LATCH(g, d)$	transparent latch
$REG(en, d)$	D-type register with enable and implied clock

2.2 Type Declarations

If a certain subcircuit appears repeatedly, it can be defined as an explicit circuit type (pattern), whereafter it can be instantiated by a simple statement. Declaration and instantiation resemble the procedure declaration and call in programming languages. Inputs appear in an explicit list of parameters. Outputs do not. Instead, they are treated like local variables, with the difference, however, that they can also be referenced in the context of the instantiation, namely by their name qualified by the instance's identification.

Of particular value is the easy *scalability* of declared types. This is achieved by supplying a declaration with numeric parameters, typically used to indicate array dimensions. This kind of parametrization embodies the most essential advantage of textual specifications over circuit schematics.

2.3 Examples

The first example is a binary adder consisting of N identical units of type *ASElement*. Input *cin* denotes the input carry, and *s* controls whether *z* is the sum of *x* and *y* or their difference (Fig. 1).

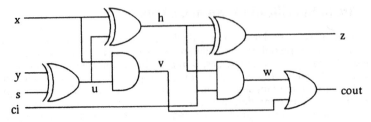

Fig. 1. Add-Subtract Element

```
TYPE ASElement;
    IN x, y, ci, s: BIT;
    OUT z, co: BIT;
    VAR u, h: BIT;
BEGIN u := y − s; h := x − u; z := h − ci; co := (x ∗ u) + (h ∗ ci)
END ASElement;

TYPE Adder(N);
    IN cin, sub: BIT;
        x, y: [N] BIT;
    OUT cout: BIT;
        z: [N] BIT;
    VAR AS: [N] ASElement;
BEGIN AS.0(x.0, y.0, sub, sub);
    FOR i := 1 .. N−1 DO AS.i(x.i, y.i, AS[i−1].co, sub); z.i := AS.i.z END ;
    cout := AS[N−1].co
END Adder
```

The second example shows a multiplier with N-bit inputs x and y and a 2N-bit output z. The circuit consists of a matrix of identical adder elements (Fig. 2). The first parameter is the product of multiplicand and multiplier.

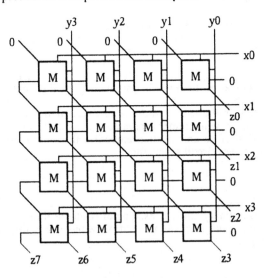

Fig. 2. Multiplier

```
TYPE AddElement;
    IN x, y, ci: BIT;
    OUT z, co: BIT;
BEGIN z := (x–y) – ci; co := (x * y) + ((x–y) * ci)
END AddElement;

TYPE Multiplier(N);
    IN x, y: [N] BIT;
    OUT z: [2*N] BIT;
    VAR M: [N][N] AddElement;
BEGIN
    FOR j := 0 .. N–1 DO M.0.j (x.0 * y.j, '0, '0) END ;
    FOR i := 1 .. N–1 DO
        M.i.0 (x.i * y.0, M[i–1].1.z, '0);
        FOR j := 1 .. N–2 DO  M.i.j (x.i * y.j, M[i–1][j+1].z, M[i][j–1].co) END ;
        M[i][N–1] (x.i * y[N–1], M[i–1][N–1].co, M[i][N–2].co)
    END ;
    FOR i := 0 .. N–2 DO z.i := M.i.0.z;  z[i+N] := M[N–1][i+1].z END ;
    z[N–1] := M[N–1].0.z;  z[2*N–1] := M[N–1][N–1].co
END Multiplier
```

Our last example is a binary up/down counter with the three control inputs *en* (enable, carry input), *clr'* (clear), and *up* (indicating the counting direction).

```
TYPE UpDownCounter(N);   (*with load, enable and clear*)
    IN ld', en, clr', up: BIT; x: [N] BIT;
    OUT Q: [N] BIT;
    VAR cu, cd: [N] BIT;
BEGIN
    Q.0 := REG(MUX(ld': x.0, Q.0 * clr' – en)); cu.0 := Q.0 * en; cd.0 := ~Q.0 * en;
    FOR i := 1 .. N–1 DO
        Q.i := REG((MUX(ld': x.i, Q.i – MUX(up: cd[i–1], cu[i–1]))) * clr');
        cu.i := Q.0 * cu[i–1]; cd.i := ~Q.i * cd[i–1]
    END
END UpDownCounter
```

2.4 The Compiler

Unlike a compiler for a programming language, which generates executable code, the Lola compiler generates a data structure representing the circuit that is most appropriate for further processing by various design tools, ideally by an automatic layout generator. Other tools are timing analyzers, fanout checkers, and simulators. In our case, the most important tool is the Checker, which verifies a given layout rather than generating one. The data structure generated by the compiler consists of a binary tree for each variable occurring in the design. Hence the compiler flattens the structured description. It also applies obvious simplification rules. They take effect, for example, at the edges of the matrix of the second example above, where some of the input parameters are zeroes.

3 The Layout Editor

A graphical editor is used to enter and modify circuit specifications implemented on an FPGA. It presents the FPGA at a low level, as close to the real hardware as possible. We first present the used FPGA architecture and then give a description of the editor's mode of operation and its implementation.

3.1 The Hardware

In our laboratory, an extension board containing an FPGA of Atmel (formerly Concurrent Logic Inc.) is used [2]. The AT6002 chip in an 84-pin package consists of a matrix of 32 by 32 identical cells. A cell implements *two functions of up to three inputs* (A, B, and L). These functions can be combinational and sequential (i.e. involving a register). Two outputs (A and B) of a cell are connected to the inputs of its four neighbors (north, south, east, and west). In addition to the *neighbor connections*, there is a bussing network connecting bus inputs and outputs of eight cells in a row or column. These so-called *local busses* are used to transport signals over longer distances between cells. They can be connected to other local busses or to additional *express busses* via *repeaters* at 8-cell boundaries. Surrounding the array of cells are 16 *programmable IO pads* on each side. These connect to the bus of the host workstation and to components on the extension board, such as an SRAM and an RS–232 line driver.

3.2 Design Representation and Modification

The editor presents the gate array in a viewer as an excerpt of the 1024 cells (Fig. 3). Every eight cells, a repeater column or row is displayed, and surrounding the array, the programmable pads are shown. Each component's contents reflect the implemented function as closely as possible - e.g. an Exclusive-Or in a cell with a constant one input is displayed as a Not-gate. To show the signal flow, connections between cells and to and from local busses, and connections with repeaters are displayed as arrows. By giving neighboring connections a different color (yellow) than local (green) and express busses (red), a visual feedback on the speed of a specific connection is suggested. Inside a cell, the same picture is displayed regardless of the source and destination direction of signals. For instance, even if signals enter a cell from below and flow to the top, the picture inside the cell suggests a flow from top to bottom. The reason for this will be explained in Sect. 3.3.2. To give signals a meaningful name – and to enable a link to a Lola description of a circuit (see Sect. 4) – textual labels can be placed at cell and pad outputs.

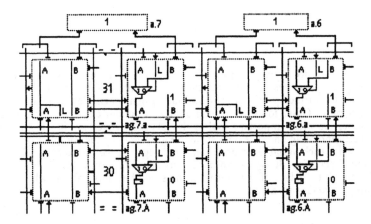

Fig. 3. Editor View with Cells, Pads, Repeaters, and Labels

The mouse is used as the primary input device to change a design. Cells, pads, and repeaters can be edited using popup menus (Figs. 4, 5). The top row of the menu in Fig. 4 shows the six different *routing modes* possible in a cell, and the four items on the left of the bottom row show the *state* of a cell [2]. The two multiplexers on the right are an often used combination of routing mode Mux and states Xor or Xor with register. Similarly, all possible configurations for repeaters (Fig. 5) and pads (not shown) are presented through a menu. The current configuration of the edited resource is highlighted in the menu with a frame. Connections between cells must be entered manually as no automatic router is provided. Thus, students learn about the problems of placement and routing in FPGAs. Fast replication of data path elements is available by selecting and copying bit slices of the layout. Cells can also be moved or copied across viewer boundaries in which a different design or a different excerpt of the same design is shown.

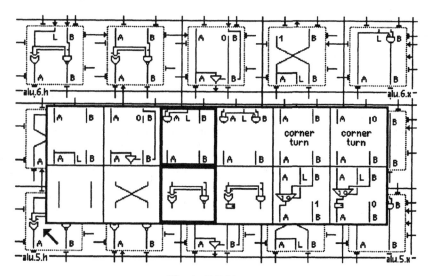

Fig. 4. Cell Menu

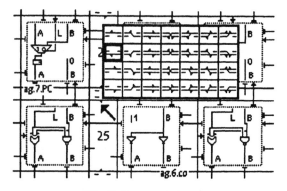

Fig. 5. Repeater Menu

3.3 Implementation

The editor consists of five modules comprising roughly 65KB of object code. The following sections discuss some of the finer points of the implementation.

3.3.1 Data Structures

We use a straight-forward data structure to represent the various resources on the FPGA. A two-dimensional array of cell records represents the matrix of cells. This allows for fast iteration over the data structure when displaying it. Similarly, the repeaters and pads are represented as arrays of records. The labels, however, are a linked list of records containing the position and caption of a label. Designs are saved to disk using a portable data format. A simple run-length encoding of empty cells, pads, and repeaters compresses typical files to 23% of their original size. Even large designs take up only 8KB, whereas smaller designs remain well under 1KB.

3.3.2 Drawing Operations

For drawing the contents of a cell, we use a special font containing only the patterns of signals flowing from top to bottom. Thereby, we get fast drawing of a design without having to distinguish between the 384 possible signal flow directions, but at the cost of a fixed aspect ratio and non-optimal print output. Making the distinction and drawing a cell's contents with multiple lines and dots slows down the performance by 50% and increases the program size by 100%. Repeaters are drawn using a font as well, but here, a special pattern exists for each possible signal flow. Despite the disadvantages when using a font, the chosen solution works well in practise. A special display option can be set where only used cells and busses are drawn. Not only does this improve display speed, but it also avoids a cluttered view.

3.3.3 Editing Operations and Undo

The problem of displaying three different menus has an elegant solution using a generic procedure. This procedure takes two procedure variables as parameters, one for displaying the contents of each menu item, and one for updating the data structure according to the chosen item. Thereby, the code for configuring cells, pads, and repeaters remains the same, only the procedure variables and the number of rows and columns in the menu change.

Each editing operation can be undone. This is accomplished by backing up the data structure before executing the operation. Then, a simple swap between the backup and the primary data structure implements the undo (and redo) operation.

3.4 Command Module and Queries

Operations that are not frequently used are provided through a command module [3]. Clock and reset lines are set with commands. Labels, cells according to their coordinates, and whole arrays according to a prefix, can be located in a design. Statistics on the design are also provided, with which different implementations of the same specification can be compared against each other (according to bus utilization and the number of cells used for routing, logic, and registers).

3.5 Downloading to the Extension Board

Once a design is finished, it can be downloaded onto the FPGA in a few milli-seconds. Only during this step, simple electrical consistency checks are performed, such as multiple sources writing to a bus unconditionally, and incompletely configured cells.

3.6 Discussion

For the intended purpose the chosen implementation worked out very well. The fast adaption of all users to our system was encouraging and the positive feedback very rewarding. In the future, we will provide configurability of the editor to support various chip sizes and IO configurations. Research-wise, we intend to develop design automation tools that support a seamless integration between the specification of a circuit and the automatically laid out design.

4 The Checker and Analysis Tools

4.1 The Checker

In a digital design laboratory, a typical design cycle might look as in Fig. 6. After initial design entry with the editor, the designer downloads the design onto the FPGA. By configuring the FPGA, the circuit is implemented and can be tested subsequently. If the test fails, the design is corrected, downloaded, and tested anew.

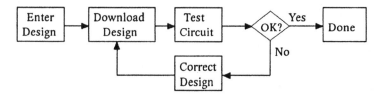

Fig. 6. Design Cycle

While downloading and testing a design is usually a matter of seconds or minutes, correcting a faulty design can be very tedious. Mostly, this comes from the fact that, while it is easy to detect an error, it is hard to find its location in the design. In traditional laboratories with electronic components being plugged together, the designer must verify manually that each component is properly wired. Our software-based approach, by contrast, offers the opportunity to construct a circuit checker program that helps the designer not only to detect, but also to locate implementation errors.

4.1.1 Representing and Checking Designs

A digital circuit is characterized by its inputs, outputs, and a set of Boolean functions combining the inputs. Each circuit output is associated with the result of such a function. The function can be represented as a *binary tree* with nodes consisting of Boolean constants, operators, variables, and units composed of several operators (e.g. multiplexers, registers). Each output forms the root of such a binary tree. A complete circuit can thus be represented as a set of trees, one for each output. Inner tree nodes represent operators with edges pointing towards the node's inputs, while leaf nodes represent constants and input variables.

Fig. 7 illustrates the equivalence between a Boolean function represented as a set of interconnected gates, a binary tree, and a Boolean formula.

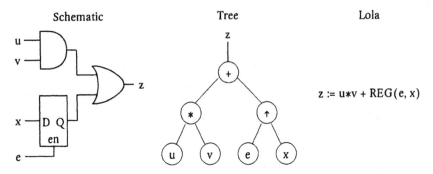

Fig. 7. Circuit Representations

Since the above representations of a circuit are equivalent, both a circuit layout and a Lola program can be transformed into a set of trees. Corresponding trees can then be compared to detect inconsistencies. Under the assumption that the Lola program describes the circuit correctly, i.e. it properly reflects a *circuit specification*, inconsistencies between corresponding pairs of trees are interpreted as errors in the layout, i.e. the *circuit implementation*.

The checker strives to find a *structural equivalence* between the specification and the implementation trees. It starts at the roots of two respective trees and, in parallel, traverses both trees from the roots towards the inputs. At each pair of nodes, the checker verifies that the two nodes match. If they match, the nodes' subtrees are checked for equivalence recursively. The procedure terminates when all nodes have been visited or a mismatch is detected.

Existing verifiers, such as automated theorem-provers [4], attempt to find an equivalence between Boolean equations by transforming them until equivalence (or its opposite) is inferred. This scheme is more flexible than matching for structural equivalence and allows for different levels of abstraction between the specification and the implementation. While such verifiers are well suited to detect inconsistencies, they typically fail in pinpointing the fault in the layout. The information needed for this purpose is either left out or lost during the transformations applied to the Boolean equations. This loss makes it impossible to locate an implementation error automatically and leaves the designer with the labor of locating it in the layout manually.

The checker, by contrast, keeps the information required to locate a part in the layout within each node. With this information available, an implementation error can not only be detected but also located in the faulty layout.

4.1.2 Using the Checker

The first step in the checking process is writing a Lola specification for the circuit. This program is compiled by the Lola compiler which generates a set of trees as its output. The trees can be viewed in a textual format as a set of Boolean equations. The output can be used as a reference in the next step when entering the design with the editor. The checker is then invoked to check the implementation for compatibility with the specification. Inconsistencies between the two are displayed textually and graphically in the layout. The checker can check complete layouts but may also be used during design entry to check partial layouts (e.g. for checking bit slices of a data path).

4.1.3 Implementation

In order to make the implementation of the checker simple and extensible, the *architecture-dependent extraction* part is decoupled from the *architecture-independent matching* part. The extractor converts the FPGA-dependent representation of circuits used by the editor into an architecture-independent set of trees. The matcher then verifies compatibility with the set of trees generated by the Lola compiler. This separation allows easy adaption to a new FPGA architecture by simply exchanging the extractor component. The extractor follows the signals from the output towards the inputs. Extraction stops at labels and constants found in the layout. When returning from the leaf nodes, the tree is constructed. Already during extraction certain checks are performed, such as detecting unconditional outputs to a tri-state bus or reading from an undefined source. The extractor also recognizes certain combinations of gates and converts them to more abstract operators, such as

$$q := MUX(en: q, x) \quad \rightarrow \quad q := LATCH(en, x)$$
$$q := {\sim}(s' * {\sim}(r' * q)) \quad \rightarrow \quad q := SR(s', r')$$

Once the trees are extracted, the matcher checks corresponding pairs of trees for compatibility. The trees generated by the Lola compiler are used as a reference while the trees extracted from the circuit are examined.

Earlier, we mentioned that the checker searches for a structural match between two corresponding trees. Demanding an exact structural match would require the designer to specify the circuit exactly the same way as it is later implemented. As this is too restrictive, the checker allows a number of *transformations* being applied to the trees. Since the goal is still to locate detected errors in the layout automatically, transformations must preserve the information needed for this purpose. The structural matching rules are relaxed and allow the following transformations:

1. *Inverters.* Architectural constraints imposed by FPGAs sometimes require the designer to connect parts of a circuit through successive inverters. For example, if an AND gate is implemented with a NAND gate, an inverter must follow the NAND gate, hence there are two inverters in series. The checker allows an arbitrary number of inverter nodes between any two nodes.
   $$x = {\sim}({\sim}x)$$

2. *DeMorgan's Laws.* The checker applies the laws of DeMorgan when necessary. For instance, the AT6002 FPGA cell lacks an OR gate. An OR gate is therefore usually implemented as a NAND gate with inverted inputs. This architecture-dependency should, however, not reflect in the specification where the OR operator is used instead.
   $$x + y = {\sim}({\sim}x * {\sim}y) \quad x * y = {\sim}({\sim}x + {\sim}y)$$

3. *Commutativity.* The representation of a dyadic Boolean operator as a node of a binary tree introduces an inherent order, by which its subtrees are compared (e.g. "compare left specification subtree with left implementation subtree"). For commutative operators (AND, OR, XOR), this order cannot be determined beforehand and the checker potentially matches both possibilities. Since the trees generated by the Lola compiler have a typical height of less than five, there is no apparent performance penalty associated with commutativity.
   $$x * y = y * x \quad x + y = y + x \quad x - y = y - x$$

4. *Associativity*. As with commutativity, associativity is an inherent property of binary trees. The checker supports only simple cases of associativity.
$$y * (x * (u + v)) = (u + v) * (x * y)$$

5. *MUX selectors*. For greater flexibility, multiplexers may be implemented with an inverted selector signal and accordingly exchanged input signals.
$$MUX(s: x, y) = MUX(\sim s: y, x)$$

6. *OR/AND with MUX*. It is sometimes more convenient to implement OR gates or AND gates using multiplexers. The checker recognizes the MUX representations as equivalent.
$$x + y = MUX(x: y, \text{'}1) \qquad x * y = MUX(x: \text{'}0, y)$$

All of these transformations can be applied to trees without losing information needed to locate errors in the layout after a mismatch.

Combined, the transformations make the checker a flexible and efficient tool for checking layouts. Its speed and its capability to check only parts of a design make it well suited for interactive use during design entry.

Design	AT6002 Cells Used	Lola Variables	Total Checking Time
UART	240	100	< 1 s
8x8 Multiplier	440	230	< 2 s
Microcontroller	770	240	< 4 s

Table 1. Checking Performance (80486, 33MHz)

4.2 The Timing Analyzer

Once a circuit is designed with the editor and its correct layout verified with the checker, the question about the circuit's performance arises. To determine the maximum operating speed of a given synchronous circuit a timing analysis tool is required. We have developed a timing analyzer which is capable of analyzing combinational and sequential circuits efficiently. It can be used interactively from within the editor during design entry but also provides a simple programming interface which can be used by future design automation tools. It provides commands to determine the maximum input delay between a given output and all of its inputs or only a specific input. If a circuit contains parts with fan-outs greater than one ("common subexpressions") their input delays are calculated only once to save computation time.

5 Conclusions

We presented an FPGA system consisting of an extension-board with an Atmel AT6002 FPGA and a set of simple and efficient software tools used to develop circuits for the board. The software consists of a compiler for the Lola language, a small hardware description language for synchronous digital circuits, an easy-to-use graphical editor with which layouts are entered with simple mouse manipulations, and a loader to configure the FPGA with layouts entered with the editor. Additionally, a circuit checker was implemented which performs a consistency check between a circuit specification in the form of a Lola program and its implementation within the editor. Inconsistencies are not only detected but also located within the layout displayed in the editor.

The software part was designed and implemented in Oberon [3] by three people in three months and consists of 13 modules containing about 6500 lines of code. Two weeks were spent developing the extension-board.

We have been using the system successfully in a laboratory for introductory courses in digital design. Due to its simplicity, the students learned to use the system quickly and were able to solve the given exercises. The exercises range from simple binary counters to a UART. At the Institute, we use the same system for experiments with programmable hardware.

All in all, we can only recommend using FPGAs in education. Their flexibility and quick reprogrammability allow interesting and diverse problem statements. By using real hardware instead of a simulator, the students also have to cope with the "real" problems of digital design such as good placement, economical routing, timing, and synchronization between components. Last, but not least, the chosen solution is an order of magnitude more cost effective than conventional laboratories using discrete MSI components and physical wiring.

Acknowledgements

We wish to thank I. Noack for implementing and testing the extension board.

References

1. B. Heeb, I. Noack, *Hardware Description of the Workstation Ceres-3*, Technical Report 168, Institute for Computer Systems, ETH Zurich, Switzerland, October 1991

2. Atmel Corporation, San Jose, CA. *Field-Programmable Gate Arrays, AT6000 Series.* 1993

3. M. Reiser. *The Oberon System – User's Guide and Programmer's Manual.* Addison-Wesley, Reading, MA. 1991.

4. R.S. Boyer, J. Strother Moore, *Proof-Checking, Therorem-Proving, and Program Verification*, Contemporary Mathematics, Vol. 29, American Math. Society, 1984, 119-132

COordinate Rotation DIgital Computer (CORDIC) Synthesis for FPGA

U. Meyer–Bäse[‡], A. Meyer–Bäse[†],W. Hilberg[‡]

[†]Institut Flugmechanik und Regelunstechnik, Technische Hochschule Darmstadt
[‡]Institut für Datentechnik, Technische Hochschule Darmstadt

Abstract. An universal CORDIC processors is able to compute a wide variety of functions, for example conversion between polar and cartesian coordinates, trigonometric (sin,cos,tan and vice versa), division, hyberbolic and exponential functions. Because CORDIC needs only simple add/subtract and shift operations, it is easy to realize it with FPGAs. We explain the CORDIC synthesis in different architecturs and of different accuracies. We examine the CORDIC synthesis for coordinate conversion from cartesian to polar $X, Y \rightarrow R, \theta$ and for computing the exponential function with the CORDIC processor supporting a former implemented artifical neural network. With our optimization the hardware effort of the CORDIC could be reduced, so that each processors may be implement each with one XC3090 FPGA from Xilinx.

1 The CORDIC Algorithm

In 1959 Volder [Vol59] developped the CORDIC algorithm (Coordinate Rotational DIgital Computer), to convert between polar and cartesian coordinates. CORDIC is an iterative algorithm to compute the coordinate of a vector rotation or to compute radius and the phase of a vector.

The method explained by Volder to compute triogonometric functions was expanded by Walther [Wal71]. With the aid of these extension, it is possible, to compute also very effective[†] hyperbolic and exponential functions using the same hardware as for the trigonometric functions [Sch74, p.162-176,181-193].

For the generalized CORDIC algorithm the iterative equations in the hyperbolic ($m = -1$), in the linear ($m = 0$), and in the circular ($m = 1$) coordinatation approach are shown in figure 1(b).

The CORDIC algorithm can be operate in either a vector "rotation" mode, or an angle accumulation mode ("vectoring"). Table 1 shows the various functions [HHB91], which can be realized with the CORDIC algorithm depending on the initial values of the register x_{in}, y_{in} and z_{in}, the coordinate system, and the two modes.

2 Examples of CORDIC–Processor Synthesis

After a brief summarize of the CORDIC algorithm and same using terminology, we will point our attention on two typical synthesis applications, and we will show that

---

[†] Insider says that HP use the CORDIC techniques in their scientific calculators [Sch74].

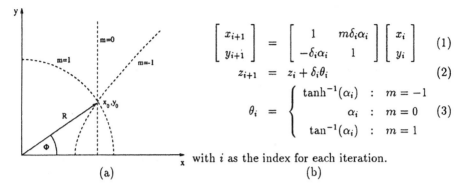

$$\begin{bmatrix} x_{i+1} \\ y_{i+1} \end{bmatrix} = \begin{bmatrix} 1 & m\delta_i\alpha_i \\ -\delta_i\alpha_i & 1 \end{bmatrix} \begin{bmatrix} x_i \\ y_i \end{bmatrix} \quad (1)$$

$$z_{i+1} = z_i + \delta_i\theta_i \quad (2)$$

$$\theta_i = \begin{cases} \tanh^{-1}(\alpha_i) & : \quad m = -1 \\ \alpha_i & : \quad m = 0 \quad (3) \\ \tan^{-1}(\alpha_i) & : \quad m = 1 \end{cases}$$

with i as the index for each iteration.

(a) (b)

Figure 1 (a) CORDIC modi (b) Definition of the rotation angle.

	hyperbolic $m = -1 \; \alpha_i = 2^{-i}$ (i=1,2,3, …, N) (repeate Iter. 4,13,40, …)	linear $m = 0 \; \alpha_i = 2^{-i}$ (i=1,2,3, …, N)	circular $m = 1 \; \alpha_i = 2^{-i}$ (i=0,1,2,3,…,N)
$y \to 0$	$\delta_i = \begin{cases} 1 & : \; x_iy_i \geq 0 \\ -1 & : \; x_iy_i \leq 0 \end{cases}$ $x_{N+1} = K_h\sqrt{x_{in}^2 - y_{in}^2}$ $z_{N+1} = z_{in} + \tanh^{-1}(y_{in}/x_{in})$ $|\tanh^{-1}(y_{in}/x_{in})| \leq 1.1182$	$\delta_i = \begin{cases} 1 & : \; x_iy_i \geq 0 \\ -1 & : \; x_iy_i \leq 0 \end{cases}$ $x_{N+1} = x_{in}$ $z_{N+1} = z_{in} + y_{in}/x_{in}$ $|y_{in}/x_{in}| \leq 1$	$\delta_i = \begin{cases} 1 & : \; y_i \geq 0 \\ -1 & : \; y_i \leq 0 \end{cases}$ $x_{N+1} = K_c\sqrt{x_{in}^2 + y_{in}^2}$ $z_{N+1} = z_{in} + \text{atan2}(y_{in}, x_{in})$ $|\text{atan2}(y_{in}, x_{in})| \leq 1.7433(99.9°)$
$z \to 0$	$\delta_i = \begin{cases} 1 & : \; z_i \leq 0 \\ -1 & : \; z_i \geq 0 \end{cases}$ $x_{N+1} = K_h[x_{in}\cosh(z_{in})$ $\qquad + y_{in}\sinh(z_{in})]$ $y_{N+1} = K_h[x_{in}\sinh(z_{in})$ $\qquad + y_{in}\cosh(z_{in})]$ $|z_{in}| \leq 1.1182$	$\delta_i = \begin{cases} 1 & : \; z_i \leq 0 \\ -1 & : \; z_i \geq 0 \end{cases}$ $x_{N+1} = x_{in}$ $y_{N+1} = y_{in} + x_{in}z_{in}$ $|z_{in}| \leq 1$	$\delta_i = \begin{cases} 1 & : \; z_i \leq 0 \\ -1 & : \; z_i \geq 0 \end{cases}$ $x_{N+1} = K_c[x_{in}\cos(z_{in})$ $\qquad - y_{in}\sin(z_{in})]$ $y_{N+1} = K_c[x_{in}\sin(z_{in})$ $\qquad + y_{in}\cos(z_{in})]$ $|z_{in}| \leq 1.7433(99.9°)$

Tabular 1 CORDIC–table with δ_i, x_{N+1}, y_{N+1}, z_{N+1} and the range of convergence (ROC).

the CORDIC FPGA solution is more efficient than the conventional signalprocessor (DSP) solution.

A conventional DSP solution is to prefer, if the function, which should be computed, may be developped in a (short) Taylor series to utilize the very fast hardware multiplier. Interesting and efficient applications of a DSP are FIR filter and auto- or cross-correlation. Unfortunately this concept will not properly work in computation of trigonometric functions. E.g., the Taylor series of the arctan function is linear convergent in dependence with the number of iterations, this means, doubling the iterations, we can get a gain of 1 bit. The CORDIC algorithm – in contrast – has a gain of 1 bit per iteration [Hah91, p.81-83].

In the following, we will show the synthesis of CORDIC–processors by explaining two applications in detail.

2.1 Demodulation of Bandpass Signals Using the CORDIC Algorithm

For an universal incoherent receiver results the inphase and quadraturphase after a mixer or sampler, whose complex sum constitute an analytic signal. The aim of a demodulator is to recover the original signal for all possible types of modulations. We will see that the CORDIC algorithm will solve this problem in an efficient manner.

2.1.1 Basis of COordinate Rotation DIgital Computer (CORDIC)

When we construct a demodulator, we use the binary CORDIC because the result of an A/D converter is normally binary (K2). Additionally we use the CORDIC-algorithm as a coordinate converter from rectangular to polar $(X, Y \to R, \theta)$, because we get directly AM- and PM-demodulation, respectively. As we have seen in the last section the conventional computation $R = \sqrt{X^2 + Y^2}$ und $\theta = \arctan(Y/X)$ is expensive and could dramatically be reduced with the CORDIC [Hah91, p.81-83].

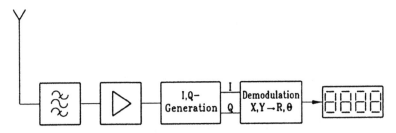

Figure 2 Hilbert sampling receiver with CORDIC demodulation.

2.1.2 Functionally Discription of the Convertion $(X, Y \to R, \theta)$

The CORDIC algorithm is an iterative procedure, which rotates a vector in the X, Y-plane by a defined angle $\pm \alpha_i$. We distinguish between *vectoring* $(X, Y \to R, \theta)$, representing demodulation, and *rotation* $(R, \theta \to X, Y)$, representing modulation.

In the following, we will only examine the vectoring mode, because we will get only demodulation (R, θ) [MB93c]. With the help of figure 3 we explain the principle. Starting with vector "1", which is received as X- and Y-coordinate, this vector will be rotated in each iteration about $\pm \alpha_i$, so that he lies finally on the X-axis. The sum of the angles $\sum_{i=1}^{i=n} \alpha_i$ represent the phase, where we are looking for. The final value of X_n is the belonging radius. It can be shown how the rotation can be reduced to simple add and shift opperations [MB91, p.30-32].

Understanding the concept of Volder [Vol59], means choosing an angle, which rotates either in positive or negative sense, see figure 3(a).

With this choose of the rotation angle, the equations for the rotations become

$$X_{i+1} = X_i \mp Y_i 2^{-(i-2)} \tag{4}$$
$$Y_{i+1} = Y_i \pm X_i 2^{-(i-2)} \tag{5}$$

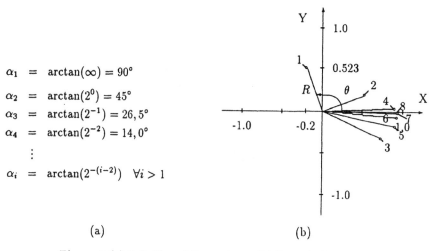

$\alpha_1 = \arctan(\infty) = 90°$

$\alpha_2 = \arctan(2^0) = 45°$

$\alpha_3 = \arctan(2^{-1}) = 26,5°$

$\alpha_4 = \arctan(2^{-2}) = 14,0°$

$\vdots$

$\alpha_i = \arctan(2^{-(i-2)}) \quad \forall i > 1$

(a)   (b)

Figure 3 (a) Definition of the angle α_i. (b) Example for vectoring.

We see, that to compute X_{i+1} and Y_{i+1}, we only need very simple arithmetic (shift and add) operations, which can be very efficient realized with FPGAs. We are free in choosing the bit width. The parameter depends on the available complexity of the FPGA and the required accuracy . The reachable accuracy for R in effective bits may be calculated with the theoretical calculation by [HHB91].

$$d_{\text{eff}} = log_2\left(\frac{1}{2^{-n+nl}|\vec{v}(0)| + 2^{-b-0.5}\left(G_m(n)/K_m + 1\right)}\right) - 1 \qquad (6)$$

This equation is shown in figure 4(a) for 8 to 20 iterations and bit width of 8 to 20 bit for the X/Y register and full scale $|\vec{v}(0)|_{\text{max}}$.

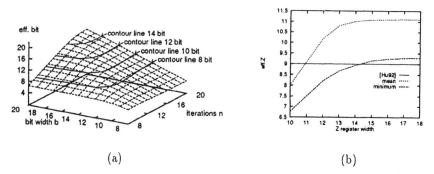

(a)   (b)

Figure 4 (a) Effective bit following (6). (b) Dependence of the Z-register width to the accuracy Z with 13 bit X/Y-resolution and 8 iterations.

The examination in [Hil92] shows, that unlike to the available literatur [Hu92], the Z accuracy (θ) depends not only on the number of iterations, radius of input, the resolution in the X/Y-path, but also on the rounding errors in the Z-path and the rounding error in Z, because of the existing rounding errors in the X/Y-path. Especially our simulation results proof, that the accuracy of Z reaches only asymptotic the radius accuracy, see figure 4(b).

Finally we point out the fact, that the resulting resolution for radius in the range 25% − 100%, which use the AM modulation of the central european radio control watch station DCF77, has a comparable accuracy as the PM/FM-signals with used LF-FAX stations like DCF37 or DCF54 with radius 100%. A portable universal LF receiver [Hil92],[BB92],[HK92] was designed with the CORDIC demodulator with accuracy as shown in table 2.

	range	minimum	mean	variance	lit. [Hu92]
radius	25% − 100%	10,160	11,160	0,470	9,648
phase	25% − 100%	9,143	11,071	1,306	9,000
radius	50% − 100%	10,160	11,121	0,453	9,648
phase	50% − 100%	9,277	11,080	1,309	9,000
radius	100%	10,161	11,061	0,414	9,648
phase	100%	9,538	10,990	1,286	9,000

Tabular 2 Effective bit width of an experimental CORDIC processor with 13 bit X/Y-path, 15 bit Z-path and 8 iterations.

2.1.3 Hardware Realization of the CORDIC Demodulator

The CORDIC algorithm may be realized as a "stage machine" or with a full pipelined processor [Hil92],[Nol91].

Both architectures can be efficiently realized with Field programmable Gate Arrays (FPGA) [‡]. If the main aim is speed, then each iteration equation may be realized in special Hardware — and we need for b bit width and $([b − 1]$-stages of the pipeline $\times 3 \times b$)-CLBs. By $b = 8$ bit we need so 168 CLBs and for $b = 12$ bit we need 396 CLBs, which are to match for the greatest FPGA (XC3090 → 320 CLB).

In figure 5 we present a state machine, which may be preferred, if the space in the FPGA is critical. An additionally reduction in complexity would be reached, if the "full parallel architecture" [Tim90, p.63] from figure 5 is reduced to the "lean structure" (only one shifter and/or one accumulator). Additionally it is possible to replace the expensive barrelshifter (BS) by a seriell right shifter (RS) or a seriell left/right shifter (LRS). For this three different architectures the resulting space for the X/Y-path in a FPGA can be seen from table 3.

As a good compromise between low latency time and space in an FPGA appears the "lean structure" with one barrelshifter and one accumulator.

In figure 6 the resulting expense is shown for a realization of the CORDIC processor in the "lean structure" (1BS+1AC) if 8 to 16 bits accuracy are nesessary. From figure

---

[‡] The specified CLB numbers (Configurable Logic Block) refere to the XC3000 of Xilinx.

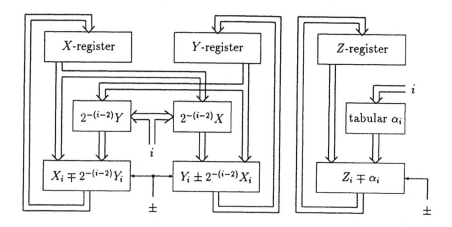

Figure 5 Realization of the CORDIC algorithm through a "full parallel" state machine.

structure	register	multiplexer	adder	shifter	∑CLB	time
2BS+2AC	2*7	0	2*14	2*19,5	81	12
2RS+2AC	2*7	0	2*14	2*6,5	55	46
2LRS+2AC	2*7	0	2*14	2*8	58	39
1BS+2AC	7	3*7	2*14	19,5	75,5	20
1RS+2AC	7	3*7	2*14	6,5	62,5	56
1LRS+2AC	7	3*7	2*14	8	64	74
1BS+1AC	3*7	2*7	14	19,5	68,5	20
1RS+1AC	3*7	2*7	14	6.5	55,5	92
1LRS+1AC	3*7	2*7	14	8	57	74

Tabular 3 CLBs for a 13 bit plus sign bit X/Y-Pfad of a CORDIC processor. (Abbreviations:AC=accumulator; BS=barrelshifter; RS=seriell right shifter; LRS=seriell left/right shifter)

6 it can be seen, that if the Z accuracy increases linearly, the expense behaves also linear. In contrast, the example of the full pipelined processor shown at the beginning of this section, which shows a square expense. The jump between 14 and 16 bit accuracy is established from the fact, that each CLB has at most 5 input variables.

The comparison between a conventional universal demodulation with a DSP and the CORDIC–FPGA realization arise as follows [MB91, p.33-37]: The direct algebraic computation of the I/Q demodulation equation with a DSP for AM, PM or FM signals is expensive for PM, less for FM and AM modulated signals. On the other side, the CORDIC algorithm needs for iterative computation for b bit accuracy only $\approx b$ iterations and it is possible to realize a pipeline structure, with which (with a

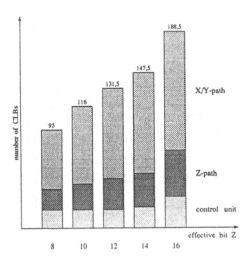

Figure 6 Expense for different requirement accuracy of Z result.

short delay) a conversion I/Q to R, θ (AM and PM demodulation simulaniously) may be performed in one cycle. For the FM demodulation only *one* differentiation must be computed after the CORDIC conversion.

Conclusion: Except for binary ASK signals the above examination shows, that the CORDIC realization with FPGA may be preferred before the direct computation with a DSP, because the CORDIC processors offer essential speed and computational advantage.

2.2 Implementation of an GPFU Neural Network with Aid of FPGA

Artificial neural networks with radial basis functions realize an universal approximator with a three layer structure [HKK90],[MB+93a], see figure 7. The hidden neurons, also called "Gaussian Potential Function Units" (GPFU) [MB+93b], compute the euclidean distance between input and reference vector.

The main computational effort is:

$$y(x) = \sum_{i=1}^{N} c_i * e^{-d(x, m_i)/2} \tag{7}$$

The aim of a hardware implementation is to increase the computational speed of the gaussian potential function through FPGAs. On these programmable logic the computation of the exponential function should be performed. In the existing DSP/FPGA realization the DSP won't waste a lot of time computing the time expensive exponential function.

To compute the exponential function the CORDIC processor must work in hyperbolic/rotation mode. For the X and Y register we get from table 1 with $m = -1$ the following equations:

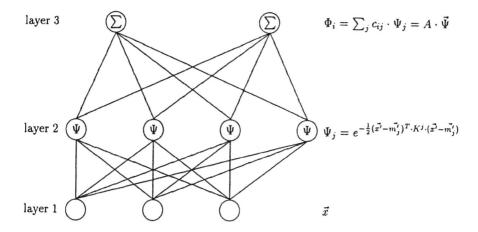

Figure 7 Three layer GPFU artificial neural network.

$$x_{n+1} = K_n\left(x_{in}\cosh(z_{in}) + y_{in}\sinh(z_{in})\right) \tag{8}$$
$$y_{n+1} = K_h\left(x_{in}\cosh(z_{in}) + y_{in}\sinh(z_{in})\right) \tag{9}$$

with the aid of the hyperbolic function the exponential function is defined as

$$e^x = \sinh(x) + \cosh(x) \tag{10}$$

To get directly e^x from the iteration, both X and Y register must be initialized with 1. Now we see, that the iteration equations for X and Y are the same. This fact, as we can see, dramatically reduce the complexity of the CORDIC architecture.

For the hyperbolic CORDIC the range of convergence (ROC) for rotation mode is

$$|z_{in}| \le \theta_{max} \approx 1,1182 \tag{11}$$

To get a greater ROC we must use negative iterations with the rotation angles $\theta_i = \tanh^{-1}(1 - 2^{-2^{-i+1}})$ instead of $\theta_i = \tanh^{-1}(2^{-i})$, because the normal rotation angle deliver for negative i complex angles.

The expanded range of convergence results in

$$\theta_{max} = \sum_{i=-M}^{0} \tanh^{-1}(1 - 2^{-2^{-i+1}}) + [\tanh^{-1}(2^{-N}) + \sum_{i=1}^{N} \tanh^{-1}(2^{-i})] \tag{12}$$

The needed ROC depends mainly on the parameter of the neural net. With $M = 2$ the ROC of the exponent x is $[-6.92631; 6.92631]$. Simulation results shows [Sch92], that the ROC of $M = 2$ is sufficient for our net configuration. With the increased ROC we get slightly different iterative equations:

$$\begin{bmatrix} x_{i+1} \\ y_{i+1} \end{bmatrix} = \begin{bmatrix} 1 & m\delta_i(1 - 2^{-2^{-i+1}}) \\ -\delta_i(1 - 2^{-2^{-i+1}}) & 1 \end{bmatrix} \begin{bmatrix} x_i \\ y_i \end{bmatrix} \tag{13}$$

With the "lean structure" the equation for Y must not be explicitly computed and it follows for $m = -1$

$$x_{i+1} = x_i - x_i \delta_i (1 - 2^{-2^{-i+1}}) \qquad (14)$$

Analyzing equation 14, we can see, that we get for δ_i two cases ($\delta = -1$ or 1) for the nonshift part of the iteration equation and a third case for the conventional iteration with positive rotation. These three possibilities changed the architecture so far as the X path needs an additionally 3:1 multiplexer before the add/sub unite [HK93], see figure 8.

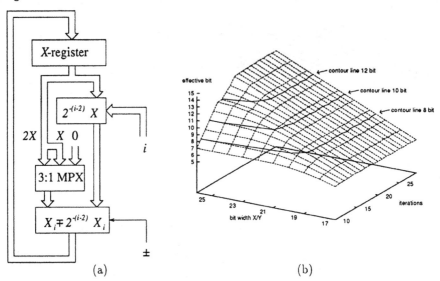

(a)   (b)

Figure 8 (a) CORDIC stage machine with ROC expansion and with a "lean structure" (θ see figure 5). (b) Error estimation of the CORDIC algorithm with 1000 test values z_{in} for each point.

2.2.1 Selection of the Appropiate CORDIC Parameters

In the former section we introduced the new efficient structure of the processor, and now we want to select an appropriate bit width and the number of the necessary iterations. These two parameters depend on the acceptable hardware complexity and the required accuracy. The simulation of the GPFU recognition rate shows that 10 bit fractional accuracy will be sufficient. The integer accuracy is 6 bit resulting in a total accuracy of 16 bit. The accuracy of the CORDIC computation depends on the quantization error of the X/Y path, the Z quantization error, the number of iterations, and the scaling. In contrast to the circular CORDIC algorithm [MB93c] the effective resolution of the hyperbolic CORDIC may not be computed analytically, because $\| B_m(j) \|$ depends mainly on the value of $z(j)$ during the iteration, see [Hu92, p.837, Lemma 2].

$$\| B_m(j) \| = \prod_{i=j}^{n-1} k_{-1}(i) e^{|z(j) - z(n)|} \qquad m = -1 \qquad (15)$$

Figure 8(b) shows the minimum accuracy with the estimation equation by [Hu92], which is computed over 1000 test values z_{in} for each combination bit width/number of the possible iterations. A comparison of the CORDIC results of the 1000 values with an exact (floating point computation) shows [HK93], that this results are ca. 1 bit better than the estimation with the equation by [Hu92]. So the estimation results of [Hu92] are quite realizable.

number of CLBs

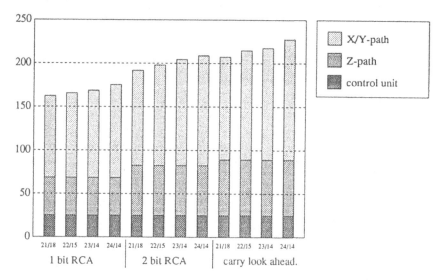

bit width X / number of iterations

Figure 9 Realization of the CORDIC processors with Xilinx XC3000 FPGA depending on different register width, number of iterations, and the architectures.

For the appropriate accuracy of 10 bit, four different solutions exist. Under use of the fastet shifter (barrelshifter) the computational time depends only on the number of iterations, and the kind of accumulator. Beside the conventional "ripple carry adder" (one CLB for each bit), we also examined double speed adder (2 bit in 3 CLB) and a "carry look ahead" adder, see figure 9. Finally we choose the "carry look ahead" adder with 22 bit width and 15 iterations, because only 320 CLBs are available and this combination would be faster than the three other ones.

3 Further Develoment

Our further examination will be a full pipelined CORDIC processor with the new XC4020 FPGA. These FPGAs have a "fast carry logic" to implement 2 bit per CLB of a fast adder/subtract realization. With this pipeline architecture, we get a great increase in speed compared with the stage machine.

Also a future aim would be an universal CORDIC stage machine, which should realize all 6 modes (3 coordinate systems each with rotation and vectoring mode).

This processor is with the existing FPGA in the required accuracy not realizable nowadays, so we hope of future devices with more complexity.

Acknowledment

The authors would like to thank to all students who have worked with us on this project. Special thanks to J. Hill, R. Schimpf, C. Brandt, R. Bach, H. Hausmann, V. Kleipa, S. Keune and T. Häuser.

References

[BB92]   R. Bach and C. Brandt. *Universeller Hilbert–Abtastempfänger mit CORDIC–Demodulation durch programmierbare Gate Arrays.* Masterthesis ST 551, Institut für Datentechnik, THD, 1992.

[Hah91]   H. Hahn. *Untersuchung und Integration von Berechnungsverfahren elementarer Funktionen auf CORDIC-Basis mit Anwendungen in der adaptiven Signalverarbeitung.* Ph.D. dissertation , VDI Verlag, Reihe 9, Nr. 125, 1991.

[HHB91]   Xiaobo Hu, Ronald G. Harber and Steven C. Bass. *Expanding the Range of Convergence of the CORDIC Algorithm. IEEE Transactions on computers,* page 13–21, 1:1991.

[Hil92]   J. Hill. *Entwurf eines optimierten COordinate Digital Computer (CORDIC) mit programmierbaren Gate Arrays.* Masterthesis DT 550, Institut für Datentechnik, THD, 1992.

[HK92]   H. Hausmann and V. Kleipa. *Aufbau eines Mikrocontroller–Systems und Entwicklung eines LCD–Graphik–Controllers auf FPGA–Basis.* Masterthesis ST 520, Institut für Datentechnik, THD, 1992.

[HK93]   T. Häuser and S. Kenne. *Entwicklung einer vollparallelen neuronalen CORDIC–Architektur mit Fließkomma–Signalprozessor und programmierbaren Gate Arrays.* Masterthesis DT 570, Institut für Datentechnik, THD, 1993.

[HKK90]   Eric J. Hartman, James D. Keeler and Jacek M. Kowalski. *Layered Neural Networks with Gaussian Hidden Units as Universal Approximations. Neural Computation,* 2:210–215, 4:1990.

[Hu92]   Yu Hen Hu. *The Quantization Effects of the CORDIC-Algorithm. IEEE Transactions on signal processing,* page 834–844, 4:1992.

[MB91]   U. Meyer-Bäse. *Entwurf und Untersuchung universeller digitaler Empfängerprinzipien.* Report 135, Institut für Datentechnik, THD, 1991.

[MB+93a] A. Meyer-Bäse et al. *Modulares Neuronales Phonemerkennungskonzept mit Radialbasisklassifikatoren. 15. DAGM-Symposium in Lübeck Workshop,* page 670–677, 9:1993.

[MB+93b] A. Meyer-Bäse et al. *Neuronale Selbstorganisation und inverse Abbildungs-eigenschaften in Netzen mit rezeptiven Feldern.* DFG-Tagung „Kognition, Repräsentation und Intentionalität in natürlichen und künstlichen Systemen, 4:1993.

[MB93c] U. Meyer-Bäse. *Universeller Hilbert–Abtastempfänger mit CORDIC–Demodulation.* In *Funkuhren Zeitsignale Normalfrequenzen,* page 65–81, 5:1993.

[Nol91] A. Noll. *Implementierung des CORDIC–Algorithmus zur Koordinatentransformation in einem programmierbaren Gate Array.* Masterthesis DT 481, Institut für Datentechnik, THD, 1991.

[Sch74] H. Schmid. *Decimal Computation.* John Wiley & Sons, 1974.

[Sch92] R. Schimpf. *Implementierung eines Neuronalen Netzes und Algorithmen zur Sprachvorverarbeitung auf einem Fließkommasignalprozessorsystem.* Masterthesis DT 553, Institut für Datentechnik, THD, 1992.

[Tim90] D. Timmermann. *CORDIC-Algorithmen, Architekturen und monolithische Realisierungen mit Anwendungen in der Bildverarbeitung.* Ph.D. dissertation, VDI Verlag, Reihe 10, Nr. 152, 1990.

[Vol59] J. E. Volder. *The CORDIC Trigonometric computing technique.* IRE Transactions on Electronics Computers, page 330–4, 9:1959.

[Wal71] J.S. Walther. *A Unified algorithm for elementary functions.* Spring Joint Computer Conference, page 379–385, 5:1971.

MARC: A Macintosh NUBUS-expansion board based reconfigurable test system for validating communication systems

Georg J. Kempa and Peter Rieger

University of Kaiserslautern, Microelectronics Centre (ZMK)
67653 Kaiserslautern, Germany

Abstract. In this paper a test system based on a Macintosh NUBUS-expansion board is presented. The test system is termed MARC. MARC is a reconfigurable test system which can be applied in the simulation loop. The primary benefit of MARC is that technology independent VHDL code can be mapped onto its reconfigurable components, thus allowing a fast real-world validation of the function described by the VHDL code instead of time consuming logic simulation. The different parts of this tool are introduced and the benefits of using it in the VHDL based design process are depicted. Using an exemplary communication system represented by a digital wireless microphone system, the aforementioned benefits are validated.

1 Introduction

The hardware description language VHDL [1] has become accepted as a viable tool for the use in the design process. In the Microelectronics Centre of the University of Kaiserslautern, VHDL is used in the design process of components required in communication systems [5]-[9]. The functionality of parts of these systems is verified by the logic simulation of automatically synthesized schematics. However, it is not possible to verify the whole system in this way for the following reason. To demonstrate and validate the performance of a communication system, usually the achievable error performance in terms of bit error rates is determined by considering large numbers of input data. By using a logic simulation tool, this validation procedure is extremely time-consuming. It would be desirable to carry out such a validation procedure in real time.

One of the benefits of using VHDL is the technology independent design [2]-[4]. For instance, it is possible to map the VHDL code on FPGAs by logic synthesis for prototyping or evaluation purposes. In order to exploit the benefits of the prototyping and evaluation capabilities by deploying FPGAs, a universal FPGA based expansion board as part of a test system for evaluating the performance of components of communication systems was developed. The test system including the developed evaluation board is termed MARC. By using MARC, on the one hand the functionality of the component under investigation can be verified in hardware and on the other hand the performance of that component can be quickly determined, e.g. in terms of bit error rates. The application of such an evaluation board leads to a significant reduction of development time.

In order to allow a flexible test of various components required in a communication system or even the whole system itself, MARC is reconfigurable. An overview of MARC and details of some hard- and softwareparts are presented in the following section. Section 3 introduces a digital wireless microphone system as an example of a communication system to be evaluated. The results of the validation procedure of this digital microphone system by deploying MARC are presented in section 4.

2 MARC

2.1 Overview

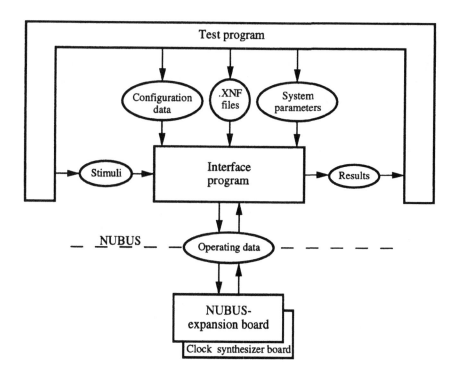

Fig. 1   MARC

Fig. 1 gives an overwiev of MARC. MARC is subdivided into a software part, containing both the "Test program" and the "Interface program" respectively, and a hardware part represented by the implemented NUBUS-expansion board and a clock synthesizer board, cf. Fig. 5 and Fig. 8. The "Test program" controls the running of the test with its

system parameters, provides the configuration data for the FPGA and the corresponding netlist file (.XNF file), creates and provides the stimulus files, simulates the environment of the system under test and interprets the results. The "Interface program"
facilitates the connection between the "Test program" and the NUBUS-expansion
board and controls the board. The NUBUS-expansion board is the platform to include
any VHDL coded and XILINX FPGA mapped system parts in the simulation process.
For this the board takes the operating data of each system part and prepares the configurable parts of the board for simulation.

2.2 Software of MARC

In Fig. 2 the petri net of the "Test program" is depicted.

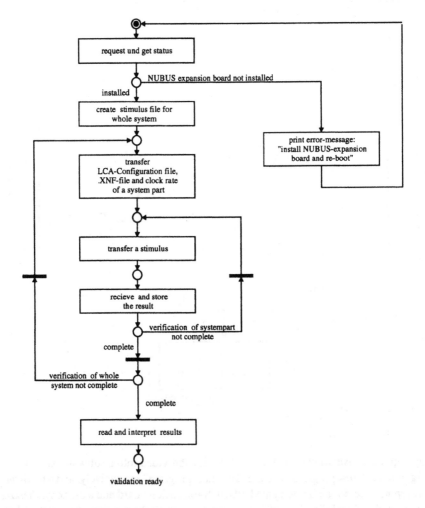

Fig. 2 Petri net of the "Test program"

In the beginning of a test run it is checked whether the hardware part of the test system, namely the NUBUS-expansion board, is installed. If the NUBUS-expansion board isn't detected, the programm puts out an error message. In this case the user has to install the NUBUS-expansion board and has to re-boot the computer.

If the NUBUS-expansion board is detected, the "Test program" creates the stimulus file for testing the system under investigation. The next task of this "Test program" is the initialization of the NUBUS-expansion board. During this initialization, the configuration of the FPGA has to be achieved once at the beginning of a test cycle if smaller systems or only parts of a complex system are to be studied. If larger systems are tested, the "Interface program" has to reconfigure the FPGA repeatedly. In this case, the FPGA's interface has to be reconfigured as well. Therefore, the "Test program" sends the LCA configuration data and the I/O data included .XNF file to the "Interface program". Additionally the "Test program" transfers some important system parameters, e.g. the clock rate of the system part to be studied, to the "Interface program". In the following the system part can be tested. The stimuli are transferred to the NUBUS-expansion board in sequential order and the computed results are received and stored. After receiving the last result it is checked whether other system parts are to be tested. If such parts are found, the described actions are repeated. Otherwise the verification of the system is completed. Finally the "Test program" interprets the stored results.

The main task of the abovementioned "Interface program" is the initialization of the NUBUS-expansion board. Therefore, the "Interface program" takes the I/O data included in the .XNF file and sends a port description to the NUBUS-expansion board. According to this port description, the "Interface program" configures the interface of the system under test. By doing so the FPGA is automatically protected against damaging by inadvertently forcing of outputs.

2.3 Hardware of MARC

2.3.1 NUBUS-expansion board

The structure of the NUBUS-expansion board is shown in detail in Fig. 3. The "NUBUS-Interface" connects the units of the board to the NUBUS. A XILINX XC4010 -5 PG191C FPGA is used as the configurable part of the NUBUS-expansion board [10]. This unit is clocked by a "NUBUS-Interface" controlled "Frequency Synthesizer". Due to this circumstance, the system is running independently of the NUBUS clock at a user defined speed. A universal interface consisting of latches provides the possibility to model different interfaces, cf. Fig. 3. The "Declaration ROM" is needed to identify the NUBUS-expansion board in order to distinguish it from other boards on the NUBUS.

In Fig. 4 the structure of the abovementioned "NUBUS-Interface" is depicted. The "NUBUS-Interface" is subdivided into four functional units. The unit "Adressdecode" monitors the NUBUS adress/data lines ADx and identifies if the "Interface program" wants to access to the NUBUS-expansion board.

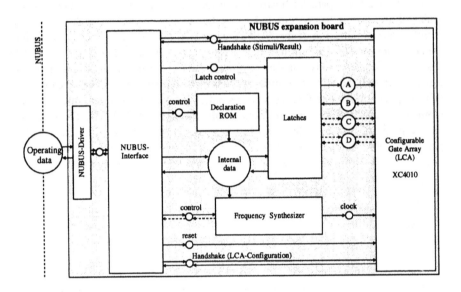

Fig. 3   Overview of the NUBUS-expansion board

The identifier /CLK is an abbreviation of the NUBUS clock signal. If a request is detected the unit "Adressdecode" activates with a control signal the "Interface control". It controls the NUBUS protocol, which is neccessary for the data transfer. Therefore, the "Interface control" can access the unit "Dataflow" in order to transfer the data between the NUBUS and the other units of the NUBUS-expansion board. Using the port description included in the .XNF file the functional unit "Latcharbiter" controls the latches which build the universal interface to the reconfigurable part of the NU-BUS-expansion board. The unit "Latcharbiter" is controlled by the units "Adressdecode" and "Interface control" as well.

The realized NUBUS-expansion board is shown in Fig. 5. The NUBUS-interface is realized as part of the FPGA on the right of the board. The NUBUS driver is realized with the parts in the foreground of Fig. 5. The "Latches" and the "Declaration ROM" are positioned in the middle. The FPGA on the left side is the aforementioned FPGA which can be reconfigured.

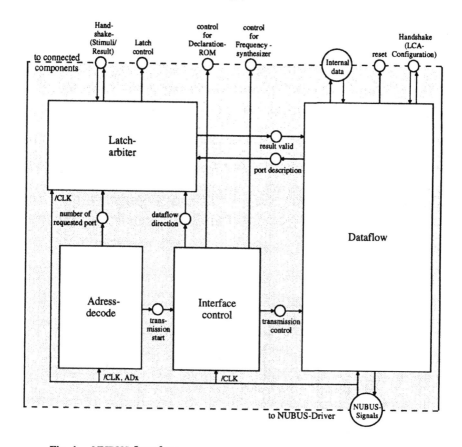

Fig. 4 NUBUS-Interface

Fig. 5 NUBUS-expansion board

2.3.2 Clock synthesizer board

To improve the performance of MARC the application on the NUBUS-expansion board can be clocked independently from the NUBUS clock. Therefore the NUBUS-expansion board can be extended by a clock synthesizer board, which allows to adjust the clock rate ranging from 1 MHz to 40 MHz in 500 kHz increments. Thus it is possible to run the system under test up to its maximum clock rate. In Fig. 6 the structure of the clock synthesizer board and its interface to the NUBUS-expansion board is depicted. On the top level the clock synthesizer board is subdivided into two functional units. One of this units, a digital frequency synthesizer circuit, is realized by an ACTEL A1020B FPGA [11]. The other functional unit contains voltage controlled oscillators (VCO) combined with low-pass filters to form a phase-locked loop (PLL) [12]. In addition the unit contains digital control logic required for this special application.

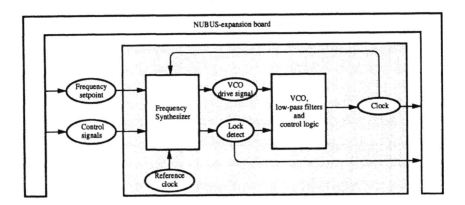

Fig. 6   Clock synthesizer board (principle design)

In the beginning of a simulation run the frequency setpoint is transferred to the clock synthesizer board. By using the control signals depicted in Fig. 6 the NUBUS-expansion board resets the clock synthesizer board and initiates the clock synthesis. The "Frequency Synthesizer" compares the frequency of the VCO output with the frequency setpoint by using a stable reference clock and performs a signal to drive the VCO. If the clock frequency matches the frequency setpoint, an active lock detect signal indicates a stable synthesized clock. If the NUBUS-expansion board receives this signal, the simulation starts.

The aforementioned frequency range cannot be realized by a single VCO. Therefore the frequency range is subdivided into smaller ranges wich are realized by several VCO circuits. The lock detect signal is used to switch between the several ranges during a particular definable time interval. In Fig. 7 an example of the functionality of the board is depicted. The frequency setpoint is shown as constant clock. The synthesized clock, depicted below the frequency setpoint trace, has to match this clock. In the beginning of the simulation any VCO circuit and therefore any base frequency is chosen, in this case a frequency which is lower than the frequency setpoint. The frequency synthesizer now drives the VCO to increase the clock frequency, cf. Fig. 7.

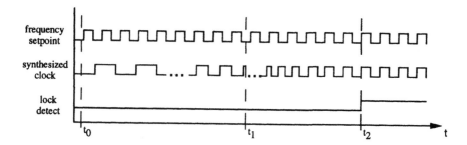

Fig. 7   Clock synthesis

The realized PLL circuits adjust the specified clock frequency after 2.53 ms at the latest. If the frequency setpoint is not within the chosen frequency range, the frequency range is automatically changed, based on the lock detect signal, see Fig. 7 at the time t_1. After changing the frequency range to a higher frequency as the setpoint, the frequency adjustment is initiated again. In the example shown in Fig. 7 the frequency of the synthesized clock is decreased to match the frequency setpoint, indicated by the rising lock detect signal at the time t_2. From time t_2 onward the simulation runs on the NUBUS-expansion board.

In Fig. 8 the realized clock synthesizer board is shown. The abovementioned "Frequency Synthesizer" is realized in the Actel FPGA on the right-hand side of the board. The timer, a predivider and the operational amplifier with the items to realize the low pass filter are positioned in the middle of the board. The parts on the left-hand side are the aforementioned VCO and the multiplexer to switch between the different frequency ranges.

Fig. 8   Clock synthesizer board

3 Digital wireless microphone system

Fig. 9 depicts the structure of a digital wireless microphone system applying frequency division multiple access (FDMA) [16].

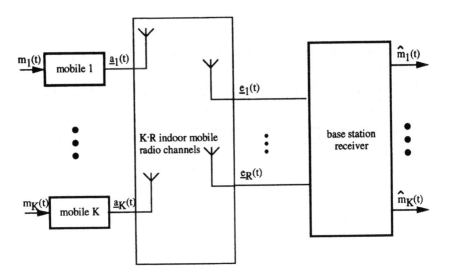

Fig. 9 Digital microphone system

The system consists of K mobiles and one immobile base station. In the system, the transmitters are always concentrated in the K mobiles whereas the base station represents the receiver. Each of the K mobiles contains a microphone. The analogue voice signals picked up by the microphones in the mobiles are converted into digital signals which are then source and channel encoded and mapped onto DQPSK (differential quadrature phase shift keying) symbols. Based on these DQPSK symbols, digital signals $\underline{m}_k$, $k = 1...K$, are generated at the K mobiles which are then modulated and converted into corresponding analog signals $\underline{a}_k(t)$, $k = 1...K$. The analog signals $\underline{a}_k(t)$, $k = 1...K$, are then transmitted over time-variant mobile radio channels. Since the considered digital wireless microphone system is designed for indoor applications, the time-variant mobile radio channels have flat fading characteristics. Besides the time-variant distortion of the transmitted signals $\underline{a}_k(t)$, $k = 1...K$, the distorted versions of $\underline{a}_k(t)$, $k = 1...K$, are corrupted by additive white Gaussian noise (AWGN) [14].

At the immobile base station, antenna diversity is applied. The distorted and corrupted versions of the transmitted signals $\underline{a}_k(t)$, $k = 1...K$, are received over R receiver antennas. Based on the R received signals $\underline{e}_r(t)$, $r = 1...R$, the base station receiver determines estimates $\hat{\underline{m}}_k$, $k = 1...K$, of the digital signals $\underline{m}_k$, $k = 1...K$, which contain DQPSK symbols by applying differentially coherent detection. The quality of the performance achieved with the base station receiver can be quantified in terms of bit error rates [15].

4 Validation

The introduced digital wireless microphone system was simulated with MARC to evaluate its digital parts. The digitally processing namely the performing of the receiver functions of the system is carried out on the NUBUS-expansion board. The other parts of the system and the indoor mobile radio channel are simulated by the "Test program" on the Macintosh PC.

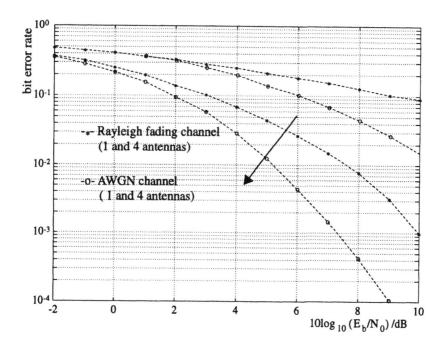

Fig. 10 Bit error rate curves

Fig. 10 depicts the simulation results in terms of bit error rates versus the average signal-to-noise ratio E_b/N_0 obtained for the considered digital wireless microphone system by various test runs deploying MARC. In order to obtain the 47 depicted bit error rate values represented by asterisk (*) and circles (o), the transmission of approximately two million DQPSK symbols per bit error rate value, i.e. 94 million DQPSK symbols in total, was simulated. The results obtained by deploying MARC are in perfect agreement with results obtained by simulations carried out on a Siemens/Fujitsu

VP100 super computer which are represented by the dashed lines [16]. Therefore, the viability of MARC is obvious.

Acknowledgement

The authors wish to thank Mr. Markus Müller for his valuable support during the layout of the boards and Dr. Peter Jung for brushing up our English. Finally the authors wish to thank Dr. Florian Graf for providing the yet unpublished simulation results obtained with the aforementioned supercomputer.

References

[1] *VHDL Language Reference Manual*. IEEE Standard 1076-1992, The Institute of Electrical and Electronics Engineers, Inc.

[2] R. Camposano, R.M. Tabet: Design representation for the synthesis of behavioral VHDL models. In Darringer, Rammig (Eds.): *Computer Hardware Description Languages and their Applications*. Amsterdam: Elsevier Science Publishers B.V. (1990) pp. 49-58.

[3] S.P. Levitan, A.R. Matello, R.M. Owens, M.J. Irwin: Using VHDL as a language for synthesis of CMOS VLSI circuits. In Darringer, Rammig (Eds.): *Computer Hardware Description Languages and their Applications*. Amsterdam: Elsevier Science Publishers B.V. (1990) pp. 331-345.

[4] D. Gajski et al.: *High-Level Synthesis*. Boston/Dordrecht/London: Kluwer Academic Publishers, 1992.

[5] P. Jung: *Entwurf und Realisierung von Viterbi-Detektoren für Mobilfunkkanäle*. Ph. D. Thesis, University of Kaiserslautern, Department of Electrical Engineering, 1993.

[6] P. Jung, J. Blanz: *Design of a Viterbi equalizer with field programmable gate arrays*. Microelectronics Journal, 24. Amsterdam: Elsevier Science Publishers Ltd.(1993) pp. 787-800.

[7] F. Berens: *Parametrisierte Verhaltensbeschreibung eines soft-output Viterbi-Entzerrers*. Masters Thesis, University of Kaiserslautern, Department of Electrical Engineering, 1993.

[8] P. Jung, J. Blanz: *Realization of a soft output Viterbi equalizer using field programmable gate arrays*. I.E.E.E VTC-93, Veh. Technol. Conf., Secaucus, New Jersey, May 18-20, 1993.

[9] G. Kempa, P. Jung: FPGA based logic synthesis of squarers using VHDL. In Grün-bacher, Hartenstein (Eds.): *FPGAs: architectures and tools for rapid prototyping*. Berlin: Springer, 1993.

[10] The Programmable Gate Array Book, XILINX, 1993.

[11] ACT Family Field Programmable Gate Array Data Book, Actel, 1990

[12] V. Manessewitsch: *Frequency synthesizers*, Theory and Design, John Wiley & Sons, 1987.

[13] M. Floyd, Ph. D. Gardner: *Phaselock Thechniques*, John Wiley & Sons, 1979.

[14] J.G. Proakis: *Digital communications*. McGraw-Hill, New York, 1993

[15] F. Adachi, K. Ohno: *BER performance of QDPSK with postdetection diversity reception in mobile radio channels*. IEEE Transactions on Vehicular Technology, Vol. VT-40, S. 237-249, 1991.

[16] F. Graf: *Digitale drahtlose Mikrofonsysteme mit Vielfachzugriff*. Ph. D. Thesis, University of Kaiserslautern, Department of Electrical Engineering, 1993.

Artificial Neural Network Implementation on a Fine-Grained FPGA

P. Lysaght, J. Stockwood, J. Law and D. Girma

Communications Division,
Department of Electronic and Electrical Engineering,
University of Strathclyde,
Glasgow G1 1XW

Abstract This paper reports on the implementation of an Artificial Neural Network (ANN) on an Atmel AT6005 Field Programmable Gate Array (FPGA). The work was carried out as an experiment in mapping a bit-level, logically intensive application onto the specific logic resources of a fine-grained FPGA. By exploiting the reconfiguration capabilities of the Atmel FPGA, individual layers of the network are time multiplexed onto the logic array. This allows a larger ANN to be implemented on a single FPGA at the expense of slower overall system operation.

1. Introduction

Artificial neural networks, or *connectionist classifiers*, are massively parallel computation systems that are based on simplified models of the human brain. Their complex classification capabilities, combined with properties such as generalisation, fault–tolerance and learning make them attractive for a range of applications that conventional computers find difficult. Examples of these include video motion detection, hand-written character recognition and complex control tasks.

Traditionally, ANNs have been simulated in software or implemented directly in special-purpose digital and analogue hardware. More recently, ANNs have been implemented with reconfigurable FPGAs. These devices combine programmability with the increased speed of operation associated with parallel hardware solutions. One of the principal restrictions of this approach, however, is the limited logic density of FPGAs resulting from the intrinsic overhead of device programmability.

This paper presents an alternative approach to previously reported neural network implementations on FPGAs [2][3][7]. The novelty of the design is achieved by exploiting several design ideas which have been reported previously in different designs and by combining them to form a new implementation. The design is based on a fine-grained FPGA implementation of an ANN in contrast to most of the FPGA implementations reported to date. It emphasises careful selection of network topology and methods of realisation to produce a circuit which maps well to the special requirements of fine-grained architectures. These include the realisation of the ANN using digital pulse–stream techniques and the choice of a feedforward network topology. It further exploits the use of run–time device reconfiguration to time–multiplex network layers to offset the logic density limitations of current devices. These topics are introduced in section two, where a reconfigurable pulse-stream ANN architecture [5][6] that is well suited for implementation on fine–grained FPGAs is described. Section three reviews some of the physical design issues that arose when mapping the ANN onto the AT6005 architecture, and in section four the performance of the network is appraised.

2. Reconfigurable ANN based on Pulse–Stream Arithmetic

2.1 Overview

ANNs employ large numbers of highly interconnected processing nodes, or *neurons*. Each neuron contains a number of synapses, which multiply each neuron input by a weight value. The weighted inputs are accumulated and passed through a non–linear *activation function* as illustrated in Fig. 1. These arithmetic–intensive operations and numerous interconnections are expensive in terms of logic and routing resources when implemented on an FPGA. Typically, as a result of these restrictions, expensive arrays of FPGAs have to be employed to implement "useful" networks [2], or alternatively, a single neuron is placed on the FPGA and used to emulate a network serially [3].

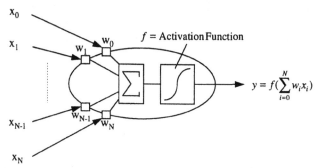

Fig. 1. The components of a simple artificial neuron.

The reported ANN incorporates three approaches to overcoming logic density limitations. First, pulse-stream arithmetic is used to provide an efficient mapping of the network onto a fine-grained FPGA. This technique is discussed in more detail in section 2.2.

Second, a reduction in the number of inter–neuron connections, which consume valuable routing resources, is made by adopting a layered, feed–forward network topology. As Fig. 2 shows, in contrast to the fully–interconnected network, the layered topology has connections only between nodes in adjacent layers. Further, supervised training is used to eliminate the need for feedback connections. This makes for easier partitioning of the network, since data flow through the network is uni–directional, from the input layer to the output layer.

Finally, by exploiting the reconfigurability of static memory–based FPGAs, the ANN can be *time-multiplexed* so that one physical layer is reconfigured to perform the function of all the other network layers. This makes it possible to implement a much larger design than would otherwise be possible on a single device. However, for this strategy to be successful it is important that the time spent reconfiguring the FPGA is relatively short, otherwise the speed of the overall network is severely degraded.

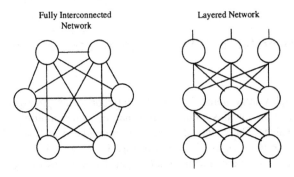

Fig. 2. Fully interconnected and layered ANN topologies

The ANN is implemented on an Atmel AT6005 (formerly Concurrent Cli6005). This is a fine-grained FPGA which is presently the only commercially available device capable of being *dynamically reconfigured*, i.e. selectively reconfigured while the logic array is active [4]. It will be shown that by exploiting this capability, and only reconfiguring those parts of the array which differ between network layers, it is possible to dramatically reduce the amount of system processing time that is lost during reconfiguration.

2.2 Pulse Stream Arithmetic

Pulse Frequency Modulation (PFM) is a coding scheme where circuit state values are represented by the frequency of narrow constant–width pulses. Fig. 3 shows an example of PFM, where the fractional value 7/16 is represented by the presence of 7 pulses in a 16–pulse window. Signals encoded in this manner can be summed and multiplied using simple logic gates. This technique, known as *pulse–stream arithmetic* [5], maps well onto fine–grained FPGAs such as the Atmel AT6005 which contain a large number of low fan–in gates.

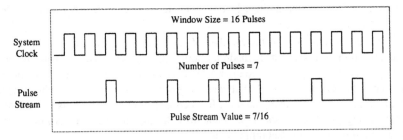

Fig. 3. Example of a Pulse Frequency Modulated signal.

2.3 Pulse Stream Neuron

In Section 1 the principal components of an artificial neuron were introduced. Here, the digital pulse–stream implementation of synaptic weight multiplication, post–synaptic summation and non–linear activation are described.

The inputs to the ANN are encoded as a constant stream of narrow pulses. Within each synapse, this pulse stream must be gated so that only a certain proportion of the pulses are allowed to pass through to the summation stage of the neuron. This proportion represents the value of the synaptic weight. A suitable gating function can be constructed by selectively ORing together a series of *chopping clocks* [5]. These are synchronous, non–overlapping binary clocks with duty cycles of 1/2, 1/4, 1/8 and so on. Fig. 4 shows a 4–bit chopping clock generator which can be used to construct weights in the range 0 to 15/16. Multiplication of the input pulse–stream by the weight value can be achieved by simply ANDing the input and the gating function, as shown in the diagram.

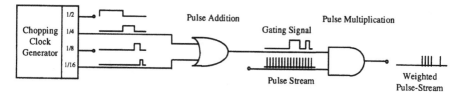

Fig. 4. Pulse arithmetic using simple logic gates

A synapse output is either *excitatory*, i.e. it increases the chance of the neuron firing, or *inhibitory*. In the pulse–stream neuron, positive and negative synaptic weights are accomplished by feeding excitatory and inhibitory synapse outputs to separate *up* and *down* inputs of a binary counter.

The neuron activation function is a simple binary step function, rather than the sigmoid function that is often used. There are two principal reasons behind this choice:

1. The sigmoid function is considerably more complex to implement, and requires neuron outputs to have a range of values rather than a simple binary output.

2. The binary step function's primary limitation applies to networks which employ back–propagation learning, which are less likely to converge on a correct solution without the smoothing effect of the sigmoid. The ANN reported here uses supervised learning, so this restriction is less relevant.

The output of the neuron is therefore calculated using a simple thresholding operation based on the most significant bit of the counter. Fig. 5 shows a block diagram of the complete digital pulse–stream neuron.

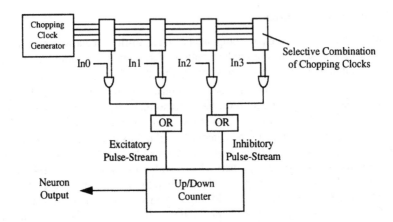

Fig. 5. Pulse–Stream Neuron

2.4 Pulse Stream Artificial Neural Network

Fig. 6 shows pulse–stream neurons connected together to form a single layer of the ANN. A layer consists of a maximum of four neurons, each with four synapses to allow full connectivity between successive network layers. The restriction to four neurons is imposed by the need to lay out the design on a single AT6005 device, and is discussed more fully in the following sections.

Inputs to the circuit are latched and encoded into non–overlapping pulse–streams so that on any given system clock cycle a pulse appears on only one input line. This ensures that pulses are processed one at a time by the neural counter. The chopping clocks are distributed to every synapse, where they are selectively combined to represent the weight value. Synaptic weights have a resolution of four bits. Higher resolution weights require more chopping clocks to be distributed to the synapses. Moreover, each additional weight bit doubles the number of pulses needed to represent circuit values and hence halves the processing speed of the network. Four bit weights were therefore chosen as a compromise between speed of operation and accuracy.

After processing of a network layer is complete, the neuron outputs are latched, and the FPGA is reconfigured to load the next layer. Any unused neurons in a layer can effectively be "switched off" by assigning them zero–valued weights. This means that the only parts of the circuit to be reconfigured are the OR gates in each synapse which are used to combine chopping clocks.

After reconfiguration, the previous layer's outputs are fed to the input latches and the next layer processed. When the final layer is completed the network outputs can be sampled.

To implement the complete circuit within the FPGA, it is important that both the input and output latches, and the FSM which controls reconfiguration, retain their state during device reprogramming. This requires dynamic reconfiguration, i.e. partial reconfiguration while the logic array of the FPGA remains active [4]. Note that in this particular system no datapath processing takes place on the logic array during reconfiguration. This limited form of dynamic reprogramming, where the logic array remains active only to maintain storage values, constitutes a sub–class of the wider class of dynamic reconfiguration. Currently, the only commercially available FPGAs capable of dynamic reconfiguration are the Atmel AT6000 series [1].

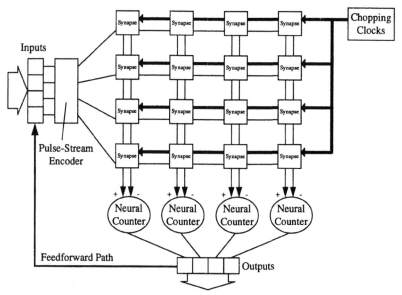

Fig. 6. Single Layer of Pulse–Stream ANN

3. Implementation on the Atmel AT6005

3.1 AT 6000 Series Architecture

The Atmel AT6005 FPGA comprises an array of 54 × 54 fine–grained cells, each of which can implement all common 2–input functions, or certain functions of 3 inputs

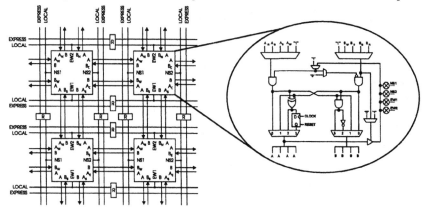

Fig. 7. Atmel AT6000 Series Architecture

along with a single storage register (see Fig. 7). Routing resources are split between slower through–cell connections and fast long–range busses, a limited number of which are available to each block of 8 × 8 cells. The equivalent gate capacity of this architecture is quoted by the manufacturer as 5000 gates.

3.2 Circuit Layout on the AT6005

The pulse–stream ANN circuitry was manually placed and routed on the AT6005 FPGA for the following reasons:

- The Atmel APR tool employs a generous placement algorithm with respect to inter–component spacing. This appears to be optimised for maximum routing flexibility, but makes it difficult to achieve the degree of macro clustering required for this design.

- Timing is critical when implementing pulse–stream circuitry, as excessive signal skew can result in errors due to two or more pulses overlapping. Subcircuits therefore have to be placed symmetrically such that delays on the signal lines which distribute pulse–streams and chopping clocks are well balanced. These special timing requirements are difficult to achieve with the current Atmel tools, which use a simple ordered list of nets to enable the designer to prioritise routing. A more advanced timing–driven layout tool such as that supplied by Xilinx would be needed to provide the necessary flexibility [8].

- The layout of the synapse circuits has to be optimised to minimise the time needed to reconfigure the device between layers of the network.

An Interactive Layout Editor is shipped with the development system, and this was used for manual design layout. Fig. 8 shows the floorplan of the FPGA with the first layer of the ANN after placement and routing.

The diagram does not fully indicate the extensive amount of long–range routing consumed by the design. Considerable areas of the logic array had cells which could not be used for logic because the adjacent routing busses were already heavily committed. The only way into and out of such areas is via through–cell routing, which is in general inappropriate for anything but short nets.

A potential shortcoming of the Atmel architecture was encountered during circuit layout. As with most designs, the ANN requires a large number of OR–gates, including a number with wide inputs. Unfortunately, both the 2–input OR–gate macros have limitations – one is slow and takes up three cells, while the single cell version is fast but has inflexible connections. Furthermore, the Atmel literature indicates that it is not possible to implement a totally glitch–free single cell OR function, due to the nature of the internal cell structure. The provision of a wired–OR capability would have been a considerable advantage for this design.

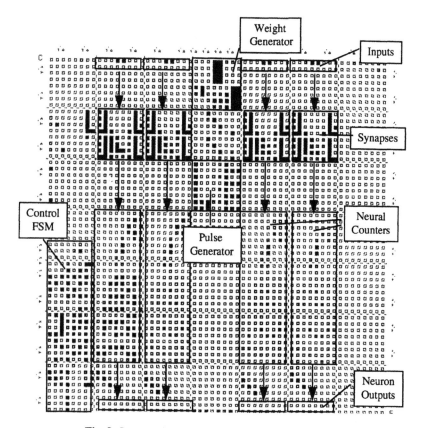

Fig. 8. Layout of a single network layer on the AT6005

The equivalent gate count calculated by the Atmel software for a single network layer (see Fig. 9) equates to a utilisation figure of 24.3%. It is clear from the table that routing forms a very significant proportion of the layout – nearly as many cells are used for routing as for logic (462 routing cells versus 476 logic cells), while 529 local and 68 express buses are also used.

Utilization Summary	Utilized
Number of Macros	273
Number of Flip-Flops	59
Number of Gates	417
Number of Turns	65
Number of Buses:	597
Local Buses:	529
Express Buses:	68
Number of IO's:	13 of 64
Number of Cells:	938 of 3136
Number of Equivalent Gates:	1216.5

Fig. 9. Design utilisation summary produced by Atmel software

3.3 Reconfiguration of the AT6005

To fully reconfigure the Atmel chip takes a minimum of 808µs, although this is only necessary when loading the first network layer. For subsequent layers partial reconfiguration can be used since the only changes to be made are to the synaptic weights.

Partial configurations are loaded into the device as a series of one or more *windows,* each of which contains programming data for a single contiguous block of cells along a row of the device. Every reconfiguration has an overhead of 10 bytes for preamble and control information, and each separate window within the bitstream carries an additional overhead of 5 bytes. In order to maximise reconfiguration speed, therefore, the following rules apply:

- The number of cells reconfigured should be the absolute minimum necessary to effect the required circuit changes.

- Configurations should be loaded at the maximum permissible rate, which in the case of the AT6005 is 10MHz. This typically requires some form of direct memory access.

- The number of configuration windows should be kept to a minimum. This has implications for circuit layout, since the reconfiguration of contiguous blocks of cells is faster than a "fragmented" reconfiguration.

It is worth noting that if any two windows are separated by less than three cells (i.e. 6 bytes of configuration data) it is faster to merge the two windows and overlay the intervening cells with an identical configuration. Experiments suggest that any stored results in these cells are unaffected by the reconfiguration operation, although this is not specified in the Atmel documentation.

4. Results and Performance

To date, the ANN has only been tested with the binary XOR function. This simple problem is non–linearly separable, which means that it requires a network with at least one hidden layer. The appropriate synaptic weights were calculated manually and subsequently incorporated into the FPGA configurations as detailed in section 2.4.

Testing of the ANN took place with the aid of an FPGA prototyping system which was developed in–house. This is based around a pair of Inmos Transputers which handle communications between the FPGA and a host computer, and also provide control over reconfiguration of the AT6005. Whilst this system is highly flexible, it is currently unable to match the maximum configuration loading rate of the AT6005, which would require a write cycle of 100ns. A mechanism to allow the FPGA to directly access fast memory to achieve full reconfiguration speed is under development at the time of writing.

With a 20MHz system clock, each layer of the ANN takes 6.5µs to produce an output. Reconfiguration between network layers for the XOR problem takes 17.6µs when a 10MHz configuration loading clock is applied. This is faster than the general case, however, since for this specific problem some weights are the same in successive network layers. The initial full configuration, for the first network layer, takes 808µs. This only takes place when the network is first initialised and so has not been included in the performance calculations.

The three–layer ANN can produce results for the XOR problem at a rate of 24kHz, when reconfiguration overhead is taken into account. This corresponds to a network performance of 0.77M CPS (Connections Per Second). In comparison, the same network implemented using full static reconfiguration, again at the maximum configuration loading rate, would produce results at a rate of only 625Hz, or 20k CPS. Thus, for this network, partial reconfiguration gives a speedup of 38 over full reconfiguration, as well as a reduction in the amount of external configuration storage needed.

The reported network is considerably slower than "static" FPGA–based ANNs such as the GANGLION, which is reported to operate at 4.48G CPS [2]. It should be borne in mind, however, that this impressive performance is achieved at considerable expense, using an array of more than 30 large Xilinx devices in a fixed configuration. Where the technique of time–multiplexing offers benefit is as a cost–effective solution to ANN implementation which uses limited logic resources.

5. Conclusions

The authors have a particular interest in investigating potential application areas for dynamically reconfigurable FPGAs. Since the only FPGAs capable of dynamic reconfiguration to date are fine–grained devices, the technology mapping of reconfigurable designs onto fine–grained FPGAs is a valuable experiment. The ANN implementation reported here has provided useful information about mapping this type of circuit onto the particular resource set of fine–grained FPGA architectures such as the Atmel AT6005. Further, the use of reconfiguration, and in particular dynamic reconfiguration, has led to the implementation of a considerably larger ANN than would otherwise be possible on a single FPGA. Whilst the current system is limited to the time–multiplexing of whole network layers, the extension of the technique to allow individual layers to be partitioned for time–sharing would offer the potential of larger networks and is currently under consideration.

The work done in developing the pulse–stream ANN has highlighted certain restrictions in both the reconfiguration mechanism of the AT6005 and the CAD tools used to produce designs on it. When compared to the system speeds possible on the logic array, reconfiguration is currently very slow. If the advantages of device reconfiguration are to be exploited in real–time applications, it is important that this situation is improved. In addition, no vendor yet provides software for the simulation of reconfigurable designs, or floorplanning tools to optimise design layouts for fast reconfiguration.

Architectural changes to the AT6005 have been identified which would increase the density and performance of the pulse–stream ANN. These include the provision of wired–OR capability, dedicated fast carry logic for counters, increased bussing resources and a faster reconfiguration mechanism.

These observations point to a possible future direction for the development of new FPGA architectures. Most design classes implemented on FPGAs would benefit in some way from having the logic and routing resources available on the device tailored to the particular application. Moreover, many designs which exploit reconfiguration contain a proportion of logic that is always static. Performance and integration levels could be further increased by providing dedicated resources to perform some of these static functions. This approach is a natural extension of the special purpose "hard macros" used in the Xilinx 4000 series devices for wide decoding functions. In the case of the pulse–stream ANN, dedicated pulse–stream and chopping clock generation could be combined with the architectural changes outlined previously to produce a Field Programmable Artificial Neural Network (FPANN). Such a device would lose the capability to implement large amounts of general–purpose logic, but would be particularly well suited to the efficient implementation of ANNs.

In general, it is conceivable that the optimisation of logic and routing resources to specific application classes could help to bridge the performance gap between FPGAs and ASICs, whilst retaining the benefits of reconfigurability.

Acknowledgements

The authors would like to thank the Nuffield Foundation, SERC and the Defence Research Agency for their support.

References

1. Atmel Corporation: Configurable Logic Design and Application Book, Atmel Corporation, San Jose, California, USA, 1994.

2. C.E. Cox, W.E. Blanz: GANGLION – A Fast Field-Programmable Gate Array Implementation of a Connectionist Classifier, IEEE Journal of Solid–State Circuits, Vol. 27, No. 3, March 1992, pp. 288-299

3. S.A. Guccione, M.J. Gonzalez: A Neural Network Implementation using Reconfigurable Architectures, In: W. Moore, W. Luk (eds.): More FPGAs, Abington, Oxford, UK, 1994, pp. 443–451

4. P. Lysaght, J. Dunlop: Dynamic Reconfiguration of FPGAs, In: W. Moore, W. Luk (eds.): More FPGAs, Abington, Oxford, UK, 1994, pp. 82–94

5. A.F. Murray: Pulse Arithmetic in VLSI Neural Networks, IEEE Micro, December 1989, pp. 64–74

6. J.E. Tomberg, K.K.K. Kaski: Pulse–Density Modulation Technique in VLSI Implementation of Neural Network Algorithms, IEEE Journal of Solid–State Circuits, Vol. 25, No. 5, October 1990, pp. 1277–1286

7. M. van Daalen, P. Jeavons, J. Shaw–Taylor: A Stochastic Neural Architecture that Exploits Dynamically Reconfigurable FPGAs, Proceedings of the IEEE Workshop on FPGAs for Custom Computing Machines, Napa, California, USA, April 1993, pp. 202–211

8. Xilinx Inc.: The Programmable Logic Data Book, Xilinx Inc., San Jose, California, USA, 1993.

Author Index

Lecture Notes in Computer Science

For information about Vols. 1–774
please contact your bookseller or Springer-Verlag